Inset - Sanikiluaq

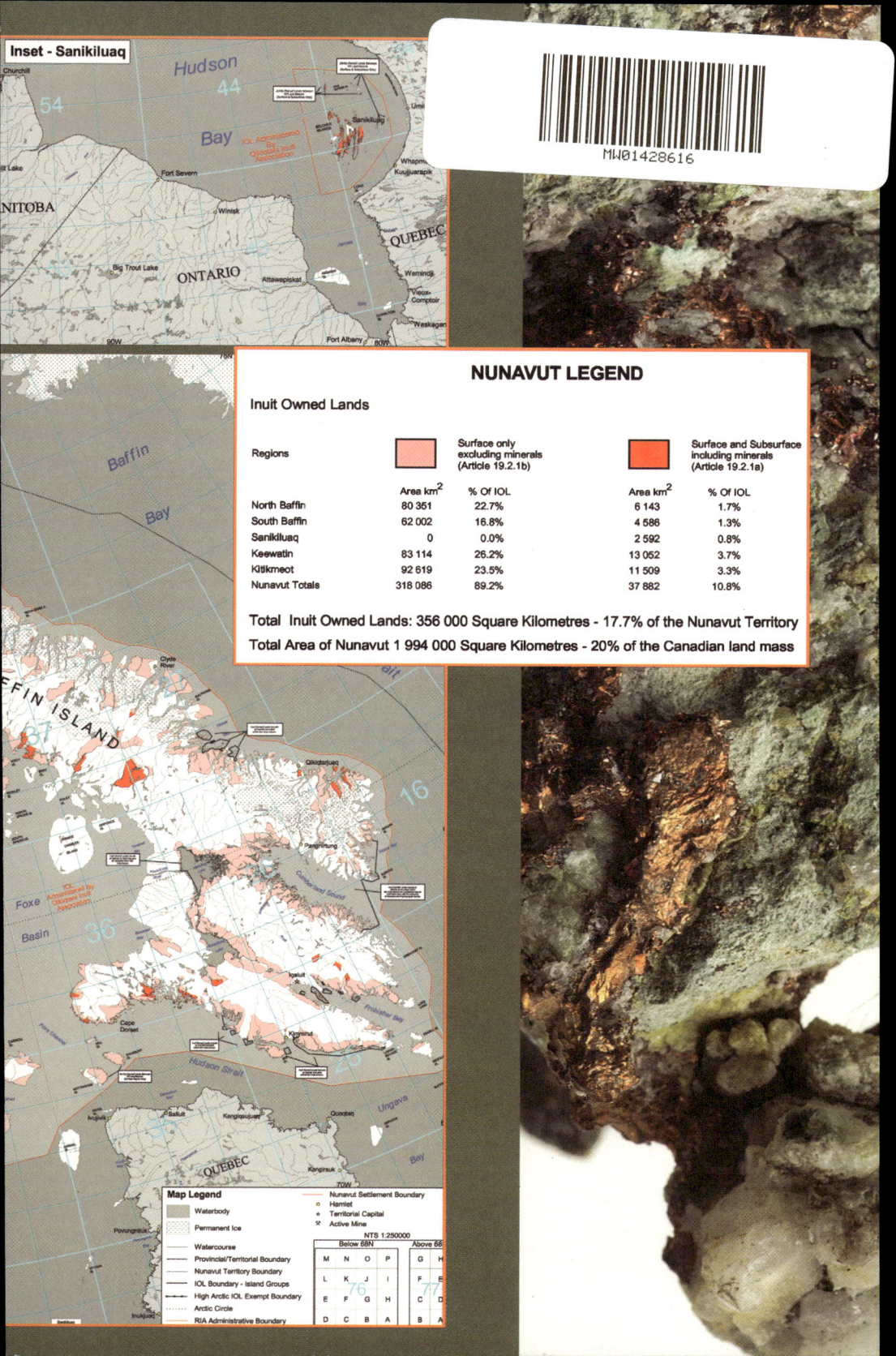

NUNAVUT LEGEND

Inuit Owned Lands

Regions		Surface only excluding minerals (Article 19.2.1b)			Surface and Subsurface including minerals (Article 19.2.1a)	
	Area km^2	% Of IOL			Area km^2	% Of IOL
North Baffin	80 351	22.7%			6 143	1.7%
South Baffin	62 002	16.8%			4 586	1.3%
Sanikiluaq	0	0.0%			2 592	0.8%
Keewatin	83 114	26.2%			13 052	3.7%
Kitikmeot	92 619	23.5%			11 509	3.3%
Nunavut Totals	318 086	89.2%			37 882	10.8%

Total Inuit Owned Lands: 356 000 Square Kilometres - 17.7% of the Nunavut Territory
Total Area of Nunavut 1 994 000 Square Kilometres - 20% of the Canadian land mass

Map Legend

- Waterbody
- Permanent Ice
- Watercourse
- Provincial/Territorial Boundary
- Nunavut Territory Boundary
- IOL Boundary - Island Groups
- High Arctic IOL Exempt Boundary
- Arctic Circle
- RIA Administrative Boundary
- Nunavut Settlement Boundary
- Hamlet
- Territorial Capital
- Active Mine

NTS 1:250000

Below 68N					Above 68	
M	N	O	P		G	H
L	K	J	I		F	E
E	F	G	H		C	D
D	C	B	A		B	A

NEW OWNERS IN THEIR OWN LAND

NORTHERN LIGHTS SERIES
Peter Schledermann, general editor

University of Calgary Press and the Arctic Institute of North America are pleased to be the publishers of the Northern Lights series. This series takes up the geographical region of the North (circumpolar regions within the zone of discontinuous permafrost) and publishes works from all areas of northern scholarship, including natural sciences, social sciences, earth sciences, and the humanities.

Other books in Northern Lights series:

Nunavik: Inuit-Controlled Education in Arctic Quebec Ann Vick-Westgate · Copublished with The Katutjiniq Regional Development Council · No. 1

Many Faces of Gender: Roles and Relationships Through Time in Northern Indigenous Communities · Edited by Lisa Frink, Rita S. Shepard, and Gregory A. Reinhardt · Copublished with University Press of Colorado · No. 2

War North of 80: The Last German Arctic Weather Station of World War II · Wilhelm Dege · Translated and edited by William Barr · No. 4

NEW OWNERS IN THEIR OWN LAND

MINERALS AND INUIT LAND CLAIMS

UNIVERSITY OF
CALGARY
PRESS

Robert McPherson

Published by the University of Calgary Press
2500 University Drive NW, Calgary, Alberta, Canada T2N 1N4
www.uofcpress.com

No part of this publication may be reproduced, stored in a retrieval system or transmitted, in any form or by any means, without the prior written consent of the publisher or a licence from The Canadian Copyright Licensing Agency (Access Copyright). For an Access Copyright licence, visit www.accesscopyright.ca or call toll free to 1-800-893-5777.

© 2003 Robert McPherson

National Library of Canada Cataloguing in Publication Data

McPherson, Robert, 1930–
  New owners in their own land : minerals and Inuit land claims / Robert McPherson.

(Northern lights series ; 3)
Includes bibliographical references and index.
ISBN 1-55238-097-1

1. Mineral rights—Nunavut—History. 2. Inuit—Nunavut—Claims—History. 3. Inuit—Land tenure—Nunavut—History. 4. Mines and mineral resources—Nunavut—History. I. Title. II. Series.

E99.E7M345 2003   346.719'50432'0899712   C2003-905527-2

The University of Calgary Press acknowledges the financial support of the Government of Canada through the Book Publishing Industry Development Program (BPIDP) for our publishing activities.

This book has been published with the help of a grant from the Canadian Federation for the Humanities and Social Sciences, through the Aid to Scholarly Publications Programme, using funds provided by the Social Sciences and Humanities Research Council of Canada.

Canada   Canada Council for the Arts   Conseil des Arts du Canada

Printed and bound in Canada by Transcontinental Printing
∞ This book is printed on acid-free Eco Book, 100% post-consumer fibre

Cover and page design and typesetting by Mieka West

www.ucalgary.ca/aina/

Dedicated to the memory of Hans Bielenstein, an inspiring geologist and people person who got me involved.

ABBREVIATIONS	:xx
PREFACE	:xiii
INTRODUCTION	:xvii
1. KEEWATIN EVENTS	:1
Economic Background	:1
Rankin Inlet Nickel	:5
The Nickel Mine's Impact	:7
Community Life	:12
Effects of the Mine Closure in 1962	:14
Central Keewatin Interaction and Leadership Growth	:16
Relocation Events and Effects	:22
Growth of Leadership and Advocacy	:28
2. EXPLORATION PRESSURES	:33
Early Geological Exploration and Prospecting	:33
Growth of Exploration in the 1960s	:38
Native Response to Explorationists and Their Allies	:50
Keewatin Exploration Boom	:54
3. INUIT POLITICAL ORGANIZATION	:57
Indian Eskimo Association (IEA)	:57
The Founding of the Inuit Tapirisat of Canada (ITC)	:61
Beginnings of COPE	:64
The Berger Inquiry	:66
Inuit Nunavut Proposal	:69

Polar Gas Pipeline Responses	:71
COPE Inuvialuit Nunangat Proposal	:77
COPE Agreement-in-Principle	:79
Gathering Sense of Unified Action	:81
Caribou Aboriginal Rights Case	:83
4. THE NANISIVIK MINE	**:89**
Introduction	:89
Early History	:89
Exploration and Pre-Development of the Mine	:92
Social Environment	:95
Feasibility Study Contacts	:96
Negotiating the Agreement	:100
Token Consultation and ITC Conflict	:103
Native Employment	:105
A Future for the Inuit	:107
Final Agreement	:108
Mine Performance	:109
Inuit Employment Study	:112
Inuit Work Record	:112
Personal Attitudes	:113
Results of the Study	:113
Ongoing Policy	:115
The Future	:116
Aftermath of Nanisivik	:118
5. THE ROAD TO NUNAVUT	**:121**
A Singular Proposal by ITC and Its Aftermath	:121
Land Claim Negotiations	:129
Emergence of the Tungavik Federation of Nunavut	:131
Some Feedback to Developers	:134
The Western Arctic Claim: The Inuvialuit Final Agreement	:138

	Land-Claim Negotiations are Stepped Up	:140
	The Coolican Report and More Negotiations	:144
	Putting the Land in Land-Claim Negotiations	:149
	At Last, an Agreement-in-Principle	:155
6.	EXPLORATION PRESSES ON	:159
	Calming the Waters	:159
	A Buoyant but Changing Industry	:160
	Some Inuit Responses	:162
	Uranium Development Projects Make News	:164
	The Effect of Land Claims on Exploration	:167
	Projects Under Native and Environmental Scrutiny	:169
	The Kiggavik Project Review Has its Detractors	:171
	Scoping Workshops and Aftermath	:176
	Gathering Opposition	:185
	EIS Fuels More Controversy	:186
	Baker Lake Rejects the Mine	:192
	UG Suspends the Kiggavik Project	:193
	Summation	:196
7.	NEW OWNERS	:203
	Agreement-in-Principle at Last	:203
	Personal Note	:204
	Organizing the Team	:206
	Prep Work and Philosophy	:207
	Thoughts on Subsurface Quantum	:208
	Meeting with Inuit in Workshops	:209
	Negotiations Begin at Pond Inlet	:213
	First Round of Cuts	:215
	Subsurface Becomes Contentious	:218
	Compromise Begins, in Part	:222
	Mineral Issue Unresolved	:225

Personal Impressions	:226
Negotiations Continue at Pangnirtung	:227
Photos	:228
Breakthrough at Resolute Bay	:237
Negotiations Continue at Coral Harbour	:243
Maps:	:244
Mineral deposits and showings in the Cornwallis lead-zinc district	
Inuit owned lands near Resolute	
Negotiations Break Down at Rankin Inlet	:249
Negotiators Try Again at Rankin Inlet	:255
Negotiations Conclude at Coppermine	:257
8. LAND BASE ACHIEVED	:269
Summation	:269
Concluding Note	:271
NOTES	:275
INDEX	:293

ABBREVIATIONS

AEC	Atomic Energy of Canada
AECB	Atomic Energy Control Board
AIME	American Institute of Mining Engineer
AIP	Agreement-in-Principle
AMCL	Associated Mining Consultants Limited
AMOK	French uranium consortium – Compagnie Française de Mokta and other French partners
BRIA	Baffin Regional Inuit Association
CAAE	Canadian Association of Adult Education
CARC	Canadian Arctic Resources Committee
CASNP	Canadian Association for the Support of Native Peoples
CCC	Concerned Citizens' Committee
CEGB	Central Electric Generating Board (U.K.)
CLINT	Community Land Identification and Negotiating Team
CMA	Canadian Mining Association
COPE	Committee for Original Peoples' Entitlement
CPR	Canadian Pacific Railway
DIAND	Department of Indian Affairs and Northern Development
DND	Department of National Defence
EARP	Environmental Assessment Review Panel
EIS	Environmental Impact Statement
EM	Electro-magnetic (in a geophysical instrument)
EMR	Department of Energy, Mines and Resources
FEARO	Federal Environmental Assessment Review Office
GNP	Gross National Product
GNWT	Government of the Northwest Territories
GSC	Geological Survey of Canada
HBC	Hudson's Bay Company
IDC	Inuit Development Corporation

IEA	Indian–Eskimo Association
IIBA	Inuit Impact and Benefits Agreement
INCO	International Nickel Company
ITC	Inuit Tapirisat of Canada
KIA	Keewatin Inuit Association
MRI	Mineral Resource International
NAME	North American Mine Explorers
NDC	Nunasi Development Corporation
NDP	New Democratic Party
NIRB	Nunavut Impact Review Board
NSO	Northern Service Officer
NAUC	Northern Anti-Uranium Coalition
NWB	Nunavut Water Board
NWT	Northwest Territories
OPEC	Organization of Petroleum Exporting Countries
OSC	Ontario Securities Commission
PCE	PCE Exploration, a Murray Watts company
PDAC	Prospectors and Developers Association of Canada
PNC	PNC Exploration (Canada) Co. Ltd.; Power Reactor and Nuclear Fuel Development Corporation (PNC) of Japan
RCMP	Royal Canadian Mounted Police
SRT	Surface Rights Tribunal
TFN	Tungavik Federation of Nunavut
UG	Urangesellschaft Canada Ltd.

PREFACE

The perspective of a bystander and peripheral participant in a momentous political process can help to shape the public's understanding, although it may differ from that of the key players. I pursued this work to enlighten myself about the background influences and motivation of Canada's Inuit in their arduous land claim struggle ending in 1993. Personal satisfaction was involved here – I did not set out to write definitive Inuit history, nor would I consider myself qualified to do so. My interpretation ought to interest northern enthusiasts and most particularly geologists, prospectors, and others connected with exploration and development projects, like myself. Others have pointed to the catalytic connection of resource development and land claim advancement, and I have confirmed its huge importance. The political maturity and determination of Inuit, of course, made it all happen.

I speak for northern would-be developers, having spent my career as an exploration geologist in the pre-Cambrian shield where much of Canada's resource wealth has originated. It has enabled me to live in the North for periods of time and to meet prospectors, trappers and Native people in the process. I met Metis activists bent on expanding Natives' role in mining, and I had occasion to employ Cree prospectors and field assistants myself. They proved to be very adept in the bush and inordinately observant, and, what is more, they were good company. Therefore it was natural for me to consider employing Inuit as my work gravitated into the Northwest Territories in the 1970s. They proved adaptable and trainable in the performance of field tasks – and saved us money as well. I learned from Simon Tookoome something of the depth in the Inuit spirit: fraternity, love of the land, competence, reverence, leadership, humour, sense of sport, value of art and beauty.

My own work in geology opened my eyes to the grandeur and beauty of the Arctic. I saw the great sweep of the Keewatin caribou-rich Barren Lands studded with volcanic buttes, great rivers, and shorelines and lakes. I saw the seemingly empty, hackly limestone reaches of Cornwallis Island, the rugged hills of Somerset Island, the endless extent of Boothia Peninsula, which was blessed with elevated beaches for Twin Otter landings and colourful granites only outshone by flowers and shrubs. I saw the coastal beauty and haunting emptiness of interior Baffin Island with glacial outpouring and seemingly endless basins carved in crustal metamorphic rocks and blazing iron formations. Not only was it a thrill to experience these areas of the land but their untapped potential for minerals as yet unseen by southerners was always with me. So I was an enthusiast and minor expert when it came to NWT geology – and this was partly responsible for the active participation of my employer in mineral exploration there. In fact, in the 1980s, Comaplex Resources was one of the largest holders of mineral claims and prospecting permits in the Territory.

In the late 1980s Inuit negotiators and their advisors established the importance of acquiring mineral rights to land since the Crown had traditionally subordinated the surface holder in the development process. It was known that substantial blocks of mineral rights were going to be granted in the land-selection phase of negotiation, and Inuit became determined to make the best of it by hiring mineral advisors. That is how I came to be involved with the TFN. I was sympathetic to the Inuit cause and took delight in preparing a mineral inventory to guide their deliberations, little knowing what the ramifications might be. In the end my job involved the briefing of community land-selection committees and participation in step-by-step negotiations with government representatives at the land claim table.

The attitude and motivation of the Native leaders I saw in action fascinated me, and after my retirement from exploration I pursued a line of research at the AINA to enlighten myself on this subject. Like most of the mining company people engaged during this dynamic period of the Native activism, I had little awareness of the issues at stake. With the benefit of hindsight, it is clear I should have – and could have – had that awareness.

Having witnessed some of the Inuit negotiations first-hand and read about the powerful background forces involved, some conclusions about the process and the future have come to mind. There was a demonstration of Inuit determination and political acumen that led me to believe they have a strong future ahead of them. Their acceptance of technical expertise and liking for technology will enable them to adopt the modern methods of land and resource management. They will have to pursue educational opportunities before they take over the necessary technical and administrative tasks from consultants and the like. Their struggle to control their fate and achieve co-management of resources has been monumental. Other Canadian aboriginal groups must be tinged with envy and longing for something like the Inuit status in our land. Perhaps the day will come when First Nations will play a co-management role in all our resource concerns as they are our people closest to the land, the forests, the fish and game. As for the Inuit, I see a unique opportunity for them to acquire mastery – and I mean mapping, GIS, and inventory and conservation skills, etc. – over the management of their own lands and resources, supported by their traditional insight. Then they will be in a position to seek the devolution of Crown resources to the Nunavut authority, at which time they will truly be the masters of the North and stewards on behalf of all Canadians.

In this work I am indebted to the Arctic Institute of North America, and to former Executive Director Mike Robinson in particular, for encouraging my peculiar blend of earth science and history. He thought it was an interesting story and offered guidance in my approaching a dual audience. Bruce Gillies and Terry Fenge gave me greater understanding of the larger picture of unfolding land negotiations. Paul Quassa and Tom Molloy provided encouragement, knowing that I was dealing with an interactive process that was undocumented except for final results. My comments brought back memories good and bad for them as key players in the negotiations, but endorsement of my opinions was neither sought or offered, naturally enough. I am grateful for family hints on the writing, endless patience from my wife Marion, and studious scholarship from my editor, Luisa Izzo. Murray Pyke generously opened his photo collection for my use.

Calgary, Alberta, September 27, 2002
Robert McPherson

INTRODUCTION

Recognition of Aboriginal title by the courts and protracted negotiation by the Inuit people with the government of Canada yielded a powerful land-claim agreement in 1993. Many issues of prime concern were settled between the parties. In addition, this agreement created a new blueprint for resource management in the eastern and central Arctic – the area that would become the Nunavut Territory. Though they were not party to the negotiations, the mining industry, and developers in general, played a significant role in the outcome. They had a motivational role that acted on the Inuit over the decades, coloring their thinking and sense of marginalization. The Inuit response to development ranged from irritation to outrage, and it spawned political organization and protest. The Inuit were also quick to understand the manifold aspects of development, including potential benefits. The government was dedicated to development and became visibly motivated and inclined to negotiate whenever it was threatened or significantly delayed. An aggressive tone and agitation for land injunctions or freezes became development-oriented tools in the Inuit struggle for self-determination, even though such issues as game and fish management, culture, and language preservation were uppermost in their minds. Their identity was associated with land, but Inuit also recognized that a traditional livelihood was not going to sustain their children, who would need to learn new skills and acquire the wage-labor jobs that developers could offer. A thriving economy based largely on nonrenewable resources would be essential for a viable new territory – the Nunavut of their dreams.

The mining industry, on the other hand, has come through years of uncertainty over access to mineral land and tenure, only to find that Inuit ownership has produced a totally new regime that goes against long-conditioned industry practices. Another layer of red tape now affects the exploration of Crown mineral lands, and prior approval is needed for prospecting on Inuit mineral lands, which have attained a sizeable dimension through the agreement. Since Inuit lands are privately owned, they can be acquired only through negotiation and at the discretion of the Inuit administrators. To the degree that developers themselves contributed to a sense of alienation, they are responsible for this new state of affairs in the Eastern Arctic. Perhaps some will lament that things might have been different if industry had changed its attitude along the way, as the following historical chapters will show. In any case, momentous changes for mining were already occurring in Canada, such as the erosion of the miner's free entry rights due to environmental concerns. Similarly, the achievement of Native self-government and control over settlement lands would have come eventually, even without the developers' catalytic influence described here.

This book documents the reaction of Inuit people to the mining industry over the three decades that preceded the land-claim settlement. It tells how they befriended and helped prospectors, guiding them to mineral occurrences; took the short-term summer jobs, unloaded ships, learned to be miners, and put up with vehement bosses; and viewed with alarm the buzz of helicopters and drills, the oil spills, and the empty drums left behind; how they regarded the cluttering of maps with mineral claims and leases and the eager presence of stakers and prospectors on their own land. Also, with growing education and worldliness, Inuit viewed the industry as a partner of government, which was never dependable or supportive of their concerns. They discerned the sham of consultation and the lack of prior notice or willingness by developers to see them as partners or genuine advisors. They opposed and yet coveted third-party rights and titles, thus clashing with government negotiators. In the end, they came to be freehold owners of mineral lands in their own right.

Mining and exploration activity, including oil and gas development, had negative overtones in the mind of Inuit people and led them to fight for management boards to regulate, impose standards, and engage in planning. The Inuit wanted development on their own terms and pushed for a new regime of regulatory bodies with co-management rights. They wanted to be first-order beneficiaries of development in Nunavut through impact and benefit agreements, ownership of mineral rights, well-placed real estate, surface ownership around mineral deposits, and royalty sharing.

Exploration people maintained a remoteness and indifference to Natives, whose sparse numbers and lack of skills made them seem superfluous to exploration and mining endeavors. Native affairs and welfare were clearly not their concern, but rather a government responsibility. Exploration people had been conditioned to take this view by operating under the free-entry principle entrenched in Canadian law. In the 1970s, when growing Inuit demands for self-determination threatened land access and development rights, industry representatives urged the government to seek a speedy resolution through the land-claim process so that they could operate with confidence and certainty. During this period, they were reluctant to engage the Inuit directly and generally shied away from bilateral arrangements and agreements.

This book does not analyze the nature of prospectors' self-confidence or their fixed attitudes about their rights and privileges. These important characteristics stem from Canadian law, which for generations ingrained "the right to explore" for Crown minerals in much of the nation's territory. From the nineteenth century, the severance of minerals (first gold and silver) was secured for the benefit of the Crown. The exploitation of minerals such as iron, coal, and gold had been regarded as a public good since early settlement times, and it was encouraged by the recognition of title and the founding of the Geological Survey of Canada to help prospectors. Over time, the details of legal title have evolved, but the concept of free entry has persisted; it is at the root of Canadian mineral administration and is responsible for the prosperity and long reach of the mining industry. The concept of the claim, introduced in the 1860s, became standard in Canadian mining acts, together with the free-entry system, around 1910. Early Canadian legislators, influenced by Gold Rush experiences in Australia and California, "unequivocally adopted the free entry system with its principles of unrestricted access to Crown minerals, free acquisition of title, and the right to develop and mine."[1] The free-entry system appeared first in British Columbia, with the British Columbia Mineral Act of 1896, and spread eastward. The Dominion Lands Act of 1898 copied the B.C. legislation, and its successors in the Yukon and Northwest Territories still show this origin. The first B.C. regulations dealt with placer gold mining, but the regulations for quartz mining made under the Dominion Lands Act were also copied from the B.C. model. In 1917, the right of the prospector to gain ownership of a claim (Crown grant) was revoked in favor of the right to obtain a renewable twenty-one-year lease. The Yukon Quartz Mining Act of 1924 deliberately made it difficult to amend the mining law, as Yukoners were suspicious of the whims of government officials! However, the Northwest Territories continued to operate under the Dominion Lands Act regulations, which have evolved into the Canada Mining Regulations of the present day. To prevent prospectors from holding on to land without exploring or developing it, the regulations gradually evolved to specify more stringent work requirements. There was no break in the underlying free-entry philosophy, although the practice has been circumscribed to respect private and municipal land. Only Alberta made a complete shift away from the free-entry system, allowing complete ministerial discretion in disposing of mineral rights.

Despite current licensing requirements, prospectors – imbued with the free-entry system for generations – feel they have an unconditional right to explore for minerals. They had better be secretive about it, too, because others have the same rights. It is not surprising, given this feeling of entitlement, their hardiness and sense of adventure, and the sometimes huge rewards for their work, that prospectors have sustained an important presence in the northern landscape, but their legal rights, secretive nature, and tendency to be interested primarily in their task have precluded congenial contact with Native communities.

Legislative solutions, due primarily to the action of environmentalists, began to affect exploration in the 1970s with the framing of land-use regulations and a permit system. At first, the application of such regulations was selective, but in time it became universal. Since the Nunavut land-claim agreement of 1993, a Native surveillance system over all lands has taken shape, in the form of co-managed boards and mandatory impact and benefit agreements for development projects. Another reality is the appearance of private (Inuit-owned) surface parcels interspersed with Crown lands: both retain Crown minerals that might interest prospectors. Special advance licensing is needed to authorize entry onto Inuit lands: the prospector's license is not enough. In addition, the private ownership of mineral rights by Inuit has become a major factor in areas with known potential. Access to these parcels, a guaranteed right of original claim holders, can be obtained by others only through negotiation with *Inuit implementation organization* Nunavut Tunngavik Inc., which uses a published schedule of terms and fees. Making these arrangements requires corporate planning and execution before the field season. Once in the field, prospecting teams cannot make ad hoc decisions that take them into or onto Inuit land.

The new regime calls for calm acceptance by industry and a fresh approach to operations that includes legal measures and public relations. Lest this sound preachy, it should be noted that the United Nations has been responding to social pressures caused by developers in the Third World with recommendations intended to forestall confrontations. The UN urges explorers to adopt the concepts of free and informed consent, participation in benefits, and compensation for and mitigation of adverse effects of their projects. The Prospectors and Developers Association of Canada (PDAC) has issued a communiqué to members that calls for the building of a new culture of exploration.[2] The exploration industry must emulate the mine-operating sector (which is more attuned to public relations) if it wants to keep working in the most prospective areas of the world.

Drawing on the experience of Canadian professionals and the outcome of a 1997 South American conference on the topic "Mining and the Community," the PDAC has urged exploration companies working overseas to develop the staff skills and the will for early-stage dialogue with local communities. The purpose is to establish a long-lasting, positive relationship that will ensure the project's "social good standing."[3] The process should start with the first geologist on the property and strive to avoid early misunderstandings that can cause polarization and entrenchment of negative attitudes, if not open conflict, putting the whole project into jeopardy. The PDAC argues that the good community relations developed by the exploration company will be appreciated by the development company later on. This social good standing should be considered value added to the property, like ore reserves, and raise its price to the developer. It is time for companies to adopt the wisdom of respecting community rights, just as they accepted environmental safekeeping. The global situation is analogous to that of the Canadian North: both call for a new culture of exploration.

There is reason to believe we are entering a new era of cooperation between the Inuit of Nunavut and mineral explorers and developers. Local participation should grow and take many forms, from employment and training to service contracting to joint-venture financing. Only then will the Inuit acquire a sure relationship to the means of production that will be fruitful in the long run.

Possibly an Inuit exploration company will eventually emerge to exploit the love of land and keen observation skills of the people on their own lands. There will be young men who do not want the desk jobs of the new Nunavut but, like their fathers, prefer a life on the land. With some serious training in mineralogy and exploration methods, they could become effective prospectors. Organized prospecting classes, effective leadership, and some cooperation from industry are all that is required. A fine tradition of Native prospecting already exists in other parts of Canada, such as Stanley Mission, Saskatchewan, where the influence of Metis prospectors Jim Brady and Malcolm Norris that began in the 1960s has touched the younger generation of the Churchill River region.[4]

Trained Inuit prospectors would be welcomed by industry for their nearness to hand and local knowledge. Their presence in the community could spur the eventual creation of heritage exploration companies. Inuit-owned lands could empower such organizations to parlay their activities into something major, possibly rivaling what the Canadian Pacific Railway land entitlement did for CPR subsidiaries PanCanadian Petroleum and Fording Coal. At that stage of development, the heritage companies would be powerful partners in the mining industry of the North.

The documentary basis or public information for this book was somewhat deficient on the side of industry, but voluminous on the side of Native history and development. The campaign for self-determination and land-claim agreements was a public process, although the actual negotiations were private. Communiqués and newspaper accounts were enough to encourage research. Indeed, academics, environmentalists, and support groups swelled the records and made the Nunavut story available to the public, including industry people, if they were interested. On the other hand, the documentation of industry affairs was often limited to economics and technical results, rather than in-depth analysis and statements of policy. Company affairs are naturally private and are made public in the guarded language of financial reports. For example, I could not find a candid industry discussion of wage rates paid to Native workers, though the charge of being parsimonious must have gotten through to them. No doubt there was a rationale for maintaining trainee rates within company wage structures, and no company would want to lure away, with overly attractive wages, the few tradesmen already employed in Arctic communities. Another event that surprisingly lacked public comment in the technical press was the veto of the Urangesellschaft uranium mine by the people of Baker Lake in 1990. The event must have sent a tremor through the halls of Bay Street, but one would hardly know it from the media.

Nevertheless, the book deals with events and policies that seem to intertwine the lives of Inuit people with those of miners and prospectors in the North. The first chapter, "Keewatin Events," covers the thirty-four-year period from about 1929 to 1963 from the standpoint of exploration history, dealing with the impact of the wage economy and contact with developers, with special reference to the Rankin Inlet nickel mine. It describes the adaptability of Inuit, their emerging leadership and sense of mixed-group communities, as well as their dependence on government intervention (sometimes with tragic results, such as relocation), and their questionable faith in whites. Chapter Two, "Exploration Pressures," describes the beginnings of exploration in 1928 and then the post-war surge brought about by new geological knowledge, the use of aircraft and geophysics, and strong metal markets. The impact of oil and gas exploration and the uranium boom precipitated Native opposition and political activism and focused attention on aboriginal rights in Canada. Chapter Three, "Inuit Political Organization," traces the emergence of Native organizations, with southern help, from the 1960s to 1977: organizations like the regional Committee for Original Peoples' Entitlement (COPE) and the Inuit Tapirisat of Canada (ITC), which was national in scope. Newly found access to legal advice enabled these organizations to focus on land claims. They participated in the Berger Inquiry and the Mahoney court case asserting aboriginal rights. The COPE agreement, the first to evolve, was followed by the Nunavut (or territorial division) proposal of the ITC. Chapter Four is a case history of the Nanisivik mine on Baffin Island. It documents the exploration and entrepreneurial events into the 1980s, the establishment of Canada's most northerly mine, government's paternalistic protection of Inuit interests and parallel expression of its own, the community's response to organized national (ITC) opposition, the mine's performance with respect to jobs, and the exploitation of jobs by Inuit for their own ends.

The fifth chapter, "The Road to Nunavut," describes the most significant part of the journey: the real achievements in land-claim negotiations from 1979 to 1990, when an agreement-in-principle was finally signed. Native motivation, sense of purpose, and achievement are portrayed using the current information of the day, available to anyone who wished to follow the progress of negotiations, or lack thereof. Chapter Six, "Exploration Presses On," is the concurrent story of mining during the land-claim process. The pressure of ongoing development injected a sense of urgency into the negotiations and sparked Native agitation for a legislated land freeze to protect their entitlement. A progressive sense of Inuit empowerment came to a head during the Baker Lake uranium controversy. The proposed mine was dealt with under the federal government's environmental protection laws calling for public input. The citizens responded to the procedures in a massive and organized manner, reflecting a high moral position that defeated the mine.

The seventh chapter, "New Owners," resumes the tale of land-claim negotiations and traces the land selection process to its concluding agreement in 1991. Whereas the previous chapters were based on public documents and interviews, this chapter is a personal history of the author's involvement as a mineral consultant for the Inuit negotiators. The material, drawn from diaries and notes, describes personal impressions. Many people know this story from their participation in the public meetings held over the two-year period, but so far it has been described only in general terms. Perhaps this account will be received as a useful contribution of a slice of history from the perspective of an observer and participant, not from that of politicians or negotiators themselves. My reception by the Inuit was warm and respectful. They had already learned the value of including outside advisors such as lawyers and environmentalists on their team. I was able to convey the scientific information necessary for mineral evaluation by plotting this on their own land-selection maps, and I could *orally* report the technical significance of federal tactics during negotiations.

The Inuit approach to these negotiations was to stonewall until they saw exactly what the federal side wanted. Their leaders understood how their land choices affected the individual communities and gave them much room to argue their cases, even halting the negotiations if each was not satisfied. The Inuit tactic was to yield very grudgingly on some matters, all the while maneuvering carefully on others that were more precious. Their sense of negotiation gamesmanship was superb, the result of years of experience at the table. All this was not surprising to the federal negotiators, who advanced the interests of Ottawa in a progressive manner. Decision-making on the Inuit side was by consensual process, and seldom did the community representatives fail to make plain their concerns, as they had to answer to their fellow citizens back home. In fact, they were a strong driving force in shaping the direction of the talks, even though the articulation of their views *on land selection* was often carried by the leaders.

Many community and personal struggles accompanied the land selection process. The Inuit "gave up" to the Crown 80 percent of their settlement lands – the lands they had occupied for ages. Of course their hunting, fishing, and traveling rights were protected by the main agreement, but the real acceptance of the whole package (with its ceding of Aboriginal rights) hinged on the quality of the 20 percent portion of traditional lands to which they would gain legal ownership. The negotiations therefore became intense bargaining sessions full of strategies and maneuvers, as each side searched for what the other really wanted while striving to achieve its own agenda. Both sides really wanted an agreement, so a rewarding compromise ultimately prevailed. The Inuit bargained hard to acquire subsurface rights where there was real mineral potential and finally gained ownership of existing claims and leases – agreeing, of course, that for these, the Canada Mining Regulations would be entrenched. In the end, there was satisfaction that Inuit had acquired their most treasured lands and secured a subsurface portfolio sufficient to enhance their economic well-being in the future – one of the principal aims of the agreement all along. The eighth chapter, "Land Base Achieved," presents a summary of the Inuit success in the acquisition of land and mineral rights.

I retain a lasting impression of how attuned the Inuit were to the value of mineral rights and how effectively they used claims and leases as bargaining chips to get what they wanted. The outcome of their efforts, and a startling realization to some, is that they are now the largest freehold owners of mineral rights in Canada. Industry must now deal with the Inuit as private owners, just as they have dealt with the CPR, whose entitlement of surface and mineral rights led to the building of the railroad and the development of some of Canada's corporate giants.

1. KEEWATIN EVENTS

A progression of events in the Arctic showed that mineral exploration played an important role in the predestination of Native land-claim settlements. It has been said, having in mind the seismic operators and oil-well drillers in the Western Arctic, that resource companies acted as catalysts in the process of Inuit self-determination. But no less impact emanated from the mining industry, most notably the uranium companies that explored the Keewatin district in the 1970s and 1980s. For about four decades, southern Canadians pushed exploration activities northward with increasing intensity. The Native residents lacked the power to influence the developers for their own benefit, or in ways that would mitigate the effects of development on their social and natural environment. Gaining this necessary power became a keystone of Inuit ambitions for self-determination. This power began to be realized by the 1980s and reached fruition in the early 1990s.

For a proper appreciation of this liberating process, we must examine the status of northern communities, the state of economic conditions, and the development of Native leadership back to the early postwar years. A good deal of this story can be drawn from the Keewatin district.

ECONOMIC BACKGROUND

With the decline of the fur trade in the 1940s came a growing realization of the decline of all wildlife populations on which the Inuit had depended. Americans left the war-effort sites like Fort Chimo, where they had been offering wage employment. Welfare considerations began to occupy the Canadian government. By the time the white-fox market collapsed in 1949, the Department of Resources and Development had engaged James Cantley, a retired fur trader, to prepare a full report.[1] Family allowance benefits began to be applied to Native as well as other Canadians.

Canadian pressure for social reform in general emerged in the 1940s. American presence in the Arctic grew dramatically during World War II, and wage employment was offered to Inuit in many places during those years. The introduction of the Family Allowances Act in 1944 and the revision of the Old

Age Security Act in 1951 had particular significance for Native Canadians. The collapse of fox-fur prices had left many Inuit with no source of cash income except welfare and family allowance payments. Attempts by the Arctic administration to address the deplorable conditions in the Eastern Arctic played a prominent role in shaping Inuit policy and revealed many dilemmas. Responsibility for Indian and Inuit health was transferred in 1945 from the Department of Mines and Resources (DMR, still responsible for Inuit affairs) to the new Department of National Health and Welfare. The next year, an X-ray machine was placed on the *Nascopie* during its annual ocean patrol of the Eastern Arctic, an indication that tuberculosis in the region had reached epidemic proportions. Epidemics of flu, typhoid, and polio also took their toll so that, despite the press accounts to the contrary, the health of Arctic residents was in crisis.[2] Their contact with southerners inflicted a cost in personal health, to say nothing of social cost, on the Natives. For example, the polio outbreak in July 1948, which caused many deaths and cases of paralysis as it spread through the Keewatin, began at the military establishment at Churchill, yet press reports tended to attribute the epidemic to the habits and lack of hygiene of the Native people.

In the 1950s, the RCMP took over the administration of northern relief, and the Arctic Patrol ship *C.D. Howe* made annual cruises to provide X-ray examinations and other medical services. Many Inuit, as many as 20 percent of the population, contracted tuberculosis. Except for those handled by church-run hospitals at Chesterfield Inlet and Pangnirtung, most of these patients had to be removed to treatment sanatoria in the South, causing much disruption of family life. Desperate living conditions and the hope of better hunting prompted Inuit families to accept relocation to new areas, serving the government's desire to assert Canadian sovereignty over the far North. Inuit would later refer to this relocation as exile. Starvation episodes resulting from the decline of natural game were highlighted by author Farley Mowat, much to the government's embarrassment.[3] The government then began to sponsor community development in a major way. The presence of northern service officers and other members of a growing Arctic bureaucracy also drew people to the northern communities where they were stationed. A growing resistance to white dominance and regulation (of game laws, for example) coincided with the emergence of settlement councils that demanded more control over community affairs. These councils laid the foundation for Inuit demands for land-claim agreements and self-determination in the 1970s. Self-confident council participants and young Inuit activists began networking to discuss the issues of the day.

The direct relief cost for 8,437 Inuit residents of the Northwest Territories and Northern Quebec rose from $10,835 in 1945–46 to $84,680 in 1949–50.[4] The total Family Allowance credits, seen as another form of relief, were $23,968 and $281,916 for the same years. Owing to backlogs, more than $300,000 was paid out in the intervening years. By this time the social, medical, and economic problems had left the administration floundering, and active intervention was seen as essential. As for Family Allowance payments, it took the RCMP, the agency responsible for administration, until 1948 to register children for eligibility, though retroactive payments were limited in 1949 to one year's benefits, lest these credits create dependency. The issuance of payment was in kind, in the form of designated food and clothing items, and credits were issued to the Hudson's Bay Company and other traders for this purpose. A welfare role was assigned to the HBC at a time when its profits from the fur trade were in serious decline. The welfare eligibility requirement that older children attend school eventually encouraged wholesale movement from camps to settlements – a trend that was encouraged by economic conditions in general. Administrators repeatedly attempted to restrict rightful Family Allowance payments to hard times or subject recipients to a form of means test. It was not until 1953 that Inuit began to receive regular cheques like other Canadians. In this way, the government limited the power that officials had exercised in doling out benefits. This reform came about thanks to the persistent complaints of Mackenzie Delta Native people, including Abraham Okpik, who played a singular role in pressing for Inuit rights.

Tester and Kulchyski, in their book *Tammarniit (Mistakes): Inuit Relocation in the Eastern Arctic 1939–63*, point out that the issue of rightful benefits was still unresolved in much of the Arctic in 1953. They cite the example of an Inukjuak resident, Jimmy Koodlooalook, whose family had narrowly escaped starvation while camped on the Belcher Islands. The family had been given niggardly payouts from their Family Allowance account at the time, but because they had helped to sustain some white prospectors who were without food, they were ultimately reduced to eating their dogs. When schoolteacher Margery Hinds questioned Jimmy's family, she wondered why white people were without their food anyway.[5]

In 1949, the federal government authorized the payment of old age pensions at the rate of eight dollars per month, with the RCMP serving as administrators and registrars. The RCMP made sure the entitlement was extended only to persons over seventy whose annual income did not exceed $400. A year later, under the administration of Citizenship and Immigration, the rate was increased to $25 per month (less relief). But there was still a reluctance to pay the pension to camp Inuit, as this would tend to encourage whole groups to move closer to the posts. The means test and proof-of-age requirements caused problems for northern bureaucrats, and they continued to delay the payment of benefits. As of January 1953, no old age pension had been paid to any Inuk in the Northwest Territories. According to Tester and Kulchyski, "The federal bureaucracy had spent thirty years trying to ensure that Inuit did not get the same benefits from social programs as did other Canadians. It persisted with its contention that these would interfere with 'self-reliance' and create 'post-Eskimos.'"[6]

By 1950, the fur trade economy of the Eastern Arctic was virtually collapsing. White fox catches at Baker Lake, for example, fell from 3,500 pelts in 1947–48 to about nine hundred in 1949–50.[7] The price per pelt had fallen from $28 in 1943 to $6.50 in 1949–50. The annual relief costs paid by the government were consequently soaring, and reached a total of $63,251 for the Eastern Arctic in 1949–50. The report that James Cantley prepared for the government in that year offered no concrete solutions for dealing with the social crisis facing the Inuit. Cantley recommended that the Arctic Services Branch supervise all services, with less reliance on the RCMP and more cooperation with the Hudson's Bay Company. Inspector Henry Larsen of the RCMP, however, had more discerning remarks: in a 1951 memo to the Commissioner, he blamed the Hudson's Bay Company, the churches, and the northern administration for the condition of the Inuit:

> The Eskimos generally have drifted into a state of lack of initiative and confusion. Conditions are generally appalling. Never has there existed so much destitution, filth and squalor as exists today, and in the opinion of some people, the conditions under which some natives live is a disgrace to Canada, surpassing the worst evils of slum areas in cities. Bad sanitary and economic conditions are gradually undermining the health of these people and if not checked will ultimately result in their extermination.
>
> Fundamentally, the trouble goes back very many years, actually to the time that traders first went into Eskimo territory and changed the whole way of life of the Eskimos that is, changed them from primarily hunters of meat to primarily trappers of fur.... Year by year the natives became more poorly-clad in storebought clothes which are quite inadequate to withstand the rigors of the far north climate and which do not even begin to compare with native skin clothing, that is, caribou skins and seal skins.... Rubber boots are being substituted for seal skin footwear. The Eskimos sell their seal skins to traders and in some cases

make up seal skin clothing in substantial quantities and donate it to the missions or dispose of it to other white residents for some small remuneration perhaps in the way of tobacco and cigarettes or some trinket ... many Eskimo habitations are made from discarded box boards, odds and ends of burlap, canvas and other waste material, and the accumulation of filth inside these habitations is indescribable.[8]

The upshot of Larsen's remarks, which were not appreciated by the administration, was a 1952 conference of all persons concerned with Inuit administration and the subsequent creation of the parliamentary committee on Eskimo affairs. Already the publication of Farley Mowat's book *People of the Deer* and its serialization in *The Atlantic* in 1952 had caused an uproar in the House of Commons.[9] Notwithstanding the scathing attacks on Mowat, it emerged that something had to be done to avert disaster in those parts of the Arctic suffering from the demise of the fur trade. Serious consideration centered on the recommendation of the 1952 conference that the government consider "the movement of Eskimos from over-populated areas to places where they can be assured of being able to make a better living."[10]

RANKIN INLET NICKEL

The surface outcropping of the nickel ore zone at the high-water mark on Hudson Bay, a rusty, sulphidic area that measured 200 by 75 feet, was discovered by R.G.O. Johnston in 1928. Johnston and three other prospectors had traveled northward along the coast from Churchill in twenty-two-foot canoes fitted with outboard motors. Though well equipped, they found this journey dangerous, especially the first two hundred miles with its very shallow water, reefs, lack of shelter, and frequent storms. Cyril W. Knight, the organizer of this expedition (and also of Murray Watts' 1931 trip to Ungava), said at a Mining Institute meeting that J.B. Tyrell's report was directly responsible, since Tyrell had identified the favorable volcanic belt thirty-five years before. Tyrell and his party had camped on Falstaff Island, less than seven miles from the Rankin Inlet discovery, according to John Dryborough.[11]

Grab samples brought out by Johnston showed the presence of nickel, and more thorough sampling the next year uncovered a section with 2.5 percent nickel and 1.25 percent copper. The Cyril Knight Prospecting Company Limited decided to carry out a drilling program in 1930, for which they shipped a motorized forty-five-foot steel launch to Churchill. The crew reached the property on July 13. Owing to deep permafrost, the freezing of drill rods became a serious problem, which was solved only by heating the seawater drill fluid using coal obtained from Term Point. In seven weeks, only 1300 feet were drilled, but the results were rewarding: the crew documented the geology and the presence of subsurface lenses of massive and disseminated ore aggregating 120,000 tons and containing 1.22 percent copper, 4.62 percent nickel, and 0.11 oz./ton platinum. These average metal grades and the massive ore running 9.25 percent nickel enabled the sponsors to talk of this as a high-grade discovery. The preliminary diamond drilling was augmented by more drilling in 1936. Then the property remained idle until 1951, when the Korean War caused nickel prices to rise, stirring interest in small and remote deposits.

The newly formed Rankin Inlet Nickel Mines Ltd. carried out geophysical surveys to trace the extension of the sulphide mineralization lying within the sill-like host body. Diamond drilling followed this work and continued in 1952, leading to a decision to explore the indicated ore body with an underground operation. The deposit was shown to contain economic concentrations of nickel, copper, and platinum group metals. A permanent mining camp was established on the 1540-acre property in 1953 under the guidance of Kelso Roberts, president; J. K. Whatnough, consulting engineer; Kenneth Wilson, mine manager; and Charles Penny, construction manager.

Exerting what J.C. Browning described as a modern and aggressive approach, the company marshaled 1500 tons of cargo in Montreal and transported it to the site on board the chartered steamer *S.S. Angus Lake*.[12] It was not known how close the vessel could approach the mine site after its voyage of 2800 miles, so a fifty-ton motorized landing craft was brought along. All equipment and supplies were landed ashore and secured above high tide in fourteen days in late August 1953. The mine's eight buildings were erected in September, and shaft sinking reached a hundred feet of depth on its way to 331 feet by the end of November. At this time fifty men, including fifteen Inuit, were working at the site under the direction of mine manager Kenneth Wilson. Underground development work and additional diamond drilling were carried out, showing that the nickel–copper ore body varied in width from five to forty feet and extended for at least 450 feet along the strike. Proven reserves to the three-hundred-foot level totaled 460,000 tons, and were graded at 3.3 percent nickel, 0.8 percent copper, 0.03 oz./ton platinum, and 0.06 oz./ton palladium.[13] At the proposed mining rate of 250 tons per day, these known reserves would last about five years.

In April 1954, Rankin Inlet Nickel Mines Ltd. was reorganized and its name changed to North Rankin Nickel Mines, but a period of dormancy ensued, obliging the Inuit to take up traditional pursuits while the company raised the capital necessary to bring the mine into production. Late in 1955, the Mogul Mining Corporation provided $2 million in the form of convertible bonds. Preliminary estimates of the total capital requirements included $2 million for plant erection and mine development, $0.3 million for supplies, and $2.7 million for working capital, for a total of $5 million.[14] Financing enabled the Company to order 2000 tons of equipment and supplies to be delivered to the property by ship in August 1956. Construction proceeded during the winter months, and under the guidance of H. Leavitt, mine superintendent, and W. Walker, milling consultant, the first nickel ore from the mine was fed to the concentrator in mid-May 1957. Metal concentrates were stored on site until the opening of saltwater navigation. In 1958, the softening of nickel prices and contract difficulties caused concern in view of the extreme working capital requirements of the mine. Supplies for the whole year had to be purchased in advance, yet concentrates had to be stored for ten months before shipment, metal settlement, and revenue flow from the smelter could take place. Dr. W.W. Webber, managing director of Mogul Mining, stated: "We are considering ways and means of effecting economies at the mine and we are endeavoring to cut costs to the bone, but we are not thinking of closing the mine for the time being."[15] The realities of operating a small mine in the Arctic plagued the company throughout its life. The first year's (1957) production of concentrates was bagged and shipped to refineries via Hudson Strait. Thereafter, the concentrate was stored in bulk and shipped to Churchill by boat and thence by rail to the Sherritt Gordon refinery at Fort Saskatchewan, Alberta. The price of nickel in 1955, the year the production decision was made, was $0.63 per pound. It rose rapidly to $0.74 in 1956 and remained there for five years. In 1961, it rose again to $0.815, but fell back to $0.79 in 1962, the year the mine closed.[16]

THE NICKEL MINE'S IMPACT

The employment of Inuit labor was a big factor in the success of the mine. Initially only a handful of Natives, rarely more than ten families, lived in the vicinity at certain seasons, relying for food on sea mammals along the coast and Marble Island, fish in the Meliadine River, and caribou inland on the tundra. These families maintained a kinship affiliation with Chesterfield Inlet people to the north. Key to the employment of Native labor on the project was the influence of Andrew Easton, who came in 1957 as a construction consultant and was later retained as mine manager.[17] Easton argued strenuously for the employment of local labor because it was so difficult to secure white employees. Some men were employed early during the unloading of cargo vessels and camp construction. In the spring of 1956, five or six heads of household were employed, and by November the number of employed Inuit men had increased to fourteen. During the spring of 1957, coinciding with the first production, the number of Inuit employees grew to twenty, and by year-end it had reached eighty. From that time through 1961, the mine provided steady employment for approximately eighty Inuit men and several women. In addition, others were seasonally or periodically employed in boat handling and mineral exploration, so as many as a hundred Inuit were receiving some form of income from the nickel mine. Many showed an aptitude for underground work and were trained to join the regular underground complement of eighteen men. Although travel to and from the mine was commonplace, the settled population at Rankin Inlet continued to rise owing to the mine's presence: in 1962, the Inuit population exceeded five hundred.[18]

To secure an adequate labor force, the company recruited prospective employees from many settlements.[19] Over the lifetime of the mine, 65 percent of workers were migrants from Chesterfield Inlet, 25 percent from Eskimo Point, seven percent from Repulse Bay, and three percent from Baker Lake. The government cooperated in the recruitment and transport of potential workers by boat and aircraft. Later on, a steady migration by dog team swelled the tide of people coming to this boom town. An important factor in recruitment was the hiring of an Inuk named Shingituk away from the Hudson's Bay Company at Chesterfield, as he could serve as an intercultural foreman.

In describing the new mine in 1957, the mine staff noted that the local labor force comprised Inuit who showed great interest and a natural quickness to learn. The company sponsored language and job training, and hiring practices tried to accommodate the Inuit's instincts and way of life.[20]

A labor pool was maintained numbering more workers than necessary at a given time so that some men could be engaged in trapping and hunting. The availability of country food was of paramount importance. Though fully employed by the mine in many cases, the Inuit men continued to hunt and fish to provide food for their families and the dog teams that remained in town. Sealing was done in the fall and winter, but it intensified in the spring, when seals were plentiful along the floe ice. Absenteeism was high during this season, especially on Monday mornings. In July and August, the favorite activity was char fishing (using gill nets) in the Bay or at the mouths of the local rivers. Still, the Inuit learned to live by the clock, more or less. In the mine, there were two shifts per day, five or six days a week; in the mill, three shifts a day, seven days a week. This was the reality of wage employment at Rankin Inlet.

The Inuit men, invariably unskilled in mining trades at the beginning, were hired on the surface labor crew for the basic $.75 per hour. When a man had "proved" himself and shown a desire for more skilled work, he underwent a period of on-the-job training, still at $.75 per hour. The skilled labor jobs paid up to $1.90 in the mine and two dollars in the mill. An elite half-dozen or so working underground shared in the production bonus, netting up to $200 per month. Inuit learned such skills as carpentry, plumbing, and steam fitting, and many held jobs as equipment operators, mill helpers, drill helpers, muckers, deck men, and cage tenders. The last job called for an aptitude in communication, while most required mechanical aptitude and strength. A few Inuit became salaried employees: these included a labor foreman, son of a Chesterfield Inlet HBC post servant, who was paid $300 per month, first for recruiting duties and later as interpreter and message carrier. A few women found work in the Eskimo cookery and received a wage of four dollars per day.[21] The Company defended these labor rates for Native workers on the grounds that their fringe benefits were significant: for example, a hot meal at noon was provided for free to Inuit earning less than one dollar per hour.

The Canadian average hourly wage for mining increased from $1.88 in 1957 to $2.13 in 1961 (from $79.35 to $89.08 per week).[22] The average weekly wages and salaries in mining were listed as $83.89 and $95.90, while those for the service industry were $45.77 and $55.38.[23] In a 1961 report to the Department of Northern Affairs and National Resources, social scientists Robert and Lois Dailey cited the mean monthly income of unskilled Inuit workers at Rankin Inlet (for March 1958) as $134.79, or $33.70 per week. This amount was 39 percent of the Canadian *average* of $86.60 per week for 1958, just as the Inuit training rate of $0.75 was 39 percent of $1.88 per hour, or 66 percent of the $1.13 average in the service industry.[24]

This minimum hourly rate was not attractive in Rankin Inlet in light of the elevated cost of food and trade goods. Consequently there was much incentive to strive for skilled jobs with higher pay. Of the seventy-two Inuit men employed in summer 1957, eighteen novices and six trainees were paid $.75 per hour while 48 (66 percent) were paid skilled rates up to two dollars. Some of the men – mostly Baker Lake or inland people – were content just to stay on the surface labor crew, while others evinced a higher ambition and motivation, even participating in the precision work of mine rescue and the use of breathing apparatus. They demonstrated a natural quickness to learn that was reported by the mine staff.[25]

Absenteeism was common among workers who were not accustomed to habitual work schedules and had a psychological and material need to hunt. Some psychosomatic illness also took a toll. Robert Williamson, in a 1974 paper entitled "Eskimo Underground: Socio-cultural Change in the Canadian Central Arctic," reported instances of men staying away from work because they feared the aggressive behavior of foremen and other whites. The normal behavior of a supervisor sometimes sharply contrasted with traditional Inuit manners of intimacy and personal interaction. High-stress situations were particularly disturbing to sensitive inland people and to others as well. The least affected were Coral Harbour men who had previously worked on Dew Line construction and were familiar with the white man's aggressive manner of address.[26]

The mine developed an ad hoc replacement system by "hiring" more men than it needed. Inuit were encouraged to report absences in advance and find replacements; otherwise, the labor foreman would be sent to find one. Replacements were usually easy to find for the unskilled jobs. The influence of wives, who most appreciated the benefits of the wage economy and opportunities at the HBC store, kept about 75 percent of the employees on a regular pattern. The other 25 percent had a somewhat patchy employment history. Delinquent or unpunctual performance led to dismissal "with a lecture," often followed by later rehiring. Only the more articulate men asked for a transfer or formally quit the job. More commonly, the practice of the disenchanted was to perform badly so as to get themselves fired.

From the spring of 1960, a more vocal attitude took shape among workers, who set about negotiating with the management over conditions of employment. Sometimes they involved the labor foreman in the process, as the mine relied heavily upon his judgment. But the foreman was distrusted by the majority of employees and considered prone to favoritism and discrimination in hiring tactics and housing allotments.[27] He was known for his obsequious attitude to management and did not stand up for the Native individual. More and more, the government welfare officer was drawn into disputes and problematical situations with the mine staff. The government was permanently represented in Rankin Inlet because of the community's potential to relieve the distress of inland Keewatin Inuit.

The wage income was welcomed by new workers because it was steady and afforded them the opportunity to buy the attractive array of goods at the HBC store. However, their traditional attitudes made it difficult to accept the objectives of the mine and the discipline required. Those in entry-level jobs on the labor gang especially were unable to identify with the mine project. Many were discouraged and would have left, had there been some other place to go.[28] Another view, and a glowing one, was expressed by Farley Mowat:

> Eskimo employees in a white enterprise were being treated as men in the full meaning of the word. Beginning with a handful of local Eskimos who asked for employment and who were given the simplest tasks on a trial basis, the Eskimos had, in a matter of two years, almost taken over the operation of the mine, and in so doing had displayed powers of adaptation and intelligence which once and for all have given the lie to the contention that the Innuit are an inferior and backward people who must be protected from the impact of the modern world.[29]

The Rankin Inlet experience was a unique one for Native people along the road to acculturation within the larger society. But there was not any sort of integration with the whites or participation in community life there: in fact, the company worked consciously to promote segregation to protect the Inuit and his culture. White mine workers housed in bunkhouse accommodation were restricted from the Native village and its social activities.

The wage income and store experiences were rather bewildering to many new residents, and there were accusations of being cheated or short-changed. The area administrator and police officer were drawn in to demonstrate that charges or deductions, though painful to the individual, were not fraudulent. Considering the low wages paid at the mine, the cost of store goods and food and the company deductions (for services such as board, housing, fuel, and medical care) were substantial. The money economy did not radically improve the material stability of most families. Approximately 65 percent of workers were in low-income brackets (between $175 and $250 per month of take-home pay). Beginning in 1960, the government operated a banking service and exercised financial control over people's lives; but it did not, according to Robert Williamson, try to teach them money management. A few families – perhaps ten at the most – were employment elites who saved some of their money or put it into purchases of bombardiers, snowmobiles, freighter canoes, and peterhead boats. Their homes were also extensively equipped in comparison to pre-industrial days.

The Rankin Inlet community was dominated by the mine and mine objectives, in what was described as a thoroughly domineering and authoritative way.[30] Most services, including medical care, were administered by the company, which always exacted a price. No individual in this company town could avoid being placed in a position of obligation to the company. With success of the mine always uppermost in his mind, the strong-willed mine manager made all the decisions. He was genuinely interested in the welfare of the Inuit – and why not, since they made an economic success of the operation? The prevailing policy of protectionism, together with segregation in housing and dining, meant that individuals were being trained to be laborers, not citizens. According to the Daileys, "Those Eskimo that are frugal, hard working, punctual, and cooperative, are in the eyes of the mine 'desirable.' Those who do not readily adjust or who do not pay attention to orders, or who malinger, are rejected and forced to leave the community."[31]

COMMUNITY LIFE

Clearly, a system was in place that restricted social contact between whites and Inuit, in the interest, it was said, of protecting the integrity of the Inuit family. Its scope included the home, dining hall, and recreation center, where drinking habits were also curtailed. Whether this boom-town life should be described, as indeed it was, as a caste system, is debatable. The Inuit were being denied social equality in Rankin Inlet, but this was scarcely different from the situation in other northern communities of the day, where Native councils were just beginning to appear. Were Inuit being prevented from developing a sense of self-direction as well? Williamson points out that during the first years of contact, when confronted with the discrimination and behavior of white bosses, Inuit tended to articulate feelings of fear and confusion and then withdraw from anything but essential involvement with whites. Long experience in this situation led to growing cynicism and lack of social commitment, modified only in the case of those whites who attempted to speak Inuktitut.[32]

The five-member Eskimo council that functioned from 1959 to 1963 in Rankin Inlet was an instrument of social organization that served the interests of the mine. The mine manager would request meetings through his Eskimo foreman. *In camera* discussions were followed by a community meeting at the school, when the foreman and one or two eloquent counselors would harangue those attending on the subject of work habits, punctuality, temperance, housekeeping, garbage, etc. After the shutdown of the mine in 1962, the council came under the direction of the government, more particularly of the area administrator. Though public meetings were popular and well attended, some regarded them as merely "playing a talking game the government always wins."[33] The interests of the white community were served by the community association that was formed in 1962 and broadened in 1963 to include a whole range of public affairs. Some half-dozen articulate Inuit demanded from the start that their people be granted full participation. They were opposed by the administrator and the school principal. The administrator expressed the government's official policy of encouraging racially separate community organizations, and the principal advocated segregation because the Inuit were not responsible or ready for participation. Very rancorous meetings took place, with the result that attempts were made to install a Baker Lake type of solution, one predicated on a larger population of whites having a vested interest in community decisions. This called for racial separation along with one umbrella organization, and Inuit spokesmen considered it complicated and unnecessary. They argued that only full participation on an equal basis would allow their people the opportunity to learn. The compromise that emerged was a combined association devoted entirely to recreational matters. In the association's various committees and at public events, more and more participation came from the inland Inuit population, who had previously been reticent and retiring. However the Inuit lacked an effective leader and spokesman. The labor foreman who might have assumed that role was a pressured and discontented man. He spoke of moving to the new resettlement community of Whale Cove.

Significant behavioral and organizational changes to Native life occurred at Rankin Inlet during the mining years, yet there were few structural changes. The major change had to do with the division of work between men and women that arose from the changed subsistence pattern. Women no longer had a busy and equal role in making clothing and preparing game food. They relinquished several of their functions to the Hudson's Bay Company and consequently became idle and bored. They remained more traditional than their men because they had no social interaction with whites. The men were separated regularly from their families and had dwindling opportunities to hunt and instruct their sons. The men let go of the equipment of traditional life, such as sleds and dogs.

Many recommendations were made by social scientists to improve the health, self-expression, education, and coordination of Native workers. They also suggested scrutiny of the company to ensure fairer wage levels and equitable benefit deductions. Crowding and bad sanitary practices had had a severe effect on Inuit health and infant mortality, and living in a larger community required huge adjustments. The Chesterfield Inlet and Repulse Bay people were kin (*Airilimiut*) and typically looked down on the people from Eskimo Point (*Padlimiut*). Social differentiation within the community was aggravated by the mine's preferences or priorities in hiring: the labor foreman was from Chesterfield Inlet.

From many sides, the Native worker and his family were urged and exhorted to save their money and plan ahead, practices which were not in their nature and caused anxiety. Inuit spending tended to concentrate on some luxury items and dispersal to the extended family. Fiscal planning and restraint had not been learned, even though some coaching was offered. The Inuit began to wonder about their future prospects and the expected life of the mine. They were told that the mine's life was predictably short, even though no planning for its demise was in sight. Obviously many people were in for a change, and observers were conscious of what the newcomers had given up in accepting settlement life. Williamson observed, "Many of the Eskimo who had worked in the Mine for more than two or three years had relinquished their dog teams, disposed of their skin clothing, and indeed some of the weapons they needed for the hunt. Relatively few ... were prepared emotionally or materially for a return to iglu life on the sparse hunting grounds of the Keewatin."[34]

Drinking facilities at the mine, which catered to the lifestyle habits of the white hard-rock miners, were irresistible to Native workers. To control absenteeism, the company imposed a nightly consumption limit of six cans of beer per man. Nevertheless, drinking became a source of anxiety to the Inuit population because it caused fights, family violence, and the squandering of paycheques. This concern became more and more acute as the end of the mine approached.

EFFECTS OF THE MINE CLOSURE IN 1962

The impending closedown prompted a thorough canvass of workers during the winter of 1961–62. Seventy-five percent of employees said they would go to work anywhere Andy Easton (the manager) went, if he would take them. Except for an abortive arctic-fox hunting project, the government did little to cushion the blow. In fact, when government workers took over town maintenance, they replaced the Inuit tradesmen who had performed that function for the mine.[35] Relocation – at least on a temporary basis – was arranged for some workers. Three mine employees went to work on the Dew Line; nine went to the Asbestos Hill operation in Ungava, following manager Easton, who came back to recruit them; seven, including a few of Easton's recruits, were hired by Sherritt Gordon Mines at Lynn Lake; nine men went into gold mining at

Yellowknife; and three went into tungsten mining at Cantung (now Tungsten, Northwest Territories).

Over the course of the next ten years, many more mine workers were relocated, including a total of twenty-five to Lynn Lake, where there was the least social disruption. The best record of employment tenure there reached sixty months. Previous experience, age, education and marital status affected the length of stay at Lynn Lake. The most stable men were competent at their work, had a rudimentary knowledge of English, were married, and had no formal education. Younger, single men with some education left earlier, either in search of mates or because they became aware of other opportunities. The former averaged about forty-eight months at Lynn Lake, while the latter averaged fifteen months – both figures compare favorably with the record of white mine workers in isolated locations. R.G. Williamson and Terry Foster, then at the Institute for Northern Studies, University of Saskatchewan, wrote a monograph entitled *Eskimo Relocation in Canada,* published in 1973. They found that while some individual experiences were painful, the overall experience in mining industry relocation was termed successful, especially the relocations to Asbestos Hill and Lynn Lake.[36] The same study noted that many relocatees had been relatively prosperous in Rankin Inlet because they were employed, but they were not the Inuit leaders in that community before their departure. On their return to Rankin Inlet, most of the relocated miners adjusted well; in fact, they prospered and gained positions of leadership and prestige in the community.[37]

In 1972, there were some nine voluntary organizations in Rankin Inlet with sixty-four executive positions, of which fifty-four were filled by Inuit residents. The fourteen Inuit who had spent some time in a Euro-Canadian environment at school or work held twenty-nine of these positions, and two wives of former relocatees held positions in formal organizations. The majority of the permanent jobholders and two entrepreneurs were persons of similar background. Williamson concluded that returning mine workers had been able to succeed socially and financially, perhaps because the quasi-urban character of Rankin Inlet prepared them for relocation and allowed the community to absorb their skills later.[38] The general experience taught many lessons in appropriate preparation, orientation, counseling, and communication – the authors came up with forty-five recommendations on planning recruitment and policies for outside work.

The Inuit population of Rankin Inlet dropped by 260 in the two years that followed the mine closure. Most of those who remained went onto social assistance, which began during the 1962 wind-down of mining operations, and engaged in sporadic hunting and seal harvesting, prices having risen by then. The government encouraged handicraft production in the form of sewing, carving, and pottery making. By 1965, when the population had risen modestly to 320, social assistance was entrenched at a level that was not excessive overall, in the opinion of Williamson, but he felt there was "a significant part of the Rankin Inlet population which had become, over almost three years, to some degree habituated to life 'on relief.'"[39] Such people, for whom the taxpayer became responsible, might be considered the true casualties of the mining experiment at Rankin Inlet.

CENTRAL KEEWATIN INTERACTION AND LEADERSHIP GROWTH

The white man's invasion of the Arctic, which had begun long before this time, had a progressive impact on the Inuit economic and social system. This has been the topic of much research, including Frank Vallee's 1962 study that centered on the Baker Lake region in the central Keewatin. Vallee noted that the Inuit's lack of English and limited education (not one Inuk in this region had completed Grade Four as of 1958) closed all but unskilled labor jobs to them. He predicted that high school graduates would be appearing in 1968; by 1970, they would be too numerous to qualify for the forty-odd skilled jobs in the region unless a mine or other industry became established.[40] He argued for a systematic exploitation of renewable resources to provide a decent standard of living on the land for about forty-five families, or about two hundred people.

Much of Vallee's comment had to do with Keewatin residents who were relocated in the late 1950s. They came chiefly from Garry Lake, the Back River, and the Kazan River and were resettled in Rankin Inlet, neighboring Itavia, or Whale Cove. The mine at Rankin Inlet had attracted the relatively mobile people from Chesterfield Inlet and Repulse Bay, and by 1958 about eighty men had full-time employment and were living in company housing. Itavia was a rehabilitation center about half a mile from the mine that comprised a number of huts, a storeroom, a workshop, a school, and two houses for Department of Northern Affairs personnel. Its primary role was to feed and care for immigrants from famine areas and others who were incapacitated or infirm. Most were welfare cases, though a few men took employment at the mine. Many relocatees, like the Kazan people, requested return to their home camps, but some responded to education and encouragement and had settled down in Itavia by 1960. The new coastal settlement of Whale Cove, about sixty miles south of Rankin Inlet, was a haven for hunters and fishermen, and it proved to be an economically viable community. It was also psychologically satisfying for its residents because there was limited supervision by and involvement of whites. Many of the Whale Cove people had lived at Rankin Inlet and were able to compare the two settlements. The consensus was that Whale Cove, with only one Department of Northern Affairs technical officer, was a better place to live.[41] There were too many white bosses at Rankin – at least one in charge of every activity except the purely domestic. One man said the white-dominated community made Inuit "feel like dogs who can't go in the tent." He was referring to the segregated social system of the mining town with its caste-like overtones.

Other relocatees who had been sent to Rankin Inlet from beyond Baker Lake apparently made the adjustment by 1960, and the majority of able-bodied men among them were working at the mine. The mine management commended these men as good workers. Their adjustment was due to their high level of skill, their youth and adaptability, and the remoteness of their former dwelling places on the Back River.

In 1958, other men were recruited from camps around the Baker Lake area to try their hand at nickel mining, and the government transported them by air – usually by RCMP patrol aircraft. By the next summer, all but one had returned home to a life of hunting and trapping. They complained of the bossiness of supervisors and mimicked the shouting and waving of arms they had supposedly endured. The returnees also found it difficult to adjust to the everyday requirements of mine scheduling and authority, which was not surprising given the nature of their upbringing on the land.[42] A few said they might try it again, admitting the attractive side of life at the mine – the parties, movies, dances, and steady income. Of the original group, one married man and his family and one single man remained at the mine; both of them had been settlement residents before. The married man lived in a company house, which he intended to purchase. As of 1960, his only concern was the possible closure of the mine during the next few years. Vallee predicted more and more natural movement of acculturated people towards localities of employment opportunity. His observations had a lot to do with the interaction of Inuit and white residents. He said that overt conflict rarely occurred, but a pattern of social segregation was common throughout the Arctic. This pattern reflected social class considerations, and not more enduring, caste-like discrimination, as the Daileys contended. A more extreme form of this pattern was the formal segregation rule imposed by the mining company at Rankin Inlet. White employees were not permitted to visit the Native village or attend dances. However, a small minority of prominent Inuit exchanged visits with white individuals, reflecting the fact that social acceptance could be gained via economic or social improvement. Some whites took on the role of socializer or paternalistic overseer. When asked what would be the ideal kind of person they were "grooming" for the future, the image they gave was that of a "steady, predictable bourgeois adult with a sentimental attachment to the Eskimo past, proud of the way in which Eskimos have manipulated nature and kept alive, but who has turned his back on features of the traditional way of life, such as the traditional religion, unusual marital arrangements, and certain features of daily living."[43] Native social skills, community consciousness, care with money, independence, and interest in arts and crafts were all prized by whites.

Conflicts arose from the fact that most Inuit were present-oriented or existentialist by nature, whereas whites were more future-oriented, purposeful, or goal-directed. Whites made sacrifices for self-fulfillment and career achievement. Vallee claimed there was a certain amount of despair among white "socializers" over these differences in lifestyle and outlook. They interpreted the behavior of the majority of Inuit as wasteful and showing an apparent lack of interest in the future of their community and race. This interpretation was used again and again as evidence that Inuit were childlike and not to be treated as full adults.[44]

In an attempt to differentiate population trends, Vallee devised the term *Kabloonamiut* (people of the whites) for the more acculturated person – the white man's Eskimo – as opposed to the term *Nunamiut* (people of the land) for the traditional, land-based person who resisted some aspects of change. The former carried the Euro-Canadian culture to others and functioned as behavior models and key figures in the interaction network. Although they had not given up all the old skills of the land, they resided permanently in the settlements, often as longtime employees of or linked with the Hudson's Bay Company, the RCMP, or a government agency. Their children tended to follow similar patterns. They had the resources to be well equipped and function as successful weekend hunters and fishermen. They were oriented to the ways of the white society and sought to better themselves, for example through education. They tended to socialize more with each other and strengthen mutual kinship bonds, rather than socializing with the *Nunamiut* or country people, who were their economic inferiors.

The *Kabloonamiut* played an active community leadership role because of their social position and tendency to speak for all Inuit in the region. In typical arrangements prevailing in communities like Baker Lake in the late 1950s, the northern service officer (NSO) was in charge, but the Eskimo council functioned at least in an advisory way. New proposals or programs were "bounced off" the *Kabloonamiut* members before being discussed with the community at large. The *Kabloonamiut* initiated routine matters and played a large part in mediating these to others, so they were truly representatives of the whole Inuit community vis à vis the white man. Their role was destined to expand when the true community development program was launched by the federal government in 1964.

The *Kabloonamiut* and their more educated offspring got the jobs as language interpreters and key people in the communication network. They communicated white ideas and instructions much better than they expressed the concerns of Inuit themselves. Shyness and pride were possible explanations for this reticent behavior, but Vallee believed that some *Kabloonamiut* were ambivalent about white society and nursed a deep-lying rejection, which they revealed by refusing to speak the whites' language.[45] While the *Kabloonamiut* were fully integrated into the Inuit community and valued for their go-between role, they were not equally accepted by whites in a social sense; that is, they were not invited to their parties, dances, and other social events.

Deviant conduct and crime began to appear in larger Arctic communities and formed an undesirable part of the assimilation process for Inuit. Baker Lake had a record of social conformity and good behavior because it was homogenous, mission-influenced, not dominated by one agency, alcohol-free, and had a diverse white population, including families. This contrasted with other centers that were dominated by construction, mining, or military activity and had many young, unattached, white males in the population. Nonfraternization and segregation policies were justified by officials facing such a situation, as in Rankin Inlet, but an unintended tension between ethnic groups built up, which itself resulted in deviant behavior.

Power and authority eluded Inuit in the 1950s because they followed traditional approaches suited to camp life, small groups, and consensus decision making. Moreover, they were a tiny minority nationally, unorganized and with limited access to resources, all of which limited their power in society. They were quick to admit that they were awed by the power, wealth, and influence of white society. Inuit tended to be complaisant and cooperative in this era, adopting two informal rules: let the white man initiate and lead, and do nothing to openly displease him.[46] Authority among the Inuit was a muted affair, limited to one's own kinship or camp resident group, where an active, successful hunter spoke for the group. Giving orders to and speaking for others was foreign to Inuit – and would cause no end of trouble for burgeoning Native organizations after the 1960s. This cultural conditioning was also responsible for the strong reaction of Inuit newcomers to the behavior of white bosses at Rankin Inlet.

Communities with mixed kinship backgrounds were springing up, where the traditional mode of authority and leadership was inadequate. Government agencies recognized the need for an expression of leadership – if only advisory in nature – and began to establish "Eskimo councils" as a means of encouragement. The emerging social class of *Kabloonamiut* filled the leadership gap. Council meetings had a strong educational value for Inuit, but the Inuit themselves did not volunteer issues and concerns of the type that preoccupied white residents. That is why parallel associations of white residents evolved (in 1960 at Baker Lake) to cope with white community problems. Membership in the white associations became a big issue, but the membership process recognized a degree of social assimilation. The association dealt with social activities; however, in effect the Inuit were still excluded from real decision making.

In Rankin Inlet, as the Daileys pointed out, the Eskimo council was a vehicle for transmitting messages and instructions from the mine or the government. This was a company town that was formally segregated along ethnic and economic grounds and integrated along authoritarian lines; it was very different from Baker Lake, where Inuit developed a larger involvement in their own destiny.[47] It was generally true that the Inuit population had a minimum of say in communities that were dominated by a single agency, such as a company. After the mine closed in Rankin Inlet, a community organization of citizens arose to pick up the pieces. Inuit reaction to this was immediate and determined: they were not going to be left out and relegated to an ineffectual Eskimo council.[48]

Rankin Inlet was unique and should be recognized for industrial adaptation and social assimilation in the short period of its mining history. It was a force for diversification in the area, but it had little effect in the form of trickle-down benefits for other communities. There was a naive hope of such benefits at the time, but Rankin Inlet generated no emerging demand for products other than a supply of labor – not even a demand for country foods for its Native workers.

Vallee believed that the Inuit were not being assimilated; on the contrary, he saw a process of social integration at work and a clarification of their ethnic identity. "The trend towards a kind of pan-Eskimoism is given impetus by the emergence and growing influence of a class of persons committed to settlement living, the kind of persons we have called *Kabloonamiut* throughout this report."[49] Articulate, English-speaking persons were emerging to speak on behalf of all – and these people were the products of settlements, residential schools, and hospitalization. As to their eventual role, Vallee foresaw (in 1962!) that the more they saw themselves as victims of discrimination and social/economic predicaments, the more likely they were to become spokesmen for protest movements and power struggles. (Traditional or *Nunamiut* people were still held in respect, and they were able to exercise their leadership skills by participating in the emerging cooperative movement, despite their lack of English.)

The social disintegration depicted by Williamson and others was being offset by a growing process of self-identification. The Inuit depicted themselves as a unique people – a pan-Northern society. Carving and other art creation, encouraged and facilitated by the government, were economically fruitful as well as satisfying and ego-building. The growth of Native-language broadcasting through the Northern Service of the CBC was another avenue of self-expression and group identification.[50] However, the demeaning aspects of social assistance continued unabated, and the prospects for employment for the growing Inuit population looked bleak.

The pattern of migration and population growth at Rankin Inlet also led to the disintegration of kinship bonds with extended families. The emerging nuclear family units with a materialistic outlook tended to discount traditional values, pursuits, and language. This cultural dysfunction tended to show itself as an alienation of the middle generation from the older folk and a progressive alienation of the schoolchildren from their parents. The craving for new experience – always strong in the Inuit culture and linked with economic security – was having a modern expression. The desire to work at the mine and to better one's position through job training affected many. The parallel desire for recognition encouraged cooperation and complaisance on the job, and the prideful creation of art works and handicrafts. Increasingly, some individuals sought to be spokesmen for the larger group in community affairs, in contrast to their restricted kin-related roles within traditional camp society. Competition in these spheres was growing, with strong implications for Inuit leadership. At the same time, Inuit were aware of and resented acts of discrimination or oppression by whites, such as attempts to exclude them from community associations. There appeared to be little room for initiative, and thus the inexperienced population traveled the road to self-government and community development rather slowly.[51]

RELOCATION EVENTS AND EFFECTS

The government's deliberate relocation of elements of the Inuit population made their condition even more desperate. This desperation affected their reaction to whites, including prospectors: it hardened their attitudes. The process of relocation and public criticism contributed to fundamental changes in northern administration policy. During the summer of 1953, the government relocated ten families, seven from Inukjuak, on the Ungava Peninsula of northern Quebec, and three from Pond Inlet, on the northern tip of Baffin Island. They were moved to two locations in the High Arctic: Craig Harbour (and two years later, to Grise Fiord, on the southern tip of Ellesmere Island), and Resolute Bay on Cornwallis Island, 250 miles to the southwest, the site of a military base and a Department of Transport weather station.[52] The government claimed the people were moved for humanitarian purposes, whereas the Inuit (much later) claimed they were deceived and moved for purposes of asserting Canadian sovereignty. More families followed, and the experiments in resettlement continued until 1956, when the Dew Line and other commercial opportunities brought about a change in policy.

Regarding potential relocatees from overpopulated areas, James Cantley, acting chief of Northern Affairs and National Resources' Arctic Division, wrote: "We must also keep in mind possible developments in the areas from which such groups may be drawn. Apart from the Mid Canada Line [radar] operations at Great Whale River, I am thinking of possible mining developments on the Belcher Islands and in Ungava Bay in northern Quebec and also at Rankin Inlet in the Keewatin. If these developments come about, as they very well may, there is a possibility that quite a few Eskimos may find wage employment which might appeal to them more than going north to hunt and trap."[53]

By the late 1950s, government officials commonly felt that the High Arctic relocation ventures had been a success – and this attitude influenced decisions taken in the Keewatin. However, Walter Rudnicki, chief of Northern Affairs and National Resources' Welfare Division, noted that Resolute Bay and Grise Fiord had experienced serious sociological problems, many of them related to kin-group rivalry and animosity, brought about by settling together families from different Arctic locations. The families were concerned about being denied the right to return home and with the strong-arm tactics of the RCMP. High Arctic resettlement had its negative side and was dead by 1963, after the Diefenbaker government came to power. The growth of Inuit self-expression and political opinion was eventually fostered by the whole process.[54]

Although the transportation and communication network enabled regular contacts by an elaborate bureaucracy, instances of famine and starvation plagued the Keewatin in 1957–58. The declining numbers of caribou and the vagaries of their migration routes caused famine among inland people at Ennadai and Garry Lakes. The fifty-five members of the Ennadai group had resisted being moved for a number of years, but were finally relocated to Henik Lake, a hundred miles to the northeast, in the summer of 1957.

The relatively traditionalist Ennadai group tended to congregate at the weather station and accept handouts to supplement their sometimes meager earnings from hunting and trapping, especially after the traders left because of falling fur prices. By 1947, when the Northern Administration Branch (Department of Northern Affairs) began distributing family allowances, yearly relief efforts were in effect. However, the government felt that this destitute group ought to be moved from such an inaccessible location for their own good and for better administration; they chose to move them in 1950 to Nueltin Lake, where the group might enjoy better hunting, participate in a commercial fishing operation (that never materialized), and be close to a trading post. By December of the same year, all forty-seven had made their way back to Ennadai Lake on foot – they were indeed an independent people who rejected being moved away from their familiar territory.[55] The government's failure to provide them with boats, fishnets, and the like may have had something to do with the failure of this "experiment." In 1955, there was an effort to study the continuing predicament of the Ennadai group. One researcher, Northern Service Officer James Houston, recommended leaving them where they were, and helping them with family allowance credits, which they had not been receiving routinely. He also encouraged the group to supplement its caribou diet by trapping and to cache more meat and fish. About moving them, he said, "They have twice avoided such attempts by returning to their original region. They would probably do so again as they are primitive (in the sense of resisting outside influence) and independent people who are not yet used to accepting everything the white man tells them as wisdom."[56] This has the ring of sound advice and ought to have been appreciated by a government that had previously sought to maintain traditional lifestyles.

In the fall of 1955, the Ennadai area caribou hunt went badly. The failure was due in part to the lack of ammunition and the people's inexperience with rifles: they were still making the transition from spear and kayak hunting. To sustain them during the winter, the Northern Service Officer flew in rations by RCAF Otter aircraft, officially as relief. In fact, they were owed wages from work around the weather station and from family allowance credits. The argument developed among officials that relocation to the area of Henik Lake would enable better supervision of their hunting efforts to be carried out from the posts at Padlei and Eskimo Point, so efforts were made through the summer of 1956 to sell the Ennadai people on the idea of another move, but without success. However, that winter proved even worse for the group: their dogs were lost to starvation, and they had to be sustained by government-supplied food and buffalo meat.

In May 1957, the group of fifty-five was flown a hundred miles to Henik Lake and supplied with tents, ammunition, fishnets, and food for a month. For whatever reason, the spring migration of caribou was late, and the people broke up into several camps around their new district. Members from one camp became familiar with the advance party at the Bray Lake project of Sherritt Gordon Mines, where they reportedly traded fish for tobacco and tea. Actually they were befriended by P. Lynn, who gave them food on several occasions and reported their hungry state to the RCMP.[57] The field men at Bray Lake secured the summer's supplies of canned food and other goods and left them ready for the arrival of company prospectors after breakup. During their absence, on about June 6, according to statements collected by Tester and Kulchyski, the Inuit broke into the company's cache and stole or wantonly destroyed $1500 worth of supplies and equipment.[58] The outraged prospectors questioned Oohootuk, an Inuit hunter who was camped nearby. He admitted that he and others had taken the supplies. Later in the summer, the RCMP arrested Oohootuk and two other participants, Mounik and Iootna, and escorted them to Eskimo Point. There they were convicted of theft, sentenced, and imprisoned until late fall. In December they were still awaiting return to their camps. It was the mining company that pressed charges, encouraged and supported by the RCMP and the Northern Service Officer, who were all bent on setting an example to discourage theft among the Inuit. Much later, geologist Jim Marshall said to me that the field party was insensitive to the true predicament of the Natives, not knowing that they were displaced and suffering extreme hunger.

Bill Kerr, the Northern Service Officer (and ex-RCMP officer) who was responsible for overseeing the Ennadai people and supplying them with food, concurred "that the instigators and ringleaders of the theft should be prosecuted as an example to other Eskimos."[59] No one seems to have been concerned in the slightest that the convicted men were group leaders and that their presence might be essential to the success of the relocation and survival of the band. Though welfare was provided at Padlei to the families of the convicted, supplies were delivered, and a technical advisor (Louis Voisey from Eskimo

Point) was left with the band until September, their situation did not improve. Belatedly, a sixteen-foot canoe and more fishnets were provided to facilitate the fish catch at Henik Lake, but no large kills of caribou took place for winter caching of meat, and the meat from the animals that happened by was sufficient only for day-to-day needs. The Inuit were in disarray. They balked at fishing and perpetrated another break-in at the Sherritt Gordon camp. They were not meeting the expectations of the government because, as Tester and Kulchyski observed, they were demoralized about finding their life out of control and by the loss of their leaders. They were a hunting people and clearly unable to adapt to a fishing lifestyle.

So much for the "official" account of this tragic event. There is another version which deserves a hearing, as related to me by R.G. Williamson, and this impacts significantly on Inuit-industry relations. The exploration crew told the Inuit to help themselves to the company cache if they needed to, but to let them know via the RCMP. In a state of starvation, the Inuit took the mining people at their word, and soon after informed the RCMP. Both Oohootuk and Nanook were naïve, simple people but kindly and long-suffering. The RCMP arrested several Inuit, confined them at the Eskimo Point detachment, and forced them to do hard labor, although no evidence of a trial seems to exist. A rock chip hit Dohukuk, the husband of Nanook, blinding him for life. The men contended that the wanton destruction of supplies and equipment at the cache was the act of barren-ground grizzly bears. Descriptions of the event and the attitude of the Inuit were alleged to be colonial-minded expressions of Voisey and the officers involved.

While the mining company in this case was relatively unconcerned about Native welfare, there was evidence that other representatives of the profession were more perceptive. In this period, the Geological Survey of Canada (GSC) was conducting helicopter-assisted reconnaissance of the Keewatin district to speed up the mapping of this area of high mineral potential. Operation Keewatin was initiated in 1952 as a trial effort. The area was chosen because of its accessibility from Churchill, its barren ground (and hence suitability for helicopter mapping), and its geological merit, as determined by Tyrell, Weeks and others. A field party of sixteen mapped an area of 57,000 square miles, employing fuel caches at Ennadai and Ferguson Lakes and ten campsites. At the end of the season, Cliff Lord drew attention to the gold, nickel, and uranium potential of the district and reported recent prospecting by Donald Cameron, Canadian Nickel Company, Kasba Explorations, Rankin Inlet Nickel Mines, and Hudson Bay Exploration and Development. Geological reconnaissance of the same type was repeated in the Baker Lake and Thelon River blocks during 1954 and 1955.[60]

Before launching this mapping program, the GSC informed the RCMP and others of its intention to fly helicopters around the Barren Lands, and apparently the police notified Inuit groups of this, warning them not to fear or shoot at the helicopters. News of the latter injunction somewhat jarred the GSC staff. In any event, they were determined to make a favorable impression. They carried extra rations in their machines, which they gave away to Native groups that they encountered. Some of the Inuit they met in this way were indeed hungry – hungry for fat, which was lacking in the fish diet they had to depend on during the caribou scarcity. Regrettably, the geologists were not in a position to satisfy that need.[61] Probably the very survival of the Inuit in the harsh Keewatin district was not possible without caribou, which supplied protein, iron, vitamins, and winter clothing, as well as fat.

As for the impending crisis near Henik Lake in late 1957, the patience of the Northern Service Officers had been exhausted. They withheld help, on the assumption that the groups of Natives were periodically visiting Padlei Post to report their troubles and secure aid. But the relocatees did not do this, nor were they visited. In the department, the debate about relocations heated up with the growing awareness that these people had simply been moved from one depressed area to another.

The Ennadai group of fifty-five had broken up into four camps within a seven-mile radius. They did not have a single, full dog team among them, because most of their dogs had died the winter before. This limited their mobility, both for hunting and for going to Padlei Post. As the winter progressed, food became scarcer and scarcer in the igloos on the shores of Henik Lake, where even jigging for fish brought a scanty return. By late January, the people were in dire circumstances, and Ootuk's son Igyaka died of starvation. Ootuk, out of desperation and envy, killed Hallow, who was sitting stolidly at his fishing hole in the ice. Subsequently Kikkik, the wife of the murdered man, fatally stabbed Ootuk. Then she set out walking to the Padlei Post with her five children, a distance of about fifty miles. On the way she had to leave behind two ailing little daughters, whom she wrapped in caribou skins and covered with blocks of snow.

Kikkik was spotted by RCMP aircraft about fifteen miles from her destination after Yahah, her brother, had gone ahead to alert the Post. Though she thought her daughters would be dead, a police constable traced her track and found one child still alive under the snow. Kikkik was taken into custody, charged with murder, and later tried, only to be acquitted. The children were flown to Eskimo Point, and shortly afterward all the Ennadai people were evacuated under the supervision of the NSO. Other members of the band had died of exposure attempting to travel to Padlei, bringing the total lost in February 1957 to seven. The tragedy of the *Ihalmiut*, as Farley Mowat called them, was vividly portrayed in his 1959 book *The Desperate People*. The book, which elicited comment in Parliament and caused outrage among the Canadian people, was based on personal interviews, but the later recounting of the same events by Tester and Kulchyski, based on government documents such as departmental memos, told the same story.

The many government reports written about these events laid responsibility on the Inuit themselves, but the tragedy illustrated the profound disruption inflicted upon these families by interference from outside and by decisions of well-meaning officials not fully aware of the implications.[62] Further interviews by Walter Rudnicki led him to conclude that the results might have been predicted: "The Ennadai Inuit were relocated twice in seven years by the outsiders, and it is vital to remember that the decision of when to move and where to move was perhaps the most important decision nomadic hunters have to make. It is clear that they had lost control of their lives. But, inexplicably, the outsiders only sporadically came with food to offer. They made decisions but did not seem to want to accept responsibility. And when Inuit took from the caches left by non-Natives, they were arrested."[63] The Ennadai Inuit, inland caribou hunters for generations, were not a happy lot in Eskimo Point and were subsequently resettled in the new community of Whale Cove. A few found their way to Rankin Inlet, where they were regarded as inferior to the dominant coastal people.

Starvation also swept through the sixty-member Garry Lake band in the winter of 1957–58, resulting in the deaths of sixteen persons. The survivors were moved out to Baker Lake and eventually dispersed to the coastal locations, Rankin Inlet and Whale Cove. The same was true of lower Kazan River people, who were also having a consistently hard time because of the decline in caribou numbers.

In 1958, government officials selected the site of a new coastal settlement at Whale Cove to resolve the desperate problem of the displaced Keewatin Inuit, to capitalize on their skills, and to exploit the fish and marine animals and possibly the inland caribou. The site was selected without consulting the Inuit. In the winter of 1958–59, about 130 individuals came to Rankin Inlet from various points, and this prompted the decision to carry on with construction at Whale Cove in 1959. Thus a settlement arose with twenty-four families made up of fifty-two adults and thirty-seven children. Of these, forty-four people were from Eskimo Point, twenty-nine from Ennadai Lake, and sixteen from Garry Lake. Less-than-successful employees at Rankin Inlet found refuge at the new settlement. They lived in rigid frame houses and tents, served by a store, a school building, and a building for visitors. The settlement proved to be successful because it was well equipped and maintained as a simple hunting base, without too many white overseers. Louis Voisey from Eskimo Point, who assisted the Henik Lake group, was involved there. Many families with dominantly traditional background and skills were relatively contented there and were able to pursue their preferred lifestyle. In 1960, even more people from other places moved in. There was some sickness in the community, and cultural distinctions made for slow integration.[64]

GROWTH OF LEADERSHIP AND ADVOCACY

Community-development personnel in this period were searching for ways to encourage Native leadership. Propelled by the Diefenbaker government's northern development policy of 1957, nineteen northern service officers were taken on staff in 1958–59 and plans were made to hire eight more in 1959–60. Peter Murdoch, a former Hudson's Bay Company manager, was among those hired during the Department's expansion. In regard to Rankin Inlet and Whale Cove, he said, "Unless we provide a catalytic influence, these leaders will not be forthcoming. Generally, camp leaders were accepted because of their amiability and lack of aggression. This type of leader was ideal in a situation where everybody knew what to do and how to do it. But in a situation as exists in Whale Cove or Rankin Inlet, this type of leader is ineffective. If the re-settlement programme is to mean anything more than survival for these people, we must begin a progressive programme of human as well as resource development."[65] The will seemed to be there, but the accepted solution, as in the relocation movement, was to call for ever-more interference by officialdom. For example, efforts were made to encourage the establishment of cooperatives dealing with crafts and services.

Much of Frank Vallee's research was a study of the fringe population and their relationships in Keewatin communities. He noted that Whale Cove was a model place to live because it had the fewest white men or bosses. Elsewhere, the whites were normally in charge of every activity and were busy trying to help the Inuit help themselves. Meanwhile, the Inuit were busy trying to save themselves from being helped! Vallee was quite impressed with the social mix occurring in Whale Cove because of the lack of supervision. The positive atmosphere there, and the "caste system" at Rankin Inlet, tended to propel people to Whale Cove.

The massive involvement of northern service officers, welfare officers, and social workers, as well as teachers, nurses, community development workers and researchers, was an attempt to address the terrible problems of the Inuit. In the main, these workers followed a policy of assimilation or integration, inspired by liberal ideas of equal rights for all Canadians. This did much to confuse Inuit cultural practices, family traditions, and idea systems, yet the expanded intervention or invasion of whites, "combined with the tenacity with which Inuit continued to express their cultural preferences, eventually contributed to the creation of new structures that gave Inuit a voice – even if it was not always a voice the state wanted to hear."[66]

In 1959, for the first time, the Government's Committee on Eskimo Affairs included invited Inuit representatives. Abraham Okpik from Aklavik, George Koneak from Fort Chimo, and John Ayaruark and Shingituk from Rankin Inlet were present at this watershed meeting, along with politicians and Department officials and Prime Minister John Deifenbaker. Even Farley Mowat was in the audience, as were Mary Panegoosho and Elijah Menarik, Inuit interpreters and advisors on Rudnicki's Welfare Division staff, which for the past year had started making family allowance payments to Inuit by cheque. Committee membership had in previous years been confined to the deputy minister of Northern affairs, the chief superintendent of the RCMP, church bishops, the head of the HBC fur trade department, and various civil servants. Language interpretation had been a problem to participants.

In 1959, the encouragement of more community councils and the gradual withdrawal of NSO control were at the core of government proposals for self-determination. Koneak was familiar with Greenland practices for the direction of self-governing institutions and hailed them as far advanced over Canada's. Okpik said existing councils had not accomplished much to date, and this sparked a discussion of aims and purposes. Councils were to be *guided* along the path of problem solving until they could exercise control; first, they had to master the white man's procedural structures, and then they could take over. They would concentrate on local initiative and action. The committee agreed with Okpik that education should be of prime importance, and that mixed councils might not be the best way to achieve this. The Rankin Inlet delegates expressed their support of the Rankin Inlet nickel mine; Andy Easton, the mine manager, also attended the meeting. But Okpik pointed to the lack of vocational training (in Aklavik and elsewhere) and wondered if the government "intended to make idlers out of the Eskimo people or to make them productive citizens."[67]

On the subject of "changing society," Ayaruark made a long and pointed speech about the right of Inuit to exercise control over their own destiny, referring to the confused and unsettled environment at Rankin Inlet, especially for inland people. He urged that an organized, cooperative attempt be made to integrate traditional and wage economies, whereby some people would hunt for the benefit of the employed and be compensated by them. If this were accomplished, workers would not be ambivalent and despondent over their inability to hunt, and employers would not consider them lazy. Ayaruark referred to other matters, such as drinking, uniting families, the disappearing caribou, and the need for better boats, that spanned the whole range of concerns of both traditional and settlement residents.

Shingituk, who had become embittered and disenchanted with white authorities and moved to Whale Cove, questioned Easton about the impact on Inuit workers when the nickel mine closed down, and was informed that trained men could be employed elsewhere. Other new mines were expected in the Northwest Territories, according to Gordon Robertson, Deputy Minister of Northern Affairs and National Resources.[68]

The invited Inuit participants at the Government Committee meeting turned out to have a clear perception of the problems facing their people and held opinions as to their solution. A process of Native advocacy was thereby launched (unintentionally), and Ayaruark and others at Rankin Inlet took immediate steps to follow up: they petitioned the minister about medical care, much to the annoyance of administrators. That they were determined to go right to the top was not surprising, since they had now met the minister – and the prime minister. The reason for the petition was not an idle one. The infirmary was run by North Rankin Nickel Mines in a building less than adequate considering the demand, for limited hours during the day, with a doctor who lacked an Inuktitut interpreter. The Indian and Northern Health Service had already been pressuring the company to upgrade the facility.[69]

Community councils operating in Inuktitut became the initial vehicles for Inuit self-advocacy in the settlements. Doug Wilkinson had reported on the operation of the Eskimo council at Baker Lake in 1957. Only one or two problems were solved during its first year of operation, but "The Eskimos took more interest in their own economic position. The Council had a disquieting effect on the non-Eskimo residents and started them thinking about their community, albeit in a negative way, as they never had before."[70] Wilkinson and other Northern Service Officers of the day confirmed that the councillors had a token consultative or advisory role and had no real power, especially to spend government money. The Northern Administration Branch policy directive stated that one local group could not make decisions about spending federal (taxpayers') funds. Nevertheless, the government admitted that community councils and local consultation were beginning to have a real purpose, as in the South.[71] Part of that purpose was the integration or assimilation of Native people into the national fabric. By 1962, there were ten councils in Arctic communities, and leadership was a question for debate. There was rivalry between the patriarchs or older hunters and Inuit of mixed blood as to who should speak for the group. Younger, better-educated Inuit were bridging the gap between the old and the new. Few of them would attempt advocacy on anything more than their own interest, with the exception of people like Abe Okpik, Simonie Michael, and John Ayaruark, who had become effective leaders in Rankin Inlet. In general, the administration frowned on the selection of a traditionally minded (*Nunamiut*) person, even though most Inuit might have confidence in him. Increasingly through the 1960s, this type of person was able to voice opinions and perspectives that went counter to the policymakers' views. In this way, the Inuit asserted that they were not prepared to be integrated or assimilated.

There was no more eloquent a spokesman in 1962 than Abraham Okpik, who published in the Inuktitut magazine a powerful essay entitled "What Does It Mean To Be An Eskimo?" He asked his fellows to treasure their heritage of survival, their culture, and their language. He concluded, "There are only very few Eskimos, but millions of whites, just like mosquitoes. It is something very special and wonderful to be an Eskimo – they are like the snow geese. If an Eskimo forgets his language and Eskimo ways, he will be nothing but just another mosquito."[72]

This was a particularly unpleasant time in Keewatin history. The mine closure at Rankin Inlet had caused dislocation and uncertainty, to say nothing of the public welfare burden. Many men were incapable of going or ill-equipped to go onto the land, and alcoholism was taking a toll. Communities along the Hudson Bay coast, including Rankin Inlet, Whale Cove, Eskimo Point and Churchill, were overcrowded and dirty. Some said conditions were appalling. In late 1962 and early 1963, an epidemic of tuberculosis swept through Eskimo Point, affecting over half the people in this crowded eighty-two-household community. It was therefore timely for Inuit residents to speak out about their lives and their future. The claim that their forefathers had provided a life as good as what modern Canadian society had to offer was easily substantiated.

In the opinion of Tester and Kulchyski, who ended their extensive review of archival material with 1963, this was a watershed period. They concluded that the voices of dissent from establishment views that came to light in the community-development era of the middle 1960s were the forerunners of the drive for self-government. The new generation of Inuit leaders would become vociferous opponents of integration or "totalization" of the Inuit people into Canadian society, with all the contradictions that the concept conveyed. The policy makers had never sought or encouraged Inuit self-definition: the people were taking that into their own hands.

2. EXPLORATION PRESSURES

EARLY GEOLOGICAL EXPLORATION AND PROSPECTING

Powerful trends of the 1920s, such as booming metal markets, ease of financing, government geological mapping, dissemination of mining news, and adoption of bush aircraft, enabled a rapid penetration of prospectors into the North to take place. This new vigor was not accompanied by a greater connection with Native people – quite the reverse was true.

In August 1927, the federal government announced that the terminus of the Hudson Bay Railway, awaited since 1912, would be constructed at Fort Churchill and issued instructions to expedite construction of the railway and port.[1] Indeed, this extension of the Canadian National Railway (CNR) system reached its destination in 1929, and the first shipments of prairie grain began in 1931. The Churchill rail development and the ease of access it provided to the North ushered in a new era of mineral exploration, since it was known from the pioneering work of the Geological Survey of Canada (GSC) that a great expanse of the Canadian shield lay immediately north and west of Churchill. Joseph Tyrell and his brother James first explored the central Keewatin by canoe in 1893, traveling the Dubawnt and Thelon Rivers through Baker Lake to Chesterfield Inlet and thence to Churchill. The following year, Joseph traversed the Kazan and Ferguson Rivers. In summarizing the work of this Canadian pair, Farley Mowat said, "The achievements of the party were spectacular. Not only did they make a tremendous journey without loss, but they brought back with them the first really accurate and detailed description of the interior of the Barren Grounds. Joseph's geological and glaciological survey of the route remained the standard by which studies of the arctic plains were measured well into the 1940s.... Between them, the Tyrell brothers contributed more to our knowledge of the arctic plains than any other men, before or after their time."[2] Their success was largely due to James's facility with the Inuktitut language, which allowed him to communicate with Natives along the route.

In 1929, L.J. Weeks and his geological party commenced a Keewatin mapping program at Mistake Bay, at the mouth of the Ferguson River. Lud Weeks was already an experienced northern mapper, having both served on the GSC patrol vessel *Arctic* in 1925 and wintered at Pangnirtung in 1926 (where, incidentally, he suffered a severe shotgun wound while hunting ducks). Weeks described his Keewatin mobilization as follows: "With completion of the Hudson Bay railway, access to Churchill at the mouth of the Churchill River is made quite easy. The distance from Churchill to Mistake Bay is about 225 miles. For this trip a forty-foot Peterhead trap boat was used, carrying the personnel and summer's supplies. The trip was made in forty-eight hours running time."[3] The party set up camp at Tavani, where Dominion Explorers Limited already had a winterized base camp. In the vicinity of the fieldwork, two large groups of well-financed and well-equipped prospectors were doing reconnaissance by canoe and airplane.

Anticipation of the completion of the railway and the likely flood of small prospectors had indeed prompted action on the part of well-heeled operators, since the general condition of the mining industry in 1928 was very prosperous. In the race to corral the mineral riches of the Subarctic, prospecting firms sailed schooners with men and supplies into Hudson Bay. Dominion Explorers brought in the *Morso* to establish winter camps and supply an aircraft, and the firm NAME (North American Mine Explorers; Jack Hammell, president) bought the *Patrick and Michael* and used it to place prospectors inland by airplane. The ships arrived after breakup, which was relatively late that year. Thus, the prospectors had to prepare camps and plan for winter survival.[4]

With an eye to small prospectors in particular, Weeks pointed out the hazards of working in this remote area and the fact that game animals (walrus and caribou) were protected by law. His advice to aspiring prospectors showed a sense of excitement over this area as a new northern frontier for exploration. Weeks noted that a belt of volcanic rocks lay along the Ferguson River as far as Kaminuriak Lake and observed mineralization in the form of pyritic quartz veins there. In a 1932 publication, he reported on three seasons of work and produced a regional map. His descriptions included the dominant rock types and mineral showings. Most important was his delineation of the Archean volcanic-sedimentary belt that extends from north of Rankin Inlet, through Kaminak Lake, and beyond to the upper reaches of the Maguse River. In fact, he noted that J.B. Tyrrell had observed similar rocks in the vicinity of Kasba and Ennadai Lakes, much farther to the southwest. Thus Weeks revealed the widespread if not continuous occurrence of these rocks and stated, "As is well known, such rock assemblages in other parts of the Canadian Shield are the home of metallic deposits and, therefore, this block of country deserves at least preliminary exploration as a possible mineral producer."[5]

In the same report, Weeks described some of the current activities of mining companies in the area. At Rankin Inlet, where Oscar Johnston had discovered nickel and copper in 1928, the Cyril Knight Prospecting Company had started diamond drilling in 1930. Under the guidance of John Dryborough, a substantial steel boat and drilling equipment had been shipped by rail to Churchill for transport to the property. The drilling was accomplished by the innovative use of an on-site water heater to help penetrate the permafrost that enveloped the rock. In this way, core samples were extracted from a body of serpentinized igneous rock two hundred to three hundred feet thick. Sulphide minerals rich in copper, nickel, and traces of platinum occurred at the base of the body. The serpentinized igneous host rock was considered to be a pyroxenite intrusive body or lense occurring within the Archean volcanic-sedimentary rocks cited above.

The Nipissing Mining Company was actively working claims on Term Island (near the later coastal community of Whale Cove), where they had discovered sheared volcanic rock bearing free gold. The trench that was excavated for sampling showed silicified and quartz-bearing greenstone accompanied by pyrite and chalcopyrite.

Considerable prospecting was done by other groups along this coastal area and along the Ferguson River. Claims were staked on many showings of silicified shear zones carrying sulphide minerals. Free gold was panned from crushed rock along the upper parts of the Ferguson River. Dominion Explorers Limited had been actively prospecting along the Maguse River system and north of Rankin Inlet.

In 1932 the same author, L.J. Weeks, ascended the Maguse River and traveled to Kaminak Lake on the Ferguson River to further document the local geology. The trading post at Padlei on Kingarvalik (Kinga) Lake was then in service, having used aircraft for the first time (rather than the Maguse River canoe route) to replenish supplies. Bush aircraft were now servicing the traders as well as the geological and prospecting parties.

Using aircraft to explore the North and set down prospecting parties in remote sites was quite new, and it played an enormous role in the mining industry. As pioneer aviator Punch Dickens noted, 1928 was a boom year in the growth of commercial aviation in Canada. The Winnipeg-based Western Canadian Airways had already completed a large-scale freighting contract into Churchill while the railway was still under construction. They bought an improved bush aircraft, the Super Universal Fokker, equipped with a four hundred-horsepower Pratt and Whitney Wasp engine.

Punch Dickens piloted this aircraft (G-CASK) on a pioneering flight over the Barrens. He left Winnipeg on August 28, 1928, with engineer Bill Naden, editor of *The Northern Miner* Richard Pearce, and mining man Colonel C.D.H. MacAlpine on board. Colonel MacAlpine, president of Dominion Explorers Limited, had chartered the flight to reconnoiter the country, inspect his prospecting parties already in the field, and apprise himself of conditions in the area. Dickens was admittedly worried about the weather and the lack of fuel and good maps in some areas west of Baker Lake. Five days after departure, the party was flying up the coast from Churchill in good weather. In two-and-a-half hours, they landed at Mistake Bay, where the Colonel's prospecting party was unloading supplies from a coastal schooner for transport ashore. The passengers stayed at their camp and the next day flew some geologists north to Rankin Inlet, after which they flew inland to look at the Barrens on their way back to camp. As the weather was good, they left on September 2 for Rankin Inlet and Chesterfield Inlet en route to Baker Lake, which was then a settlement of two trading posts, an RCMP detachment, one missionary, and two Inuit families – about twenty people. The rest of the flight included 550 miles of unexplored territory along the Thelon River and Dubawnt Lake. Using an extra couple of ten-gallon kegs of aviation gasoline – and one tea-break to load it – the party easily reached Stoney Rapids in northern Saskatchewan. While there, they searched for and found the Dominion Explorers prospecting team that had made an arduous eighty-mile, two-week canoe trip from Stoney Rapids. Punch Dickens flew back from their camp in forty-five minutes, "and the Colonel was thoroughly sold on the bush airplane." After visiting Fort Smith and Cold Lake, where the Colonel inspected a Sherritt Gordon mining property as Company director, they were back in Winnipeg on September 9, having flown 3,956 miles in forty air-hours.

Fellow passenger Richard Pearce described this flight with lengthy comment and photographs to show how "in touch" *The Northern Miner* was with events of the day.[6] His headline was "Northern Miner First to Fly Across Canada's Barren Lands." It was a newsworthy item, but Pearce was not able to report any significant findings by Dominion Explorers, North American Mine Explorers (NAME), Cyril Knight Prospecting, Nipissing Mines, or other organizations working in the area. This did not dampen the ardor of explorationists, who recognized the need to go over the area thoroughly and systematically using base camps and aircraft. They foresaw many years of activity in this large Subarctic region. In arguing for this kind of organized prospecting, Pearce said, "The experienced prospector will find his bush sense of little use to him, for when he gets beyond the tree line he might just as well leave his axe behind. His gravest difficulty will be to get fuel, and he will probably have to use a primus stove, which burns gasoline, not a very satisfactory substitute for the bush fire he is accustomed to." Unless he is prepared to "live like an Eskimo" he is likely to have trouble. Supplies are hard to come by, and "traveling the coast in small boats is none too safe, wind and tide being troublesome. In contrast, the Dominion Explorers' schooner had one-third of its cargo as coal, and elaborate insulated-wall buildings were put up at Mistake Bay, where a crew of thirty men were wintering. Fortunately their bush aircraft, now fitted with skis, brought back fresh supplies at Christmas time."[7]

As a footnote to the extraordinary 1928 season, J.B. Tyrell, who was by then a successful mining executive, said he could see his hope fulfilled. He was awaiting results from the summer's exploration of the Barren Lands with interest. In his geological mapping of the area thirty years earlier, he had noted sulphide mineralization in volcanic rocks, similar to that in Quebec, but it was largely copper sulphide. He stressed the copper potential of the area – and it so happened that copper developments elsewhere (e.g., at Noranda in Quebec and Flin Flon and Sherridon in Manitoba) were much in the news.

The use of bush aircraft for prospecting, now an established practice, introduced an important change in exploration. No longer did prospectors make contact with Native people or trappers as they entered the bush: they could now overleap and pretty much ignore local residents in areas they selected to explore, especially when they were well organized and financed. This had the effect of depriving prospectors of valuable advice and help that local people might offer. Of course, there are many counterexamples; there was also a positive side to this trend of diminished contact. Whites were sometimes carriers of infectious diseases, to which the Natives had little or no resistance. A case of this occurred in 1928, when the Dominion Explorers schooner *Morso* picked up four Inuit families at Wakeham Bay in northern Ungava to act as guides. All of them fell sick with influenza and survived only through the diligent efforts of G.H. Blanchett, head of the expedition. The same epidemic was reported to have caused forty deaths at Wakeham Bay.[8]

Although the mining boom of the 1920s stimulated unprecedented exploration and interest in frontier areas, the stock market collapse of 1929 and subsequent Depression had the opposite effect. Some of the programs that had already begun, such as the diamond drilling at Rankin Inlet, carried over to 1930 and were briefly resumed in 1936, but in general, exploration in remote areas like the Keewatin stagnated through the Depression and World War II.

GROWTH OF EXPLORATION IN THE 1960s

Through the 1960s and beyond, economic conditions, technology, and government policy controlled the type and pace of prospecting activity in the Northwest Territories. On the whole, the industry was on a growth trend, but there were some setbacks in the form of recessions. One such setback was the downturn of 1961, which caused falling nickel prices and resulted in the closure of the North Rankin Nickel operation in 1962. Other recessions that put the brakes on industry expansion were recorded in 1970, 1975, 1981, 1983, and 1991. The following review of prospecting and mineral highlights from an industry perspective was drawn from such sources as the *Northern Miner*[9] and the *Canadian Mining Journal*.[10] Much of this activity, which afforded little or no local benefit, provided a backdrop to growing Native restlessness and self-expression as their traditional land base appeared to be declining. Inuit reactions to exploration and development are discussed in the following section.

The industry's drive to stake claims, survey, and occupy the land for mineral exploration and development was vigorous and irrepressible. Why? Economic conditions and capital formation, notably stock market activity, were buoyant. Belief in the yet undiscovered mineral wealth on the Canadian frontier was a powerful magnet. The laws of the provinces and territories provided a legal framework for exploration that was open, straightforward, democratic, and encouraging. Rooted in the ethos of "free entry," the regulations allowed prospectors to go anywhere, secretly if necessary, to explore Crown land. Governments encouraged such exploration and even provided incentives, in the belief that economic progress would result. Lastly, the newer aircraft and advancing technology made exploration immensely easier and more thorough, but also more costly. Technology could reveal a great deal about huge areas of rock, rapidly and remotely.

In 1961, in response to the global iron-ore boom, the Belcher Mining Corporation and others found noteworthy deposits at Great Whale River. Exploration in Ungava was especially upbeat, as the Province of Quebec had reopened this vast terrain to ordinary claim-staking. (The area had previously been closed to staking and was made subject to block licenses in 1955.) Particular areas of interest were the Le Moyne nickel find and the asbestos discoveries (Asbestos Hill) in the Wakeham Bay–Cape Smith belt, the former having received about $1 million worth of work by Asarco. The active asbestos property of Murray Mining, named after discoverer and famous Arctic prospector Murray Watts, passed into the hands of Raglan Nickel Mines. The Asbestos Hill project was soon to hire unemployed Inuit miners from Rankin

Inlet. However, the industry's interest was turning increasingly to gold and silver, since oversupply and low prices tended to plague the market for base metals and uranium. Because of the low uranium demand, producers participated in "stretch out" programs, which slowed production down to extend the life of the operation. Oceanic Iron Ore engaged in ambitious exploration at Payne Bay Ungava in northern Quebec. Arduous dog sledding from Justin Lake to Payne River with local men was necessary to serve extended winter programs in difficult conditions.

The use of geophysical surveys, especially the airborne variety, for mine finding was becoming popular in the light of the successful system developed by McPhar Geophysics for International Nickel Company (INCO). Industry was pleading with government for more ready acceptance of geophysical surveys. Proposed revisions of assessment regulations in both Saskatchewan and the Northwest Territories would give credit for airborne geophysical surveys conducted before claim staking. This acceptance of work done without being physically on the ground would increase the separation of the explorer from the land. The new 1961 Canada Mining Regulations (applicable to the Northwest Territories) were meant to discourage foreign ownership but encourage exploration. The regulations were drafted after wide consultation with industry over a two-year period. They included generous claim tenure and lease provisions. In addition, a new form of large-scale land acquisition was created under a prospecting permit system for areas not being actively explored. Also of great interest was the mining incentive offered: a three-year "royalty holiday" upon the start of production.

New radio communications systems were put in place, much to the delight of mining companies operating in remote areas. There was talk of the critical role of transportation in mine development and the importance of a government-funded, 430-mile railway to the Pine Point lead deposits of Cominco. Inuit were later employed in construction and operation of the publically funded line (whose life span, as it turned out, was a mere two decades).

In 1962, the upswing in activity was widespread, and every part of the country participated. Within the northern frontier this applied particularly to Ungava, where activity was spurred on by base-metal finds made in the previous year in the Fort McKenzie area. INCO resumed work at Contwoyto Lake, where a discovery (eventually to become the Northwest Territories' most important gold mine) had been made the year before. Companies with geophysical capability were targeting base metals or nickel because of their strong magnetic signature, but on-the-ground field observations were turning up iron formations bearing gold mineralization, as at Contwoyto Lake. Giant

Yellowknife Gold Mines sponsored an active program in the territories. Selco Exploration, a British firm with geophysical equipment mounted in a Canso flying boat, was spreading its surveys northward from Manitoba.

Lest the modern techniques and sustained effort by major companies displace the little operators, independent prospectors were encouraged by a new assistance plan offered by the Department of Northern Affairs and Natural Resources in the Yukon and Northwest Territories. In the first year, the territories supported thirty-five two-man prospecting parties, twenty-three in the Northwest Territories and twelve in the Yukon, by paying half the field expense for sixty days of fieldwork, to a maximum of $2000. The GSC conducted helicopter-supported mapping on the Boothia Peninsula and King William, Somerset, and Prince of Wales Islands, as well as on Axel Heiberg and Ellesmere Islands.

In 1963, a rich iron discovery in northwest Baffin Island was announced by Murray Watts, president of British Ungava Exploration, a consortium of large companies. This was a fitting discovery for Watts, who had persevered as a prospector in the remote North since first wintering in northern Ungava in the early 1930s. He had learned much from local Inuit and quickly hired them to assist his field crew. The Mary River iron ore deposits were described as large and very high grade, running as much as 71 percent iron. Elsewhere, scientists concentrated their efforts on the Muskox Intrusion, an igneous rock with presumed metal potential, located about fifty miles south of Coppermine. There, the government contracted to drill very deep scientific core holes, perhaps the deepest ever, to promote research and development for nickel and platinum. Airborne magnetic surveys sponsored by governments were gaining favor and attracting technical interest in their interpretation. INCO returned to do further drilling on its promising Lupin gold discovery at Contwoyto Lake and also to explore the nickel copper occurrence at Ferguson Lake in the Keewatin district. The gold discovery caused a rush of staking interest by such operators as Rayrock Mines, Giant Yellowknife Mines, Falconbridge Nickel Mines, and Cominco. As an added incentive, J.A. Fraser of the GSC reported on bedrock geological studies made by helicopter in the Bathurst Inlet region. Roberts Mining was to carry out multi-year explorations, which would lead to the Hope Bay silver and gold discoveries.

The Pine Point lead deposits south of Great Slave Lake were now being developed for projected production at 5000 tons per day upon completion of the railroad in 1966. These deposits, known since the 1890s, were first appreciated in the 1920s from the geological work of MacIntosh Bell, who said they were similar to the important Tri-State or Mississippi type of bedded lead deposits. They were explored further after 1946, when geologist Neil Campbell shrewdly hypothesized that faulting from the Yellowknife area underlay them. Canadian uranium reserves were rated as the largest in the world (though much of the ore was low grade), but operating mines were still having to cut back because of pinching markets. Nevertheless, predictions of a vigorous nuclear power industry in the 1970s, especially in Europe, served to arouse interest in the accumulation of dedicated uranium reserves through exploration.

In 1964, Texasgulf Inc., a large sulfur and base-metal concern, confirmed a major zinc deposit on Baffin Island, where it had been drilling for the previous two seasons. This deposit was destined to become the Nanisivik mine. The huge sedimentary iron deposit of Crest Exploration at the Snake River, Yukon, was also capturing the imagination of the industry. Though considered lowgrade at 46 percent iron, the deposit was 323 miles long and up to 330 feet thick. The Vangorda Creek base metal deposit was being explored by Kerr Addison Mines, and Yukon silver properties in general were active. Baffinland Iron Mines (formerly British Ungava Explorations) raised finances to carry out drilling at Mary River, which required moving supplies seventy miles inland from the Baffin Island coast. Selco continued to explore in its sophisticated and quiet way (now termed "scientific prospecting"), searching for gold in the Barren Lands. Much of the current exploration work was technical and well financed.

Meanwhile, the Prospectors and Developers Association of Canada (PDAC) was studying why individual prospecting was declining, especially in the remote and expensive North. They honored Albert Zeemel, whose discovery in 1953 of the Gunnar uranium ore body was the last major discovery by an individual with simple tools (a pick and a Geiger counter). Government-financed airborne magnetometer surveys were gaining industry acceptance and inching northward. Jointly financed by the province of Ontario and the GSC, the latest aeromagnetic survey was to cover a tract of 32,000 square miles bordering on James Bay. Though geophysical anomalies might emerge from this and other surveys, the painstaking follow-up of prospectors and field workers was still required to pinpoint mineralization on the ground.

In 1965, the new CNR railway and promising geology at Pine Point made this area look interesting to Conwest Exploration, holder of seven hundred claims. Pyramid Mining Company was also active, and in fact succeeded in making a discovery that would eventually be sold to Cominco and justify a mill expansion (in 1967). Iron production increased with the start-up of the big Wabush mine in Labrador. Work by Baffinland Iron continued with the support of a new partner, Hudson Bay Mining & Smelting. Reserves at Mary

River were put at 187 million tons, averaging 69 percent iron, and attention was turning to the feasibility of shipping the ore by sea, with the aid of icebreakers. Consulting geologist J.D. Bateman, who had just died in 1965, was honored for his part in organizing the Panarctic consortium to explore for oil in the Arctic. The consortium drilled the first well – a dry one – on Bathurst Island in 1964, and set about to accelerate the program.

For the first time, an Inuk made the mining news, as a consequence of the booming activity around Contwoyto Lake. Noel Avadluk, now a retired resident of Coppermine, was one of many Inuit who staked claims and transferred them to various companies in the staking rush. He spoke English and had worked for the RCMP as a special constable. He was self-taught in minerals and geology, which enabled him to carry out technical and physical tasks appropriate to the work, aided by his traditional land skills and the help of his wife and partner, Angela Omigluk. Avadluk engaged in winter prospecting, which was usually not attempted by white field workers, sometimes along with George Turner, under grubstake arrangement with private companies. In the winter of 1964, Avadluk had made several gold finds for Roberts Mining Company in the Bathurst Inlet area, and follow-up work on claims was being undertaken. Individual prospectors' interest in the North was actually at a low ebb as a result of alternative employment. From 1964 to 1965, memberships in the PDAC declined from 1,060 to about 400, and only a hundred attended the annual meeting in March 1965. Viola MacMillan resigned after many years as the association's president, and shortly afterward became legally embroiled in the Windfall affair, a notorious instance of stock manipulation. The mine-financing business, the lifeblood of junior exploration companies, has its unseemly side. A new securities act was passed by the Ontario legislature in 1966, and the Ontario Securities Commission (OSC) began its overseer role in 1967. The GSC continued its Arctic mapping, which included the northeastern district of Keewatin and the Melville Peninsula.

The biggest 1966 news was the start-up of the Pine Point lead mine, endorsed by government ore and concentrate export licenses and federal financing of the new railroad. The government also financed a feasibility study by Canadian Bechtel Ltd. of a possible lead smelter for the same area. So as not to ignore the "little man" in its drive to boost mining in the Yukon and the Northwest Territories, the government offered to provide forgivable cash loans of up to 40 percent of approved exploration expense. Focusing on the risk and high cost of northern exploration, and despite the high level of activity and the buoyant economy, the government broached this loan plan without consulting industry. It was proposed as a substitute for income tax measures available to other members of the industry, who appropriately would not qualify for these loans. Mining was clearly seen as a revenue source and means of bolstering the deficit economy of the territories.

The original discovery of Pine Point mineral galena was reportedly made by Natives of the area, who learned, possibly from traders, how to smelt this mineral for homemade buckshot. A tribute was also voiced to Ojibway guide Moses Fisher, who in 1930 took Ontario government geologist J.E. Thompson to an area of "burned ground" or rusty sulphide gossan that would become the Manitouwadge area's Geco base-metal mine. Fisher staked the Geco mineral showing in 1943, but he had to let the claims lapse because he could not get anyone interested.

By the 1960s, space-age metal demand for columbium and molybdenum was broadening the targets for mineral exploration, in an already buoyant global market picture. The British Columbia copper boom was on. The Cantung tungsten mine on the Yukon/Northwest Territories line, to which seven Rankin Inlet men had journeyed for employment, started up production. By now, such was the reputation of Rankin's Inuit miners that companies, including Cantung, hired men with no mining experience. Loneliness for family drew them all home in about nine months. Prospectors' lectures of the day included technical training and information on territorial developments at Pine Point, Mary River, and Dynasty, as well as new information on GSC reconnaissance mapping in the northern district of Keewatin. Iron expert G.A. Gross said the high-grade Mary River deposits were of major significance.[11]

By 1967, uranium exploration activity was accelerating. A modest rush in the St. Simeon area of Quebec was based on radioactive pegmatite occurrences along the contact between granite and gneiss. Airborne scintillometer surveys were among the tools used to identify uranium concentrations. The rush spread quickly all along the north shore of the lower St. Lawrence, with a hot spot at Johan Beetz that involved many participants. In line with the renewed interest in uranium, Gunnex (the wholly owned subsidiary of the successful Gunnar Mining company) made plans for a new exploration program in the previously worked Beaverlodge area of Saskatchewan. Geophysical survey contractors were fully engaged, and there were calls for more radiometric survey capability. Canadian Aero Service and McPhar Geophysics, for example, were purchasing scintillation counter equipment. The Panarctic Oil consortium, organized and directed by J.C. Sproule & Associates of Calgary, added two mining companies, Cominco and Bankeno, to their group. The current private financing of the group totaled $15 million, but an expenditure of $30 million directed at the High Arctic over a five-year period was proposed. Nationally, there was growing awareness of the manpower shortage. The Canadian Mining Association (CMA) became a strong advocate of more immigration. It was noted that the Canadian "million dollar club" of exploration company budgets now numbered more than a dozen. Nevertheless, these major participants recognized the crucial role played by junior companies, syndicates, and grubstakes in the search for new mines.

The Northwest Territories experienced considerable activity. Selco Explorations continued its substantial program of gold exploration at Cullaton Lake in the Keewatin, and also pondered the past season's prospecting along the Arctic coast. The belts of volcanic rock north of Great Slave Lake were being targeted for base metals, and those programs were taking precedence over gold exploration. With the encouragement provided by improved transportation facilities, prospectors were moving in earnest into remote areas like Coppermine and Baffin Island.

The 1967 report of the Royal Commission on Taxation, led by Kenneth Carter, caused consternation in the mining industry. Its recommendations to eliminate the royalty holidays and capital-gains exemptions enjoyed by the mining industry had the potential to dampen activity.[12] In addition, independent prospectors recognized that the times were changing. The "old time" prospector, who was limited to scratching the exposed outcrop of the land (about 10 percent), was an endangered species; he could not afford to chase the new frontiers (for example, in the Arctic) or employ geophysics to penetrate the overburden. Shortly, however, there would be help for the prospector in the form of a one-man electromagnetic device, the EM-16, invented by Valno Ronka and manufactured by his Toronto firm, Geonics. The device used the global submarine low-frequency signal system of the U.S. military. The new airborne proton magnetometer was another effective instrument to scan beneath the overburden. A popular Canadian application was marketed by Barringer Research of Toronto, whose helicopter system was being deployed to the James Bay and Hudson Bay areas. Helicopters such as the Bell 204 were being used in the Yukon and elsewhere as field workhorses, lifting drilling rigs and drums of fuel over stretches of impassible muskeg and tundra. Exploration for minerals had become high-tech and costly, and needed well-heeled companies.

The Geological Survey of Canada *Report of Activities* was eagerly read for geological and mineral hints that might form the basis for prospecting programs. Details were accumulating on the Ennadai-Rankin greenstone volcanic belt, and the Hurwitz formation conglomerate in the Keewatin was being studied and sampled as an analogue of the Witwatersrand (South African host rock for gold) and the Elliot Lake (uranium-hosting) conglomerates. The

Coppermine River basalt flows (lavas) were explored for copper-rich veinlets amid enthusiastic claim staking. The Itchen Lake to Contwoyto Lake map area was investigated. The federal government adhered to a philosophy of greater growth in the North and administered this under the resource and economic development group of the Department of Indian Affairs and Northern Development (DIAND), formed in 1966. Programs stressed were the Roads to Resources policy, which helped finance roads in the Yukon and Northwest Territories, and the Northern Airport Policy, which provided similar aid for airstrip construction and production airport facilities. In addition, the Northern Mineral Assistance Regulations offered exploration assistance grants covering up to 40 percent of expenses to Canadian-controlled ventures by individual prospectors or corporations not eligible for incentives under the Income Tax Act. In fiscal 1966–67, this program attracted twenty-seven applications, and a total $340,000 was granted to the six projects approved.[13]

Economic studies of various types and development analyses were conducted by DIAND's Economic Development Branch, and the department continued its role in land management, claim recording, and mine safety. The prospecting fraternity had become quite used to its privileged status: with the lone exception of Toronto mining engineer Larry Labow, who thought windfall profits should be taxed, prospectors defended and justified their tax incentives and supported immigration policies that would help provide needed labor.[14] For the first time, representatives of major mining companies joined the PDAC, previously the province of the "little guy," and their influence was to change the character of the Association and revitalize it. Optimism pervaded uranium circles, as commitments to nuclear energy were growing – nearly half the twenty-one American orders for new steam plants were nuclear, according to Robert Nininger of the Atomic Energy Commission. In an address to the 1967 PDAC convention entitled "The Outlook for Uranium and Atomic Energy," he had urged unflagging persistence by prospectors to determine the true value of uranium areas.[15] Geologist Stanley Davidson, who died in 1967, was hailed as the father of airborne electro-magnetics (EM), having first experimented at McPhar Geophysics in 1948. His invention led to the INCO electromagnetic instrument that was credited with the base-metal discoveries of Heath Steele in New Brunswick and Thompson in northern Manitoba. Canadian geophysicists were becoming world leaders in the field. The Anvil lead–zinc–silver mine at Faro, Yukon, received the production go-ahead following a feasibility study. The Discovery gold mine north of Yellowknife, small by comparison, was doing well.

In 1968, the growing demand for minerals dominated the American Institute of Mining Engineers (AIME) convention in the United States, where it was well known that domestic ores were being depleted. Obviously the Canadian industry would like to serve a global market, especially the American market, which consumed around 25 percent of the world's mineral commodities, but this would be difficult if the Carter Commission were to cut tax incentives, contending that Canada had enough mineral production to meet its own needs. One scientist at the AIME meeting, Donald Horning, called for more substitutions like plastics and urged more sophisticated management policies, because societies had overestimated the world's capacity to absorb industrialization. Horning predicted that by the year 2000, the carbon dioxide content of the atmosphere would increase by 25 percent, possibly causing serious climatic changes. He also said that waste control must be accepted as an integral cost of doing business. Horning's statement was the first strong environmental message coming from a major mining convention.[16] At the same meeting, Roger Pemberton of Scintrex Ltd. discussed another timely topic: uranium analysis using new radioactive spectrometers. His talk foreshadowed the coming boom of airborne uranium exploration that would reach many remote places in the world. Market demand for nuclear fuel in the 1970s was predicted to rise significantly.

Also in 1968, Canadian exploration, fueled in part by the participation of oil companies such as Superior Oil and Imperial Oil, was headed for a record year of investment. The previous year's success in the Coppermine district had stimulated wide interest: Coppermine River Ltd., a subsidiary of PCE Exploration, a Murray Watts company, had outlined 3 million tons of high-grade (3.48 percent) ore. Huntec Ltd. intended to fly airborne surveys on behalf of Coppermine River Ltd. and about fifty other companies, focusing attention along the twenty-four-mile Teshierpi fault cutting basalt rocks. The biggest year yet was happening in the Coppermine River play, assisted by a 3000-foot landing strip south of town, with over fifty claim holders and some 30,000 claims in existence. Cooperative endeavors led by PCE were planned. They included a regional geophysical survey of this vast area, and an airlift by Hercules aircraft from Norman Wells and Echo Bay on Great Bear Lake, to bridge the gap from the end of the overland freight haul. Exploration for minerals – copper, in this case – had become a large-scale venture. A rush unparalleled in Canadian mining history was developing; it was spreading to Victoria Island and quashing the belief that Arctic exploration imposed prohibitive problems. Little concern seemed to be voiced over the transportation difficulties that would face a potential Coronation Gulf copper producer.

Companies like Dome Mines, INCO, Noranda, Cominco, Bankeno, Campbell Red Lake, and others were participating in Panarctic Oil's 44 million acre program of mapping and seismic surveys in the Canadian Arctic Archipelago, aided by $9 million from the government. Brinex had a large program that included the drilling of the Kitts uranium project in Labrador. The Iron Ore Company of Canada, which transported its production over the 370-mile Labrador railway, had expanded its capacity to 10.5 million tons of concentrates and another 10 million tons of pellets. This huge complex was an indication of the vigor of the iron business. Buoyant prices were able to support the always-critical transportation costs for this bulk mineral commodity.

The 1968 Prospectors and Developers Association of Canada convention had an upbeat theme: "Canada's mining future – a new era?" Much exploration interest was focused on uranium: examples included the GSC mapping program, lake-water geochemical sampling by Eldorado Nuclear, and airborne radioactivity measurement by Roger Pemberton. Geochemical uranium tests, the result of GSC work at Bancroft, Ontario, promised another tool to prospectors. The GSC described the mineral potential of the Tavani-Kaminak Lake region and Project Winisk, the reconnaissance mapping of the Hudson Bay Lowlands. Murray Watts gave a talk about prospecting in the Coppermine River area. Clearly, the industry was in an aggressive mode, and many of the nation's exploration hot spots were in the northern frontier areas.

International companies exploring for uranium were Metallgesellschaft A.G., in Labrador; Motka (Canada) Ltée, in eastern Canada and the Uranium City/Fond-du-Lac belt of Saskatchewan; and Federal Resources, at Faraday, Ontario and Beaverlodge, Saskatchewan. In another significant step, oil and gas companies, particularly the large integrated companies, joined in the search for minerals. Such companies had the resources to pursue low-grade, remote deposits such as columbium in the James Bay Lowlands (Imperial Oil), iron in the Yukon (Standard Oil and Crest Exploration), uranium at Beaverlodge (Numac Oil and Gas), uranium in Saskatchewan (BA Oil and Gulf Oil), and copper in the Yukon (Central Del Rio Oils). Chemical and paper companies were also involved. Selco Exploration, a ten-year pioneer of input electromagnetic surveying in Canada, continued its program of gold exploration, which included Northwest Territories sites like the iron formations at Cullaton Lake.

The reigning exploration philosophy, as expressed by INCO, was verging on missionary purpose: to provide minerals for a hungry world struggling to provide a better life and eliminate poverty. Mineral producers predicted that raw-material demand would rise geometrically in this era of progress. This would require "big thinking" on the part of industry and remodeled government policies geared to the dynamism of the future.[17] Visionary mining leaders had emerged in Canada, like Thayer Lindsey, who at the peak of his career was president of ten mining companies, director of fifteen others, and in control of forty to fifty others. The crown of his empire, of course, was Falconbridge Nickel Mines. Upon retirement, Lindsey was honored for his vision and geological sense by other mining men, such as Murray Watts. Gordon Gibson of the Northwest Territories Legislative Council advocated a visionary change in transportation: provision of icebreakers or ice-free channels to serve Coppermine, Baffin Island, and Arctic oil and gas needs.

Multitudes of small companies were exploring for minerals all over Canada, but the raising of risk capital for their operations reached a crisis point in 1968. The Ontario Securities Commission (OSC) had imposed new regulations that severely restricted junior capital formation on the Toronto Stock Exchange. PDAC spokesmen Joe Rankin and Murray Watts deplored the bureaucratic overkill and red tape, applied in the interest of protecting the public from crooks. They called for the establishment of a "junior exchange" for primary distribution of shares and the raising of speculative risk capital, on which the industry depended. Without it, they said, the junior firms would face extinction. The OSC crackdown heralded the rise of vigorous stock exchanges in other jurisdictions, notably Vancouver.[18] Meanwhile, Ray Price, a dynamic personality and Baptist minister, had published his often inaccurate history book about Yellowknife, which had gone from mining camp to capital of the Northwest Territories in thirty years. Speaking of the 1930s, after the Duck Lake gold discovery near Yellowknife and Gilbert Labine's uranium/silver discovery at Great Bear Lake, Price said:: "Prospectors prowled through the bush, planes skittered onto lakes, geologists collected samples and developed theories, and the local Indians quietly observed the invasion."[19] The remarks were also appropriate thirty years later. Never before had so much exploration attention been focused on the Arctic frontier.

In 1968, the Canadian areas of aggressive uranium exploration were Bancroft, Elliot Lake, Beaverlodge, Johan Beetz, and Makkovik. Together, they illustrated a wide variety of geological environments that host this metal: pegmatitic, sedimentary, vein, and volcanic. The same year, the federal government lifted its restrictions and allowed the Crown-owned Eldorado Nuclear to conduct exploration outside Saskatchewan for the first time, notwithstanding that the former Eldorado mine on Great Bear Lake had shut down in 1960. Beaverlodge, Saskatchewan, would be its first target, but the Crown corporation expressed interest in the Northwest Territories as well. Competition for properties with merit had accelerated. In fact, the French-financed Amok (Canada) Ltée. had jumped the gun, surrounded the Beaverlodge properties with claims, and even spread its control toward Stoney Rapids by acquiring claim blocks. In the feverish atmosphere of uranium resurgence, no promising ground was left for newcomers.

Dynamic Mining of Calgary found itself in this position in 1968. Led by coal engineer Al Swanson, Dynamic chose Saskatchewan as the most favorable geological environment, but decided to do wildcatting rather than buy up other people's claims. Swanson was aware of the importance of sedimentary basins for the occurrence of uranium – after all, the Elliot Lake and many Southwest American pitchblende deposits were found in sedimentary rocks. He decided, therefore, to fly the Athabasca sandstone basin, which borders the more ancient Beaverlodge geological area, using the new spectrometer technology. Scintrex Ltd., a Canadian geophysical contractor, carried out this extensive survey. With encouraging results, the company acquired nineteen exploration permits to cover its prominent spectrometer anomalies. This was a bold step for a small exploration firm, but within months a deal was made with Gulf Minerals to finance a large program of field prospecting and shallow drilling. The work led to the Rabbit Lake discovery, announced in December of 1968, which launched a new era of uranium development near Wollaston Lake. The Saskatchewan mineral resources department was soon inundated with applications for permits.

In 1969, fresh from its successful airborne radiometric survey in Saskatchewan, Dynamic Mining looked farther afield to the Dubawnt sandstone basin of the Keewatin region. Geologically speaking, Swanson saw this as a sister basin to the Athabasca (it had been so identified during the GSC mapping) and again employed the Scintrex airborne survey system. The results showed positive uranium anomalies, which were followed up on the ground. Late in the 1969 season, the ground crew literally walked onto pitchblende veins on the shore of Christopher Island, at the eastern end of Baker Lake. Claim staking commenced in that area and at Kazan Falls. Large holdings of exploration permits were later acquired, which fueled a new and major staking rush to the area. In the Keewatin, the Dynamic group of companies conducted ground exploration independently, rather than seeking partners as they had done in Saskatchewan. The GSC continued its mapping work in the southern Keewatin, encouraged by the economic potential of the area.

NATIVE RESPONSE TO EXPLORATIONISTS AND THEIR ALLIES

By 1970, the Mary River iron project on Baffin Island was in advanced stages of exploration, with nearly enough data to back up the evaluation of its feasibility for production. Baffinland Iron Mines was always sympathetic to local labor, inasmuch as visionary geologist Murray Watts had known and trusted the Inuit since the 1930s in northern Ungava. He knew they were successful miners at Rankin Inlet, and he hired many for the summer field programs at Mary River. Watts spoke in glowing terms of the advantages of a local labor force, but he never had the opportunity for large-scale hirings. Author Frank Rasky wrote, "He puts his faith in the Eskimos of Baffin Island. They are conditioned to the environment, and the Eskimos among the sixty workers Watts employed to do the drilling, sampling and road building at Mary River were very adept. He wants very much to give the Eskimos a stake in their community. Indeed, two decades from now Watts envisions Mary River as a metropolis of 7000 people and his iron mountain will be responsible for the first Eskimo mining and technology college of the north."[20] In a 1966 technical article, Watts and D.K. Megill described two exceptional Pond Inlet employees: Simonee was photographed driving a bombardier tractor and drilling for samples with a Copco percussion drill, while Mosesee was operating a bulldozer and flying a Cessna 206 aircraft.[21] In 1968, Graham Rowley wrote an article entitled "Baffin Island enters the Modern Age." He noted the remarkable high-grade iron deposit at Mary River – in fact, the very ore body of massive hematite was a prominent landmark named Nuluijak Hill. This hill guided Inuit traveling north to Milne Inlet, the site of Baffinland Iron's proposed deepwater port, which would require construction of sixty-five miles of railway. Rowley further stated, "Only one problem stands in the way of the Baffinland development and that is the short shipping season. It is a formidable problem, for even in a good year Milne Inlet freezes in October and does not open until late July. In only a few short weeks the output of the whole year, stockpiled at Milne Inlet, would have to be moved from Baffin Island to the world's markets." As for the Inuit, he said they had the intelligence, sense of humor, and cheerfulness to adapt to new conditions and different standards: "No people have shown more understanding of strangers or been more helpful to them. They deserve similar treatment in return."[22]

The purely exploration-minded uranium companies working in the Keewatin at first had little contact with Native residents. A few Inuit were employed indirectly by white expediters at Baker Lake to haul fuel or handle supplies. The first task of these exploration companies was prospecting and geological mapping, for which a technical staff from the south was employed. As the work grew in scale and complexity, some Inuit were hired at Baker Lake and Arviat to work as camp laborers, drillers' helpers, and soil samplers. Helicopters were used extensively from the start in 1970 and became a familiar sight to local communities. The operators had no sense that helicopter movement was affecting the behavior of the commonly sighted caribou herds. In rare instances, Inuit employees were dropped off for some hunting before their scheduled return to town.

Unprecedented exploration activity was taking place in the North, not only for minerals but also for oil and gas. What sort of side effects or reactions were these developments causing? In his 1972 volume *The Violated Vision*, James Woodford revealed that "the current rush to exploit northern resources has touched off a wave of concern among Original Peoples, and new leaders are emerging. They are educated, eloquent and determined to seek a solution once and for all to their claims to the land."[23] The first outcry was voiced by Western Arctic Indians, notably the Indian Brotherhood of the Northwest Territories, who called on the minister in 1970 to stop all setting-aside of treaty lands. Roy Daniels, president of the Indian Brotherhood of the Northwest Territories, pointed out that "negotiations and submissions in reference to such claims (Treaties 8 and 11) will take time to make ... but time is running out in much of the area covered by the treaties. Permits to explore for oil and gas have been issued and many mineral claims have been staked."[24] Lloyd Barber, named Indian claims commissioner in 1969, was asked to inquire into such urgent issues. The State of Alaska, moreover, had declared a two-year land freeze on further disposition of public lands in 1966, pending a comprehensive claim settlement (in 1969). Clearly the explosion of oil and gas exploration in the state had created the pregnant conditions for a settlement. Similar circumstances prevailed in Arctic Canada. It was not simply the "look of the map" that caused alarm (these were subsurface dispositions); the actual field practices of developers were offensive.

The 1970 Coppermine conference of Inuit and other Native representatives was an urgent response to development pressures. Most notable of these was the disruptive seismic activity on Banks Island, known as a bountiful land for its white fox trapping, following the issuance by the Department of 308 oil and gas permits.[25] The Sachs Harbour representatives, Peter Esau and Peter Sidney, framed a conference resolution with supporting signatures from twenty-two Inuit communities that was transmitted to the prime minister and to the Indian Affairs minister, Jean Chrétien.

This story reached the *Globe and Mail* and caused a furor. Professor Peter Cumming, acting as legal advisor to the Inuit, said, "The discussion at the conference clearly indicated the incompetence and indifference of the Department of Northern Development (charged by statute with looking after the welfare of the Eskimo people) in administering the affairs of the native people of the North."[26] This was a momentous conference – the first national conference of Inuit and other Natives from across the North. "Everyone sensed the magnitude of the forces being thrust upon the far north and that for the first time the *land base*, the integral element of aboriginal culture, was significantly threatened."[27] The conference gave impetus to the recently formed Committee for Original Peoples' Entitlement (COPE) with Agnes Semmler, a Metis from Inuvik, as president, and Victor Allan (an Inuk) and James Koe (an Indian) as vice presidents. In this way, the struggle to assert aboriginal rights – a struggle long pursued by the Nisga'a nation in British Columbia – erupted in the far North.

Added to this surge of Native demands was the emerging environmental or conservation movement. After all, Professor Jim Lotz had written in *The Myth of the Rich North*[28] that none of the Arctic minerals were actually needed (as if local need were ever a concern in the global mineral arena) and the known deposits required subsidies, to be paid for by the Canadian public. Between 1966 and 1969, the federal government spent $100 million on mine development in addition to funding other programs, such as prospectors' assistance. The Canadian Wildlife Federation (CWF), the Canadian Commission for UNESCO, the Conservation Council of Ontario, and W.A. Fuller, a University of Alberta zoologist, called for a moratorium – or at least a slowing of development – until environmental assurances, oil-spill technology, and Native questions could be addressed. In the wake of the blowout of Panarctic gas wells, barge accidents, and seismic blasting, etc., ecologists and conservationists called for a national environmental policy and a new department of the environment, to which all the conservation responsibilities of DIAND would be transferred.[29] Soon, the practical and realistic Canadian Arctic Resources Committee (CARC) would be founded to support northern Native groups in the Berger Inquiry. Powerful forces were being assembled at this time to bring pressure upon government, to say nothing of the insistence of developers and economic pressures caused by the OPEC oil crisis. The Inuit were determined to use outside allies, especially ecologists and environmentalists, to bolster their case, and the idea of a land freeze had real appeal to them, but the strident, anti-technology assertions of writers like Woodford (that, on ecological grounds, the white man and his machines should be excluded from the Arctic altogether) hardly reflected their sentiments. Woodford expounded, "The reverence for life and land of the Indian and Eskimo has been replaced by the arrogance of the technocrat and engineer."[30] This was a utopian approach by an academic who, like the government, had failed to ask the Inuit what they wanted.

Environmental degradation was not seriously considered at the Sixth National Northern Development Conference held in Edmonton in the fall of 1973. Its theme was "Mining and Canada's North," and its preoccupation was government mineral policy. Conference resolutions made the following recommendations: increase incentives to maximize mineral exploration and development; establish rules to clarify mineral policy; increase funding for research on mineral potential; establish a transportation coordinating agency; improve navigational aids; conduct studies of the energy potential of the North for northern use, including mining and smelting; expand northern research centers and staff them with Northerners; and achieve an early, just, and equitable settlement of Native land claims in northern Canada.

Social responsibility was discussed by Pat Carney, a Yellowknife consultant to the Mackenzie gas pipeline study group. Her last point dealt with the widening credibility gap between the government and residents of communities. Despite a policy of consulting with local people on proposed developments, decisions went against the people because "22 million Canadians and 220 million Americans ... require energy, and their views will inevitably be considered also."[31] Carney cited examples: the Inuit community of Tuktoyaktuk, which objected to the implementation of seismic programs on nearby Cape Bathurst, and the people of Wrigley, who had taken a firm stand against the Mackenzie Highway.

The conference sponsored the attendance of three Native delegates: Meeka Wilson, secretary-treasurer of the Inuit Tapirisat of Canada, from Frobisher Bay (Iqaluit); Isadore Yukon, from Fort Franklin; and François Paulette, from Fort Smith. Chief Paulette, addressing the closing session, encouraged Native people to become educated and participate in the control of development in the North. He noted that Canada had finally recognized aboriginal rights the previous year by empowering the Commissioner of Indian Claims to conduct negotiations, adding that he hoped the government would "give us full satisfaction and fulfillment of our Native land claims in the North. We are willing and we want to assist and participate in the development of the North."[32] The Conference chairman received a telegram from Tagak Curley, president of the newly formed Inuit Tapirisat of Canada (National Eskimo Brotherhood), who was clearly miffed at not being invited to attend. Curley said social problems would be perpetuated if Native people were not involved, and that ITC was "seeking to develop a spirit of cooperation between the Inuit people, the mining industry and the government which will be important to all parties."[33] The Conference chairman mentioned Meeka Wilson's presence and invited Curley to come to the seventh conference the following year.

KEEWATIN EXPLORATION BOOM

As the pace of uranium exploration increased, large, competitive companies emerged – companies that could afford sophisticated techniques and high levels of logistic support.[34] Mineral claims and prospecting permits covered much of the Dubawnt geological basin around Baker Lake. In April 1974, Inuit from Baker Lake stated their concern about the effects of large-scale exploration on the land, water, and wildlife. They asked Minister Judd Buchanan to declare Baker Lake a land-management zone under the territorial land-use regulations and to consult settlement councils fully before granting exploration permits under the Canada Mining Regulations. In due course, their plea was refused, so, in order to forestall 1975 exploration programs, the council urged through ITC that a land freeze be obtained for five blocks of land around Baker Lake – critical areas for wildlife and cultural significance. This request was also refused by the minister.

In 1975, Pan Ocean Oil Ltd. added eight new prospecting permits to the existing one and worked on them from its base camp near Kazan Falls. As field geologist with this company, I had one of my most memorable experiences with Inuit employees at Kazan Falls. Other explorers were Uranerz, Rio Alto Exploration, and Noranda – as well as Cominco, which had budgeted $700,000 for drilling and other work on 2300 square miles of property. The Geological Survey of Canada was also conducting fieldwork at Thirty Mile and McQuoid Lakes. W. Shilts of the GSC observed that over a hundred personnel were engaged in exploration activities during the summer of 1975.[35] Yet the district was not designated a land management zone, so the land-use regulations did not apply. As a precaution, however, the game management division of the GNWT requested that mining company helicopter pilots observe a minimum altitude of 1500 feet over caribou calving and post-calving areas from May 25 to June 20. Company helicopters were indispensable for field operations, and they made frequent flights to Baker Lake to transport personnel and supplies. The settlement of Baker Lake remained the logistical center for all the exploration activity in the district, as it provided an airstrip with navigational aids, airline service, float-plane charter service, and expediting. In the 1976 season, Cominco expanded its uranium drilling program after discovering a good mineralized zone 373 ft. thick, which averaged 2.6 pounds of uranium oxide per ton. The company operated a thirty-man camp, which was prudently located away from the sensitive Kazan Falls area. Cominco was working lands optioned from Pan Ocean Oil, which considered stepping farther south in its search, to Yathkyed Lake.

The Falls on the Kazan River are a scenic wonder and a fisherman's dream. Geologically speaking, they were caused by cross-cutting black diabase dykes that buttressed the Dubawnt sandstones. Northern canoeists might negotiate the upstream rapids, but they always portaged around the Falls. The Pan Ocean camp, which I knew well, was situated a mile downstream on a gravel spit. The spit hid a small lagoon, where Native kayak paddlers had gathered in the past to launch surprise attacks on caribou herds that were crossing the Kazan River. A straight, open stretch of water afforded a perfect landing place for pontoon aircraft that could nudge onto the steep, sandy shore of the spit – that is why we chose this site as a base camp for our work.

In 1975, a number of Inuit were hired as field assistants to perform frost boil sampling for geochemical analysis. These included Simon Tookoome of Baker Lake, the well-known artist and traditional athlete. Tookoome and the others were amiable companions and capable workers for the short few weeks of our residence there. They taught us games, made fish spears, and carved bird figures from the abundant caribou antler lying about. Best of all, they taught us how to fish for giant trout in the racing water below the Falls. The summer's experience taught us respect for these men but in no way implanted any idea of their concerns and aspirations. They never conveyed any hostility or concern about our presence in that traditionally cherished area. Perhaps they responded to our own respect for burial and game storage sites along the Kazan River, which were visible from our helicopter though hidden from passing canoeists. Neither did they question the activity of the exploration companies or expound on their land-claim ambitions, but we could all see the special significance of the area: the Kazan Falls were a treasure to local residents, and we understood why the government prohibited use of the area to outsiders later on. Neither we nor others prospecting the area found any mineralization of economic importance; regrettably, we left behind a legacy of errant oil drums and burned-out diamond drilling equipment.

Revisions to the territorial land-use regulations declared all parts of the Northwest Territories to be land management zones as of March 4, 1977. At that time, the minister elected to defer for one year five new prospecting permits and four new land-use permits in the Baker Lake area, on the grounds that increasing levels of exploration activity might have side effects on wildlife resources. Helicopter flights and drilling operations were suspect. An environmental study was commissioned to review this situation and report back within a year's time. Although the minister's action did not affect existing permits and programs, both industry and the media interpreted it as a "land freeze."[36]

3. INUIT POLITICAL ORGANIZATION

INDIAN ESKIMO ASSOCIATION (IEA)

A voluntary organization in the South exerted a powerful influence on Canadian Native affairs in the 1960s and 1970s. This was the Indian Eskimo Association (IEA), whose objectives were to promote concern for the total well-being of Canadians of Indian and Eskimo background, encourage their acceptance and participation, stimulate understanding and cooperation, foster cultural expression, advertise services and benefits, strengthen local and regional action, disseminate information and provide education and consultative services, carry on comprehensive research, and develop policies on issues of the day. This was a grassroots organization sponsored by educators, public servants, and religious leaders who had a deep concern for the plight of Native Canadians. The Association gained favor and secured the support of individuals, corporations, and volunteer groups.

The movement began in 1955, with an Ottawa group organized by Father André Renaud to study the topic "The Indian in Town." That subject was later placed on the conference agenda of the Canadian Association for Adult Education (CAAE), which expressed its deep concern by setting up a standing committee, boldly called the National Commission on the Canadian Indian. The commission attracted wide interest in the form of briefings from government officials and opinions from delegates at meetings and conferences. The CAAE became convinced of the need for a national organization, and in 1958 hired John Melling to be its first director. Melling traveled nation-wide to consult with voluntary organizations and Indian communities. The Commission drafted its own newsletter, distributed an enormous amount of printed material to its 1400 members and some fifty study groups, and also submitted a brief to the new parliamentary committee on Indian affairs. Valuable advice from its many contacts enabled an organizational committee to draft a constitution, which led to the incorporation in January 1960 of the Indian Eskimo Association.[1] The new association, an autonomous national organization, was welcomed enthusiastically.

The first executive board of the Indian Eskimo Association included members with broad experience. President Clare Clark, who had been a staff member of CAAE and executive secretary of the Commission, was also on the board of the YWCA and a member of the Canadian Youth Commission. Father André Renaud, director general of the Oblate Fathers Indian and Eskimo Welfare Commission, had served parishes in the Yukon and done graduate research in education and anthropology. Dr. J.A. Munro served on the Board of Missions of the Presbyterian Church. Dr. Gilbert Monture, a distinguished geologist and mining engineer, public servant, and international advisor, was himself a registered Indian from the Six Nations Reserve. Director John Melling was a British academic interested in intercultural relations. Bernard Neary was president of an educational publishing firm, a former teacher in the Indian schools of British Columbia, and superintendent of education in the Indian Affairs Branch. The research committee of the association was headed by Frank Vallee, an anthropologist from McMaster University, and included Vic Valentine, Department researcher, and Frank Scott, McGill law professor, among others.

The Indian Eskimo Association's parliamentary brief stressed the need for Native Canadians to be free to choose their own goals and receive an education, in order to achieve growth of independence, cultural advance, and confidence in the future. In Toronto, the IEA sponsored conferences on Eskimos, with a view to informing the public and arousing interest in their condition. Frank Vallee warned against oversimplification of the problems facing both Indians and Eskimos, and speakers like Abraham Okpik and R.A.J. Phillips of the Department of Northern Affairs (DNA) gave informed opinions. Phillips remarked, "To find the Eskimo a place in the sun, we must strive to see that it is not in our shadow."[2] Conference conclusions about services and projects called for the distribution of study materials on the status of Native people to stimulate discussion. It was firmly resolved that the initiative and request for projects and services should come from the Eskimos themselves, who should become involved in a cooperative effort. After a busy first year, president Clare Clark stated how appropriate a voluntary citizens' organization was to meet such a need and how theirs was cheered and supported across the country. There was a notable awakening of Canadians. Clark said that the careful launching and apprenticeship of the IEA had enabled it to work amicably and cooperatively with government agencies. She emphasized the need to overcome an almost total lack of communication between Native and other Canadians and invited all Canadians "to join with us in a great crusade to remove the barriers, and to create conditions favorable to advancement, whereby the Native people *by their own efforts* can take their rightful place in the life of our country."[3]

Typically, the annual meeting of the Indian Eskimo Association was followed by a conference covering many topics, such as enlightened relationships, community development, education, communication, and economics. Elijah Menarik from Ungava described successful Inuit cooperatives involving a fishery, sawmill, freezer, and retail stores. Elliott Moses of the Six Nations Reserve in Ontario summed up the value of the conference and stressed that attention should be directed at youth: the future was in the hands of the young people, who must find a way to achieve a full life. Moses favored integration without assimilation.

The IEA continued its strong support of conferences and regional meetings and disseminated conference proceedings widely, thereby drawing attention to Native views and contributions. For example Frank Calder, Nisga'a activist and member of the British Columbia legislature, spoke in Winnipeg about self-government on reserves and leadership training. At a Saskatchewan conference, Metis representative Malcolm Norris argued that treaty Indians and Metis must adopt a united front to accomplish anything. Frank Vallee addressed the Northwest Territories Legislative Council on the subject of community development and regional planning, of which he was a strong advocate. In his opinion, the then-current government policy of placing its hopes on a development or private capital scenario (to recover minerals such as nickel, gold, oil and iron ore) was not good enough.[4] Helen Buckley from the Centre for Community Studies, University of Saskatchewan, echoed Vallee's view in regard to Saskatchewan: "The superiority of private capital as an instrument of progress makes no sense when we have only to look at the record – how few jobs it has created for Indians in the past."[5] In 1966, Abe Okpik, now an appointed Member of the Northwest Territories Council, chaired a panel of Inuit to discuss problems of northern communities. Panel members included George Koneak, Minnie Moore (later Freeman), Anne Padlo Lidbetter, and Elijah Menarik. They discussed the disturbance of game by aircraft, the shortage of housing, the absence of intersettlement communication to share experiences, the need to retain language and teach children in Inuktitut for the first four grades, and cultural survival.

In 1969, the Indian-Eskimo Association announced: "To further its policy of supporting Indian organizations, the IEA hopes to provide the research they need on treaty and aboriginal rights." The committee struck for this purpose was aware of the woeful lack of such research and how much it was needed for an informed response to the upcoming Indian Claims Bill. There had been discussions during an Ottawa human rights conference with representatives of the National Indian Brotherhood and Metis organizations. Subsequently, committee chairman Ernest McEwen (now on the IEA board of directors) made field trips to the four western provinces to consult local Indian and Metis groups. It was decided that the project should begin with hearings in the Northwest Territories (Treaties 8 and 11); in the Prairies, to review the legal status of Metis; and in Alberta (Treaty 7). The basic questions to be investigated were unsurrendered lands, alienation of reserve land, misuse of trust funds, unfulfilled treaty promises, and failure to act honorably.[6] The association received an anonymous grant of $25,000 for the project and appointed Douglas Saunders, former legal associate of Thomas Berger, to head it. Many lawyers and other academics were already represented on the research and aboriginal rights committees, including Walter Currie, who was now the national president of the Indian Eskimo Association.

The long-awaited government White Paper on Native Policy, tabled by Minister Jean Chrétien in June 1969, was given wide publicity in order to illustrate the vehement Native opposition. This opposition spurred on the Association's legal rights project: its report was immediately used as ammunition against the White Paper policies, especially the extinguishment of aboriginal rights. The subsequent government appointment of Lloyd Barber as the claims commissioner, whose only power was to discuss treaties, was also vigorously opposed – an understandable reaction, considering that half of Canadian Indian people were not under any treaty! Within a year, the Indian Chiefs of Alberta presented their own "Red Paper," entitled *Citizens Plus*.

The Eskimo Conference of thirty-three delegates held in Coppermine in July 1970 was co-sponsored by the Indian Eskimo Association. A major outcome was the work of legal consultant Peter Cumming from the University of Toronto. Cumming eloquently described the indifferent government policy and the denial of Inuit rights inherent in allowing seismic exploration crews to overrun the traplines of Banks Island and permitting a prospector to stake and control a traditional soapstone quarry used by the Coppermine people. Cumming continued to write on the subject of Inuit legal rights and did much to encourage the building of "Arctic power."[7] This was a period of great legal ferment, but the Association did not forget other issues, such as the teaching of Native children in Canada (a special project of president Walter Currie, who was assistant superintendent with the Ontario Department of Education). Communication difficulty in the Arctic was another huge concern, and to this end the IEA hired twenty-six-year-old Tagak Curley as a field worker to undertake liaison with Inuit communities. Based in the Association's Edmonton office, Curley was already "expected to become a staff leader for a self-organized Eskimo association speaking for all Eskimo people."[8] The same news item in the Association's bulletin spoke of the already rising "parka power" and the urgent need for a single voice in the Arctic, foreshadowing the emergence of the Inuit Tapirisat of Canada (ITC).

The Indian Eskimo Association by this time was generously supported (86 percent) by private sector giving and fund-raising walkathons. There were 200 corporations and 227 organizations supporting its work, in addition to two thousand paying members, not including 197 libraries. Its busy programs in these years were managed by director Allan Clark and president Walter Currie. The past ten years witnessed many achievements, including the establishment of ITC, but the original organization was beginning to sense that its seminal work was done. In 1973, the IEA responded to reality by metamorphosing into the Canadian Association in Support of the Native Peoples (CASNP), with T.H.B. Symons as its first president and J.E. Gasson as executive director. In this form, the association continued its publishing, fund-raising, and promotional roles until 1978, when it was dissolved.[9] By that time, Native organizations in Canada were managing confidently and sustaining themselves with government funding. Their debt to the eighteen-year support and vitality of the Indian Eskimo Association, formed and sustained by a collection of well-meaning Canadian citizens, will long be remembered.

THE FOUNDING OF THE INUIT TAPIRISAT OF CANADA (ITC)

Born and educated in Coral Harbour, Tagak Curley had two years of education in Ottawa and working experience at the GNWT adult education department in Fort Churchill. Based in the Edmonton office, he would endeavor to meet with community councils to promote a "self-organized Eskimo Association speaking for all Eskimo people."[10] The forces that southerners termed "parka power" were mobilizing to speak with a single voice, so that before the decade was over, few Canadians would not know the wants, desires, and demands of the Inuit.

Laying the groundwork and overcoming communication problems associated with a far-flung population were Curley's first tasks. The Indian Eskimo Association had sponsored the Tundra Conference in 1969 and the Coppermine Conference in 1970, referred to above, which led to the formation of the Committee for Original Peoples' Entitlement (COPE). In 1971, the IEA declared a policy change: it would facilitate and coordinate Native and

Metis efforts to form their own organizations, thereby working itself out of a job. Norman Dunne, a former chemical engineer with Shell Oil, was the IEA staff person in charge. His expertise and contacts were in the Western Arctic, whereas Curley's were in the Keewatin. As a result of Curley's work, six Inuit community leaders were invited to attend, at IEA expense, a Board meeting in Peterborough in February 1971 to exchange views. These leaders were Josiah Kablutsiak, Noah Qumak, Jacob Owetaltuk – note that Nunavik people were included – C. Makpah, Mary Cousins, and Ipeele Kilabuk. They felt it was urgent to learn their legal rights if they were to be part of northern development, rather than its casualties: "We must review our own situation in regard to preservation of our culture, economics, and our political development," Curley said. They announced an organizing committee to form the Inuit Tapirisat of Canada (Eskimo National Brotherhood), with Noah Qumak of Sugluk (Quebec) as chairman, Josiah Kablutsiak of Igloolik as vice-chairman, and Tagak Curley as secretary-treasurer. They proposed a founding conference in Ottawa, and requested that the IEA help them to secure the government funding that had been accorded Indian and Metis groups. Curley hoped that, as a result of the conference, the Inuit would be better prepared to set their own priorities. He felt sure that if the government continued to introduce programs without consulting the people, it would not meet its own goals. "To be successful in development of the North, it needs to involve the people, and enable them to have control over their own lives."[11] It is interesting that questions of development, as well as the desire to save their culture and independence, figured in the very first statement of concerns by the founders of ITC.

Thus began the formal enterprise of Native education, organization, and advocacy in the far North. Not that other voices were not joining the rising chorus: the Eskimo University at Frobisher Bay, an Inuit group organized by Inuit to prevent the loss of their traditional culture, began to criticize low-flying military jets because of their effect on caribou, especially in the winter, when the exertion of stampeding could damage the animals' lungs.[12] The Voice of Women, a national women's organization, echoed this criticism. Mary Cousins of Frobisher Bay wrote a challenge to the mining industry and government over lack of communication about resources and development. Her goal was to raise the question at a Toronto forum (called "The Canadian North: Eskimos or Natural Resources?") sponsored by The Canadian Friends Service Committee and the IEA. The ITC sounded the alarm about nationalism and language exclusion in Quebec, where the province was negotiating with the federal government to take over the administration of Native peoples. The Tapirisat's major emphasis, however, was to make its people aware of the organization's goals and their own legal rights. The IEA's name change, from Indian Eskimo Association to Canadian Association in Support of Native Peoples (CASNP), reflected its changing character: the Association was assuming more of a service role for Natives in Canada now that their own organizations were thriving. Peter Cumming, professor of law and IEA Board member, headed a fifteen-member legal research committee. Aided by federal funding, they produced a manual entitled *Native Rights in Canada* and a booklet entitled *Inuit and the Law*.[13] Efforts at general fund-raising were also beginning to bear fruit. The World Council of Churches program to combat racism provided a $2500 grant for ITC public relations and meetings. Peter Cumming's initial work proved so valuable a reference document for Native leaders, students, and lawyers that it was revised and enlarged into a second edition.

The founding conference of the new Inuit Tapirisat of Canada organization (ITC) was held at Carleton University in Ottawa on August 18–27, 1971. In attendance were twenty-three representatives from the western and eastern Arctic and northern Quebec. The meeting was a natural sequel to the Coppermine Conference of July 1970, when it was recognized that a national organization was needed to represent all 17,000 Inuit in Canada. The educated and energetic Tagak Curley became the first president of ITC. By 1972, northern environmentalism was gaining momentum. The Canadian Arctic Resources Committee (CARC) sponsored its "North of 60" conference in Ottawa, involving a cross-section of concerned Canadians. Natives from a wide spectrum were developing plans for their land. In 1973, the ITC announced a plan for "sane progress" that included a temporary freeze on exploration and development.[14] Among their demands were a fair and reasonable settlement of land claims, appropriate ecological protection, new mechanisms for participation in decision making, comprehensive agreements on land and water use, and economic plans.

Up to this point, their energies had been spent on trying to get federal recognition of the ITC. The government acquiesced and agreed to fund the Tapirisat's land-claim project. In regard to a land freeze, the Inuit refused to accept the government assertion that Native claims would not be prejudiced by prior and continuing exploration and development. They were worried about very real environmental and social impacts, and they did not suppose the government would make any fundamental changes in policy. Would the government return Atkinson Point, the location of Imperial Oil's gas well, to the people of nearby Tuktoyaktuk? The Inuit said they wished to participate in development, but they believed that a temporary freeze would be in the best interests of the people, the Northwest Territories, and orderly process.[15] This statement was made in response to the government's new policy, announced by Indian Affairs minister Jean Chrétien on August 8, 1973, of willingness to negotiate with claimants on the basis of their traditional interests.

Peter Cumming was giving talks about Native land rights and northern development. He contended that the discovery of oil at Prudhoe Bay, Alaska, meant the industrial economy was just around the corner, and this would stimulate exploration in the Mackenzie Delta and the Canadian Arctic Islands. This incipient industrial era would be "as traumatic a force as the collective impact of all those forces and changes experienced by the Native state to date." Unlike other cultural or economic forces to act upon the North, this one threatened the very land base of the Native people.[16] Cumming believed that by means of an equitable settlement, Canada and the Canadian people could take a position of enlightened stewardship not yet achieved with Indians or Inuit. He stressed the crucial importance of land ownership and control to the Inuit people. Land was an integral part of their culture and identity, and they would not simply accept money in return for the lands they had traditionally occupied. He expressed his belief that "an acceptable mix of all the necessary elements to a fair settlement can be arrived at."[17]

The ideas and concerns of Inuit were now being spread across the Arctic by conference gatherings and local meetings and publications. People were being sensitized to the issues, whether or not those issues affected their own area. For example, the 1973 annual meeting of the ITC was held in Baker Lake. This was not all a bed of roses for the ITC: relating to the local people and moving at a pace acceptable to them was not easy. The ITC's 1972 clash of words with the community of Arctic Bay over the Nanisivik mine in Baffin Island had illustrated the independent opinion rooted in local communities (see Chapter 4). In response to the urge for local representation, the Keewatin Inuit Association (KIA) was founded in 1975 as an affiliate of the ITC. The Inuit Cultural Institute dedicated to the preservation of *Inuktitut* was formed in 1973, under Tagak Curley's tutelage, with its headquarters in the Keewatin community of Eskimo Point (Arviat).

BEGINNINGS OF COPE

In January 1970, a group of Inuvik residents, concerned about their future in the face of booming exploration, met to form the first northern Native organization, the Committee for Original Peoples' Entitlement (COPE), most especially to present a strong voice at the forthcoming Coppermine Conference of Arctic Native People. Oil had just been found at Atkinson Point, and they felt it was time to speak up about the encroachment, or Native people would really lose out. They knew of the growing Native movement in Alaska, where activists were holding up oil development to get a fair land-claim settlement. What COPE people sought in order to establish and realize their rights was a united voice for all aboriginal peoples of the Northwest Territories. Recording the early history of the group, consultant Peter Usher stated, "The purpose of the committee was above all to find out about the proper entitlement of the aboriginal people, and to make sure they got it."[18] Though government maps showed the territory granted by Treaty 11, signed with the Mackenzie Valley Indians in 1921, extending right to the Arctic coast, the Indian signatories had never understood that to be the case, since the coast was beyond their territory – and Inuit had never signed the treaty. The committee therefore believed that the Natives had rights to the land that comprised their territory, and the main problem facing them was to get control of it.

In the three years following its inception, COPE recruited members, consulted widely, and initiated collective protests against the government. For example, they protested the incursion of oil companies onto Banks Island, where seismic surveys were disturbing traplines, and subsequently onto Cape Bathurst near Tuktoyaktuk. They became affiliated with the Inuit Tapirisat of Canada and thereby secured a voice in Ottawa for direct action. The government did not recognize COPE, or provide it any funding; in fact, government worked to oppose the Committee, since it would only recognize one Inuit group, the ITC, apparently expressing a divide-and-conquer mentality.

COPE wanted to be a regional Native organization for the Mackenzie Delta and Western Arctic, recognizing that the NWT Indian Brotherhood spoke for all Indian people about their treaty rights and the ITC for all Inuit. The ITC agreed to support the effort of COPE to get core funding and recognized its important regional work on local problems of land use and social abuses. In turn, COPE expected that the ITC would research and conduct negotiations with the government on a land settlement for all its members. In 1973, it was reported that "Inuit Tapirisat is starting a land use and occupancy study to document all the lands belonging to Eskimo people, and how important these lands are to them. Inuit Tapirisat is hiring the best technical and legal advisors for this project. Native people everywhere will be fully involved."[19] In fact, Peter Usher did the Western Arctic land-use and occupancy study, under the direction of Dr. Milton Freeman, with the assistance of Bertram Pokiak, Victor Allan, Peter Thrasher, and the hunters and trappers' associations in the communities.[20] COPE busied itself as a watchdog organization to defend

Native interests and protect the integrity of their lands in the face of oil and gas activity in the Mackenzie Delta and Beaufort Sea areas.

With the help of the Canadian Arctic Resources Committee, they hired environmentalist Douglas Pimlott in 1973. His chief contribution was to investigate the widespread offshore drilling scheduled for 1975 and ways of eliciting preparatory studies by government. His report to COPE released in 1974 provided technical support and mobilized the membership. News releases drew headlines that accused the government of backing hazardous oil drilling. COPE's board, under Sam Raddi, its elegant and effective leader, announced, "Balanced, long-term development ... is being sacrificed for immediate profits and a panic reaction to the energy crisis .... It is appalling that neither COPE, Inuit Tapirisat of Canada, Settlement Councils, nor hunters' and trappers' associations in the region have been consulted."[21] Digby Hunt of the Department of Indian and Northern Affairs promised consultation with Native residents, but it was never forthcoming.

THE BERGER INQUIRY

The Mackenzie Valley Pipeline Inquiry of 1974–76 produced an authoritative statement of Native Northerners' aspirations based on the evidence of three hundred experts at formal hearings in Yellowknife and nearly a thousand individuals at meetings at thirty-five northern communities. Commissioner Thomas R. Berger found that the pipeline development would militate against a just and equitable settlement of Native claims, and he recommended a postponement of ten years. This was a momentous decision for the northern economy and the urgent needs of northern peoples. Berger said that the postponement would protect the environment and ultimately bring about a more diversified economy that was fulfilling to all.

The Inquiry sought in its deliberations to understand the historical context and the essence of the modern demands of Native Northerners. This process was inspiring to the Native participants because it enhanced their credibility and confidence. Negotiated settlements and environmental assessments could now take center stage in an unhurried manner, without the threat of development. The traditional alignment of government and developer would, for a time, be less intimidating to Native groups.

The pressure of incoming settlers and prospectors had originally served to motivate the federal government to press for treaties with Native peoples in Canada. That pressure edged into the territories at the turn of the twentiethth century, with the signing of Treaty 8. Many years then elapsed before development activity reached the lower Mackenzie River. It was the discovery of oil at Fort Norman in 1920 that prompted the next action. Indeed, the Dominion land surveyor issued a warning that, until a treaty was made, the right to dispose of oil and gas resources was open to debate. In 1921, the Treaty Commission travelled down the Mackenzie River and negotiated Treaty 11 with the Dene and Metis to the north of the Treaty 8 region.[22]

The Inquiry was forcefully told that both Treaty 11 and Treaty 8, signed twenty years earlier, were agreements of peace and friendship, not the selling off of land – so their Indian homeland was still intact. Native people were now troubled by the encroachment of the oil industry. The Inuit (including Inuvialuit) experience was somewhat different, but their modern attitude about treaties was the same. Inuit of the Mackenzie Delta had been invited in 1929 to join Treaty 11, but they had refused, thinking the whole land was already theirs. Over the years, the actions of whalers, traders, and governments led to general distrust.[23] Then came the seismic exploration companies that "ploughed up their land in the summertime and ploughed up their traps in the wintertime," according to the testimony of Vince Steen of Tuktoyaktuk. The oil companies deserved more respect – more than the whalers, the fur traders and white trappers – but it was too late: the people would not take the white man's word at face value, and they now opposed the pipeline. Steen concluded that the Inuit (Inuvialuit) were asking for a land settlement because they no longer trusted the white man to handle the land that they owned, and figured they had owned for years and years. Thomas Berger summarized the Native position as follows: that before any development proceeds, "their right to their land and their right to self-determination as a people must be recognized."[24]

Having listened in his patient way to the Native articulation of views, Berger reported at length on what it was that northern Native people really wanted. They did not want to make way for industry as western Native people had made way for settlement. They did not want to repeat history, but instead wanted to take a new direction. Their right to the land and political authority over the land must be recognized. As James Wah-Shee said, "What is at issue is land, not money .... What we are seriously considering is not the surrender of our rights 'once and for all,' but the formalization of our rights and ongoing negotiation and dialogue." He called it a developmental model that could be "a source of pride to all Canadians and not an expensive tax burden."[25] Indeed, a comprehensive land-claim policy was emerging in government in 1976, with categories of land and resource management among its elements.

To survive as distinct peoples, northern Native people wanted to govern their own land and resources, participate directly in resource development, and benefit from royalties. An eloquent statement on this was made before the Inquiry by Robert André, on behalf of the Dene. Thomas Berger concluded that the claims of Inuit (Inuvialuit) really coincided in principle with those of the Dene.[26]

Thomas Berger referred to the Alaska Native Claims Settlement of 1971 as an assimilation model: the Natives allowed extinguishment of their claim (to 375 million acres of land) in return for 40 million acres (10.6 percent) and close to $1 billion. He wrote, "The Dene and Inuit oppose any settlement that offers to pay Native people for their land and to assimilate them into the larger society without any special rights or guarantees for them or their land."[27] Clearly land would be a key component of any agreement, and it would have to be held through collective ownership so as to remain the birthright of future generations. Both Dene and Inuit sought to retain the immense tracts of land that supported them and the animal populations they depended on. The Inuit of the Western Arctic (including the Inuvialuit) used nearly 100,000 square miles, while the Dene occupied 450,000 square miles to support themselves.

Berger noted in his report, "The native people maintain that the use they make of the land requires them to control vast tracts of it. They reject a land settlement that would give them title only to discrete blocks of land around their villages. They reject any suggestion, therefore, of an extension of the reserve system to Northern Canada. For this reason, also, they reject the model of the James Bay Agreement as a means of settling their land claims."[28] In that agreement, the Cree and Inuit of Northern Quebec had surrendered aboriginal rights over their traditional territory (410,000 square miles) in return for cash and specific land rights. In Category 1 lands, in and around villages, they gained exclusive use of the surface: 2100 square miles (0.5 percent) for the Cree and 3250 square miles (0.8 percent) for the Inuit. In Category 2 lands, they acquired exclusive hunting and fishing rights: 25,000 square miles for Cree and 35,000 square miles for Inuit. In Category 3 lands, that is, Quebec public lands, they could hunt and fish also. In no case are these rights entrenched in ownership, and mineral development by others is fully allowed.

Northerners rejected this model, since they saw "ownership and control of the land itself as the only means of safeguarding their traditional economy."[29] They demanded that their rights not be secondary or subservient to alternative uses, that they be free to protect the environment and to derive benefits from (or even participate in) developments.

The crucial role of mineral rights came to light in the testimony of Andrew Thompson, a law professor from the University of British Columbia. Thompson said that ownership of the surface of the land, without ownership of subsurface rights, is often of little value. Ownership of mineral rights usually carries with it a right of access: the surface owner has to give way when the owner of subsurface resources wants to exploit them. For example, the James Bay Agreement even failed to remove this priority right in the case of Category 1 lands; Quebec Natives have to allow access to mineral holders on, or even adjacent to, their Category 1 lands. Thompson recommended that the settlement of Native claims should confer management over minerals, either by legislation or through ownership.[30]

The dissertations of the Berger Inquiry provided some strong enlightenment to northern Canadian Natives yet to achieve a settlement of their claim. The inquiry also boosted their credibility and their morale. For once an authority had come to them and listened to their opinions and enabled them to launch the necessary research to forge potent arguments. They could reasonably expect that a similar process would lie ahead.

INUIT NUNAVUT PROPOSAL

The Ottawa-based Inuit Tapirisat of Canada (ITC) meanwhile had grown and evolved to become a strong national voice. In 1973, the group founded the Inuit Cultural Institute, which selected Tagak Curley as executive director in 1975. By 1976, James Arvaluk had become president of the ITC, and its thirteen-member board included representatives from COPE (Sam Raddi of Inuvik), from regional Inuit associations in the Northwest Territories, and from Inuit associations in Quebec and Labrador. A land-claim proposal, approved by a hundred-delegate meeting in Pond Inlet, was presented to the prime minister. Many Inuit and their advisors had worked on the declaration, including former claims director Tagak Curley, professor Peter Cumming, and John Amagoalik, who had left his job with the Government of the Northwest Territories. The Tapirisat worked on three land freezes, at Tuktoyaktuk, Spence Bay, and Baker Lake. It helped to assess community views on the Polar Gas pipeline project, the Nanisivik mine project, and the NWT advisory council on game management, and it founded the Inuit Development Corporation (IDC). Its affiliated organization COPE had played a major role at the Berger Inquiry into the Mackenzie pipeline, and the Northern Quebec Inuit Association had (under intense pressure) signed the James Bay agreement.

The ITC announced its land policy in 1976, but this "Nunavut" proposal to government was later withdrawn for reconsideration "on the grounds that it did not reflect the wishes of the Inuit in the villages and contained too many ideas of non-Inuit consultants."[31] The Inuit (including Inuvialuit) claimed 250,000 square miles in fee simple, including subsurface rights down to 1500 feet of depth. They wanted the right to select 50,000 square miles from that traditional territory and render this selection unencumbered by cancellation of the existing oil and gas dispositions at government expense. Then, exploration and development would proceed with the consent of the communities and subject to agreements covering economic participation, employment, co-management, and fixed royalties, as well as social and environmental impact benefits. The ITC made allowance for cases that were deemed of national importance, proposing that parliament might expropriate such lands. Thomas Berger observed that this land ownership model had the merit of attempting to accommodate both Native and national interests. The follow-up by ITC of the 1976 Nunavut proposal contained revisions that concentrated on territorial division and the structure of Nunavut.

Thomas Berger's recommendation on April 15, 1977 to postpone the Mackenzie Valley pipeline project for ten years would allow time to negotiate and settle Native claims in the North. This was a major advance for the affected Native groups, and others could also see that aboriginal concerns were now being seriously listened to.

Unquestionably, the ITC wanted the final Nunavut proposal to reflect the wishes of all Inuit. To this end, John Amagoalik moved to Frobisher Bay and initiated a stepped-up program of community workshops. The newly published reports on Inuit use and occupancy of the 1.23 million-square-mile settlement area and the renewable resources inventory were a help. COPE, pressured by the intense oil and gas development in the Beaufort Sea and the Mackenzie Valley pipeline proposal, withdrew from the ITC land-claim initiatives to proceed with an independent land claim. Yet ITC also felt the need to protect Inuit lands and waters from intense development pressures and directed briefings to the minister about government policy on exploration and the Polar Gas pipeline proposal.

Part of the work of the Baffin Regional Inuit Association (BRIA) was to assist the Arctic Bay hamlet in its dealings with Nanisivik Mines over the mine tailings disposal system. For this work, BRIA hired a lawyer and field worker. The Keewatin Inuit Association (KIA) got more involved in the Polar Gas pipeline evaluation. Its executive director, André Tautu, worked with Calgary social researcher Frank Tester to plan a detailed study of all aspects of the pipeline passing through the Keewatin district. They submitted their proposal to the federal government for supportive funding.

POLAR GAS PIPELINE RESPONSES

The Polar Gas Pipeline was proposed and researched by a 1977 consortium including Trans Canada Pipelines, Panarctic Oils and Petro-Canada. The proposal was to build a forty-two-inch pipeline to deliver natural gas from Melville Island. The route from Melville Island crossed to Cornwallis Island, then to Somerset Island, southward along Boothia Peninsula and through the Keewatin district, passing west of Baker Lake and crossing the Kazan River at Thirty Mile Lake.

Few explorationists knew the extent of the Inuit reaction or the content of the Polar Gas impact studies directed in 1977 by Frank Tester. The first volume, *Reflections of the People*, was published in 1978. It described not only the traditional knowledge and experience of Keewatin Inuit but also their sensitization to development issues, particularly the short-lived Rankin Inlet nickel mine.[32] Tester pointed out the overwhelming importance of caribou. "Caribou were the backbone of almost every aspect of Inuit society; they determined life, death, relationships between people, hunting partners, were the substance of songs and stories, determined one's warmth [by providing indispensable winter garments], and along with fish were a very major component of the diet. This almost total dependence on caribou, which was true as recently as twenty years ago, explains the very obvious concern voiced for hunting areas and caribou by Keewatin residents. It must be emphasized that while limited alternatives to caribou are available in Keewatin communities, they are unsatisfactory, and caribou continue to occupy a central role in the total well being of the Inuit of the region."[33] Community statistics, local history, and a map of the proposed pipeline routing were provided in the work. The southbound pipeline route from the Arctic Island gas supply would pass along the Meadowbank River, cross the Thelon River northwest of Baker Lake, and cross the Kazan River at the west end of Thirty Mile Lake. In other words, the pipeline would traverse many of the well-used hunting areas near the town of Baker Lake.

Tester conducted lengthy interviews with many residents of several Keewatin communities, aided by translation services and a menu of relevant topics. A complete English-language translation of the statements of all Inuit interviewed was provided in the report without comment. The statements were almost unanimous in their rejection of the pipeline as proposed, after much study, by Polar Gas.

Michael Amarook said he was concerned about the environmental aspects, such as leaks and spills from trucks and other vehicles, and the social impact, especially from alcohol, which had ruined families at Rankin Inlet. He preferred to fight for a better deal: for the Inuit to secure royalties from oil and gas and minerals and to control their future. Amarook, a genuine activist, later became president of ITC.

John Avaala did not want the pipeline to be built around the area, especially along the Thelon, where he hunted. Already the survey helicopters were causing annoyance to people along the proposed route. The Hunters and Trappers Association, of which Avaala was president, had already complained about low-flying aircraft. He believed exploration company employees carried hidden guns to take game and fish, and he predicted the same from pipeline construction workers. The lure of pipeline jobs for young people would mean the loss of the traditional way of life.

Harold Etegoyok said it would be a lot better if the pipeline were not built. Jobs would be short-lived, and food prices would be forced up. Polar Gas had already failed to hire as many men as they had promised at Baker Lake.

David Ikuutaq said the construction of the pipeline would cause a lot of noise from machines and dynamite. It would scare the animals away, and the people needed the animals.

Toona Iqulik called Rankin Inlet an example of how equipment, planes, and prospectors drive the caribou away; the proposed pipeline would disturb the hunting area. In the Kazan River area, the drilling (by Cominco) drove the caribou away. Most of the western part of Baker Lake is where the people go hunting – the pipeline should be built farther away. Also, traveling eastward on the lake could be affected by supply barges.

Matthew Innakatsik also referred to the Rankin Inlet experience. He said the smell of the pipeline – from oil or fuel or gas itself – would drive many animals away, and there would not be any near it at all.

Bill Kashla doubted that many pipeline jobs would come to Inuit. By the time many young people were trained, in three to five years, the project would be finished. Most of the work would go to Whites. And if the pipeline were to break or explode, what would become of communities or the white foxes? He had little faith in using "papers" (i.e., soil tests and mapping) to understand the land. He said that ever since the helicopters and the exploration crews and those people looking for resources began going around our communities, there have been hardly any caribou near our communities any more. The same will happen with construction activity and blasting. He was aware that exploration for more gas was going on to fill the pipeline but he wanted to create Nunavut first, because after they find the gas, they would be saying that this is their land. He added that all development would have an adverse effect on animals and would bring social problems – that is why the people did not want the pipeline. If it were built, there would need to be more involvement and regulation by council. Prices would go up, and the people not working would suffer, as the prices for fox and wolf skins do not go up. Kashla favored moving the pipeline farther west. When this was put to the company, they said the proposed route was the best route. Kashla said that if industry was not going to consult the Inuit, then the Inuit would oppose the development. He wondered whether the opinions of Inuit were even reported back to head office.

The same kind of critical comment was repeated time after time in every community, by people familiar with the locally sensitive areas and the social vulnerability of their people. Reverend Armand Tagoona of Baker Lake was also aware of migratory cycles and natural events, but he stressed the social disruption and loss experienced during the Rankin Inlet nickel mine episode.

David Tagoona, entrepreneur and former councillor, was more positive about the pipeline project. He expressed a need for work and realized that Inuit could not live by hunting any more, since there were not enough animals left. He thought the ITC and KIA and the settlement council should spend more time looking for jobs, not just opposing Polar Gas: right now there were no opportunities. He had little or no education and deplored welfare, so he was happy with the work he had been doing, hauling fuel and expediting for Polar Gas. He was against the construction staging area proposed on the lake. This would be across from the community, near active fishing sites, and it would encourage alcohol coming into town. He was skeptical that pipeline leaks could be fixed promptly because of the severe weather. The Thelon River crossing would be vulnerable. He did think that the project would have an influence on local prices and that men would quit $6.00 per hour jobs to work there.

In contrast, David's brother William Tagoona thought the pipeline would be the most destructive thing anyone could put on the land – it would destroy land and therefore animals. He said they might be able to clean up mining, but a pipeline would be there for all time. He favored controlled development and regulations that would give the Inuit some say in writing, so that the money would stay in the North. He feared the rough, tough construction workers who had no respect for the Inuit or the animals. He was a strong supporter of Nunavut, because he felt it would give Native people a chance to participate as equals in Canadian society. He said that the South has a passion to overdevelop, but Northerners would go more slowly: they were getting organized and could do a lot of planning. The uranium land freeze had given them the benefit of not having many exploration and development people there that summer, but they would be back, and that was disturbing, because their attitude is different from that of the Inuit. Tagoona had heard stories from his brother of the contempt shown by drilling crews.

Luke Tunguaq made a point about Inuit avoiding garbage that might "spook" the caribou. He had worked hauling drums of oil for mining exploration crews, but he quit because he heard that drums were being left all over the land. The crews' misdeeds made him feel guilty. Older people had long been making similar complaints, but nobody ever listened to them. Just as white people had made laws about caribou, the Inuit wanted to make laws concerning the helicopters that were disturbing their land and animals. Game officers had also paid no attention. He felt the same way about the Thelon Game sanctuary and the pipeline – promises made and not kept.

Tagak Curley of the Inuit Cultural Institute gave Tester a long statement in Rankin Inlet. He said Polar Gas was trying to avoid government red tape by its consultation process, but was now learning about Native red tape! In such public relations efforts, he felt, the oil and resource companies were merely advancing their own ideas. They did not listen to what the people had to say, and the people had no access to corporate decision-makers. Curley thought it was not a proper process of planning for land use, with notices of proposal after the fact. He said the Inuit took the commonsense approach: if it is going to be harmful to Natives, do not do it. The companies offered benefits like jobs and training skills and opportunities for young people – and were trying to "tame the Native people." He said, "There is no other way they can get to us, to the youth, but there will be a backlash."[34] He believed the big rush was because the Inuit wanted to establish a bargaining position, and the companies wanted to get the pipeline there before that happened. He said the industry must pursue real open involvement and participation rather than merely seducing people into supporting something, tactics he described as "sickening." The middle-aged people, the hunters and trappers, were the ones who would suffer from intensified development, losing their livelihood, values, and self-respect. If the Inuit rejected the pipeline proposal, then politicians should take that seriously. At the moment, the only way politicians were trying to understand Native concerns was with regard to land-claim settlements. To get the whole spectrum of Northern values and concerns accepted, the Inuit would have to start political lobbying. It was not good enough for one company, like Polar Gas, to try to interpret Inuit concerns. The Canadian people would have to recognize that this pipeline was going to be more dangerous and harmful than the one proposed in the Mackenzie Valley. Inuit might have the same right to oppose development and express dissatisfaction as other Canadians, but they would have to use all avenues to get their concerns out.

Norman Ford, a person with employment connections to the Hudson's Bay Company and the RCMP, said there was a lot of alarmist publicity about what would happen. People were afraid the caribou would disappear; the truth of the matter, he maintained, was that wolves had been killing off a lot of game over the years. He said there was nothing wrong with a pipeline.

Anthyme Kadjuk said that mostly young people would go and work on the pipeline, but they would forget how to hunt.

Pierre Karlik said old people always said yes to the *Kabloona,* and it was good for them, but now people should sometimes say no. They had been told what could happen. People were left behind in Rankin after the mine closed, and the miners never helped the people after they left. The people were left to live on welfare.

Joe Paterk said the pipeline was not going to affect him in any way, so he had no fear about it. He had had a good experience at the mine, although he had broken some bones. He said he could not survive any more without a job.

Pissuk said there were more caribou around Rankin than when he was young. He thought the caribou would adapt to noise and smells, but the pipeline would need people to police it, as they policed the mine when it was operating.

Donat Milortuk of Repulse Bay had a more skeptical view of rules and regulations: that they are often ignored. Inuit should have a hand in enforcement, especially when it came to garbage, and the gas and oil spills from equipment and barges. In the past, people looking for minerals had been told by Inuit not to leave anything behind. But the Inuit had found all kinds of things abandoned, even from small crews, so the people from Polar Gas would surely ruin the land, too.

Louis Orsokitik had similar thoughts about exploration garbage. He mentioned that Nanisivik people had said "yes" to a mining company, and then they had found that their animals were gone. They had asked Repulse Bay for help. The same thing could happen in the Keewatin.

Tommy Misheralak said he did not want this pipeline. When the mine at Rankin Inlet closed down, the people were left with nothing, and there were no animals around the settlement. Right now, the animals were going back to Rankin after ten years, but if the pipeline were built the animals would not come back. He felt he was paid low wages at the mine. He had worked under ground for years in that mine, but after finding out he was not making a lot of money, he had quit. He was concerned that the pipeline would be built right where the Whale Cove people trapped white foxes.

Norman Voisey thought the caribou would be scared away and also wondered about the fish. He said there would be trouble because young people who worked would not feel like hunting anymore.

The foregoing selected and paraphrased statements, as well as others from Frank Tester's report, showed a strong opinion trend: they revealed a people heavily conditioned by their past experiences of contact with whites at the mine and elsewhere. The people believed that their lifestyle would be threatened by development and did not trust developers as a group to act in their interest. There was an abiding belief that the caribou on which they depended would be adversely affected in some way – the animals were already skittish and unreliable. The contributors worried about the welfare of their youth, but had doubts about the lasting benefits of working on the pipeline. The company had not consulted the Inuit in the early stages and learned their basic concerns, nor did they welcome suggestions that might have proved mutually beneficial, such as locating roads for easier access to hunting areas. Quite the contrary: the company labored away with field studies and route analysis and presented only a final package. The Baker Lake people's suggestion to move the line farther west, beyond the Thelon game preserve, was out of the question: the best route had been selected. Inuit were quite aware that they were not being seriously consulted. On the other hand, they had a growing sense of their power to control development. Establishing such power had already become the chief ambition of land-claim discussions, so the citizens elected pretty well en masse to reject the Polar Gas pipeline proposal, interjecting into the consultation process what Tagak Curley called "Native red tape." The weight of Inuit opinion and the sense of purpose evoked by Polar Gas were not appreciated or even known to mining companies of the day. In the minds of local people, mining companies were grouped, along with Polar Gas, as would-be developers and polluters of the land. North Rankin Nickel Mines was blamed for all the social impact and the disruptive aftermath of its existence. Yet, the exploration community was scarcely aware of this impact, or of the hostile feeling in the Keewatin communities. The Inuit remained, at least for the time being, polite and reasonably receptive to exploration companies, taking up the few direct and indirect jobs that were offered.

COPE INUVIALUIT NUNANGAT PROPOSAL

After the Inuit withdrew the "Nunavut" claim in 1976, it was taken up by the Inuvialuit-backed organization, the Committee for Original Peoples' Entitlement (COPE). Their "Inuvialuit Nunangat" claim was then hurriedly put forward because of the Mackenzie Valley pipeline timetable.[35] Prompted by the concerns of the Berger Inquiry and the onset of offshore drilling in the Beaufort Sea in 1976, the 1977 Inuvialuit proposal for a land-rights settlement was a determined and reasoned document. It dealt only with the defined Western Arctic region that they traditionally occupied: 100,900 square miles of land and 53,500 square miles of sea, nurtured by another 11,000 square miles of wildlife breeding areas – or 165,400 square miles altogether. The Inuvialuit numbered about 2,500, residing chiefly in the communities of Inuvik, Aklavik, Tuktoyaktuk, Sachs Harbour, Paulatuk, and Holman Island. The hundred-page document dealt with many issues, among them land ownership, control of development, and royalties.

The Inuvialuit recognized the national importance of oil and gas development and the impending Mackenzie Valley pipeline, which they opposed on cultural and environmental grounds. Thus they took the time to delineate, with the help of the Geological Survey of Canada, those portions of the region with petroleum discoveries and high petroleum potential. They then chose lands outside those areas, with low potential, to be designated as Inuvialuit land. They were seeking to negotiate full ownership of the designated lands with the government. The Inuvialuit wished to claim 50,000 square miles, or approximately one-half of the lands they occupied, through a land selection process.

The Inuvialuit Nunangat specified three classes of land. In Class I lands, the Inuvialuit would own surface and subsurface in fee simple, excepting oil and gas rights. Private owners or lessees of oil and gas rights on these lands would lose those rights under the new agreement. The Inuvialuit would have to approve any new dispositions and developments on these lands; essentially, new developments were not wanted. Class II lands would convey surface and subsurface ownership in fee simple except for oil and gas, but existing permits would remain intact. Development could take place, but under the control of the Inuvialuit. Class III, or public, lands would remain in Crown hands; new development would proceed on these lands, but would be influenced by community corporations and hunters' and trappers' groups. The Inuvialuit, a thoroughly land-based people, saw land ownership and participation in land-use decisions as vital to their culture and economic well-being. Yet they sought in negotiations to avoid conflict over lands with oil potential, and to minimize their influence on development, by selecting lands where little development was in progress.

Inuvialuit lands shown in Schedule "D" of the proposal would comprise the bulk of selections, but an additional three thousand square miles would be selected from the Western Arctic Regions generally. It was proposed that COPE would select blocks from time to time, subject to negotiations. It was proposed to select approximately four hundred square miles of land surrounding each existing community. Each community would administer one square mile as municipal land, and community corporations would administer the surface rights on adjacent lands. The document included a detailed treatment of alienations. While the discovery areas of Atkinson Point and Parsons Lake were not included in the Schedule "D" lands, the Inuvialuit desired the critical areas of Cape Bathurst and Husky Lakes, where no discoveries had yet been made. Moreover, they were asking the government to cancel existing oil and gas rights in those areas, as they did not want any development there for the foreseeable future. The government was urged not to grant any further rights for oil, gas, coal, or minerals on any Schedule "D" lands until a final agreement was reached. After signing this agreement (or agreement-in-principle), the government would provide a six-month notice of any intention to grant new permits or licenses, allowing the Inuvialuit to initiate the land selection process on the affected lands.

As part of the land settlement, the Inuvialuit requested a royalty of three percent on oil and gas produced after May 13, 1977 on all lands in the Western Arctic Region, including subsea lands. The royalty rate was deliberately low in light of its wide application (justified, it was said, by the traditional offshore lifestyle of the people). Royalties would be payable to the Inuvialuit Investment Corporation (75 percent) and the Inuvialuit Development Corporation (25 percent). A royalty cap was set at $5 million plus one-half the per capita GNP for the previous year multiplied by the total Inuvialuit population for the same year.[36]

The COPE negotiators understood that aboriginal rights were multidimensional. In recognizing that extinguishment would be part of a final agreement, they specified that Inuvialuit were striving for fair compensation and benefits for extinguishment of land rights only.[37] They realized, of course, that the aim of the federal government was to extinguish *all* rights, and there was apprehension that unextinguished rights were not specified in the 1977 proposal. Indeed, the ITC was advised that the Inuvialuit Nunangat was a sellout because its legal language did not specify exclusions.[38]

Subsequent to the Inuvialuit claim, negotiations had to be undertaken with a new and unfamiliar minister of Indian and Northern Affairs, J. Hugh Faulkner. The negotiations with Faulkner and staff did not go well. It was reported that "the government offered very small amounts of land and royalties and their offer was considered unacceptable by COPE negotiators as a fair settlement of Inuvialuit land rights. The COPE negotiators felt that the government didn't want to give any more because they want our settlement to be similar to James Bay and the government does not believe the Inuvialuit or any Native people have any legal rights to their land."[39] The James Bay Agreement of 1975 granted the Crees surface ownership of 2,140 square miles of land without any subsurface rights. This amounted to a paltry 1.3 percent of their traditional lands, or 0.32 square miles per person. Though there appeared to be "insurmountable divergence between the positions of the two parties," a breakdown (and possible litigation) were avoided by establishing a working group of COPE and government representatives to seek a common position.[40] Inuvialuit negotiators kept up their efforts by concentrating on other priorities expressed by their people, such as the protection of whaling, sealing, hunting, and trapping regions from encroaching oil and gas exploration. In regard to royalties and compensation matters, they hired private consultants to help them calculate the optimum cash and land compensation that could be negotiated in a settlement.[41]

The year 1978 was significant. Pursuant to the Berger report, the North Slope of the Yukon was removed from public-land status and categorized as a future national park for preservation's sake. Northern people viewed this change as a positive step toward land preservation. The working group agreed that one of the basic goals of a settlement was the provision of fair compensation or benefits in exchange for the extinguishment of Inuvialuit land rights. The position paper finally submitted to Cabinet in May 1978 contained the phrase "specific rights, benefits and compensation to the Inuvialuit in exchange for any Inuvialuit land rights that now exist."[42] Discussion of the paper ensued at the COPE general assembly, after which COPE signed the "Inuvialuit Land Rights Settlement Agreement-in-Principle" in October 1978.

COPE AGREEMENT-IN-PRINCIPLE

The uniqueness of the Agreement-in-Principle (AIP) was a result of the vision and perseverance of the Native people and of national pressures in relation to northern petroleum development. Sam Raddi, the president of COPE, described it this way: "This settlement cannot be thought of as a model. COPE worked hard over a long period of time to develop our own claim. We made over 14,000 interviews with individual Inuvialuit to find out what they wanted. We worked hard during negotiations to avoid certain parts of the James Bay model and the Alaska model which did not suit our needs. There cannot be a model. Each claim must be developed and settled on its own merits. That is what the government policy is for comprehensive claims. We hope that other groups use whatever parts of the Inuvialuit, James Bay, and Alaska settlements that are good for them."[43]

The Position Paper and subsequent Agreement-in-Principle specified that a settlement would provide the Inuvialuit with the following lands: five thousand square miles in fee simple with surface and subsurface rights, comprising seven hundred square miles around each of six communities and eight hundred square miles in the specially prized Cape Bathurst area, where an exploration moratorium was in effect and dispositions (alienations) were to be terminated by the government. In addition, the Inuvialuit would acquire surface rights to 32,000 square miles of the settlement area, subject to existing alienations. The total 37,000 square miles would legally become "Inuvialuit lands" with the completion of the final agreement – a generous allotment under modern settlements, according to legal expert Constance Hunt.[44] The total allotment amounted to 22 percent of the Inuvialuit traditional area, compared to 11 percent won by Alaska Natives and 1.3 percent by the James Bay Crees. Keith Crowe of the Office of Native Claims pointed out that this was the first Canadian agreement that conveyed both surface and subsurface rights, including the beds of lakes. The surface lands (30,000 square miles) amounted to 12.5 square miles per person, and the subsurface lands (5,000 square miles) to 1.56 square miles per person.[45]

The COPE Agreement-in-Principle specified a land selection process that excluded privately owned land, existing subsurface dispositions, and oil and gas reserves. Surface ownership allowed right of entry to present and future Crown mineral-rights holders. In return, the Inuvialuit would gain the right to negotiate with developers for rentals, fair compensation, employment, service contracts, and other business arrangements on Inuvialuit lands. Royalties were not negotiable, but the Inuvialuit would receive royalties from oil, gas, mineral, and quarrying rights in existence on their lands at the time of the final agreement.

GATHERING SENSE OF UNIFIED ACTION

Propelled by the bullish uranium market, exploration flourished from 1969 onward in favorable portions of the Northwest Territories, especially the Keewatin region. H. Herchmer remarked that by 1969, the various activities of mining companies had become a "major subject for concern in the community," but there was no forum to express that concern until the debate over aboriginal rights materialized in the 1970s.[46] Herchmer's remark may have referred to criticism about Rankin Inlet's nickel mine experience, or to the Indian Eskimo Association's discussions with activist Tagak Curley and the subsequent Coppermine conference in 1970. Eight Keewatin residents attended that conference, including Peter Ernerk of Churchill and Lucien Taparti of Rankin Inlet. Ernerk, who served as a conference translator, said, "How many of us realize that this vast land of the NWT is our land? You and I may think so. How many in Whale Cove or Repulse Bay or Rankin Inlet think about this? They think of the spot where their house is as theirs but do not stop to think about the region around their settlement." Agnes Semmler of Inuvik responded, "I don't know, do they?" Ernerk replied, "No, they don't recognize the situation."[47] As an Inuktitut-language CBC commentator, Ernerk was determined to spread the essential message of this conference. Noah Qumak from Sugluk, Quebec, one of the most inspired contributors, spoke of the damage and destruction of wildlife and the environment by white man's activities – and the lack of compensation. Possibly the nearby Asbestos Hill mine was on his mind. He was a strong supporter of unified action and of resistance to Quebec's takeover in Ungava, and his words drew applause at the conference.[48] The influence of Western Arctic representatives, whose focus was on oil and gas exploration, and of Peter Cumming's legal advice, had a profound effect on other Inuit attendees. The conference resolutions called for a halt to certain activities and consultation with local Native councils on development activities.

In 1974, the residents of Baker Lake, concerned for their caribou resource, sent a petition to the federal government requesting a freeze on new mining activities in their area. This plea was turned down by the department, so the community sought the assistance of the Inuit Tapirisat of Canada. The ITC's formal proposal of 1975 to Minister Judd Buchanan was also turned down, on the grounds that the territorial land-use regulations were sufficient to control abuses. It was not until 1977 that the new minister, Warren Allmand, agreed to a one-year cessation of mineral rights dispositions and land-use permits in the area. Existing permit holders, however, were allowed to continue their activities. Meanwhile, the government commissioned a study by Interdisciplinary Systems Ltd. of Winnipeg to investigate the impact of exploration and development activities on Kaminuriak caribou herds.

In his 1976 annual report, ITC president James Arvaluk described 1976 as a year of steady pressure from exploration and development and quiet progress on the main project of Inuit land claims.[49] The Tapirisat's comprehensive proposal (the Nunavut proposal) had to be withdrawn in September 1976 for more study and consultation with community groups. Arvaluk hoped to move as quickly as possible with a resubmission "to protect our land and resources while there is still something left to claim, but at the same time it is vital that our decisions are the right ones, because our children and their children will have to live with them forever." He welcomed the appointment of Warren Allmand, who had promised to review government policies concerning oil, gas, and mineral exploration permits and might order public hearings into the effects of the Polar Gas pipeline on Inuit communities. The project committee had representatives right across the Arctic. Those from the Keewatin region were Michael Amarook (Baker Lake), Thomas Tiktak, Peter Tapatai (Rankin Inlet), and Michael Mautarinaaq (Arviat).

When land-claim negotiations began again in 1978, there was some expectation that the land freeze would be renewed – but, just in case, the citizens of Baker Lake sent a petition signed by 80 percent of the population requesting that the freeze be continued until a settlement was reached.[50] The new minister, Hugh Faulkner, balked at this request, since the Interdisciplinary Systems consulting report questioned whether exploration activity had any significant effect on caribou behavior.[51] The companies argued that the caribou had been overhunted, an opinion that was endorsed by some Canadian Wildlife Service biologists, but the CWS had documented much higher kill-rates in the 1950s, and this fact gave the Inuit a basis for rejecting the notion of overhunting.

The minister announced that the land freeze would end, except for certain restraints at times when, and places where, caribou are especially vulnerable. He did not mention the land-claim process. Eric Tagoona, the new ITC president, responded, "We were told for so many years that the federal government was our protector; but now it seems as though they just can't wait to give our land away to the mining companies."[52] Baker Lake representatives made last-ditch attempts via the courts to find a solution to the problems allegedly caused by mineral exploration on the 30,000 square miles of land used for caribou calving and migration. Four restrictions were requested: a total freeze on mining activity within twenty-five miles of caribou crossings, community referrals in the exploration approval processes, environmental and socioeconomic studies before issuance of licenses, and proof that minerals could not be found elsewhere. The community sought a permanent court injunction against mineral exploration, but this action might have far-reaching effects. Gordon Wray, the settlement secretary-manager, said, "If we go to court the case will be fought on the issue of aboriginal rights. It could have a dramatic effect on land claims in the NWT."[53]

CARIBOU ABORIGINAL RIGHTS CASE

Thus began the Baker Lake court case, the first legal attempt to argue aboriginal rights in the Northwest Territories. Eric Tagoona was successful in launching the case as a joint action by the hamlet of Baker Lake, the Baker Lake Hunters and Trappers' Association, and the Inuit Tapirisat of Canada. The defendants were representatives of the federal government and six mining companies – Urangesellschaft, Noranda, Pan Ocean, Cominco, Western Mines, and Essex Minerals – that collectively had spent over $11 million on uranium exploration in the Keewatin. More than a hundred residents of the hamlet supported the action, though many were apprehensive about this bold step. On April 18, 1978, the lawyers for the plaintiffs asked Justice Patrick Mahoney of the Federal Court for an interim injunction, which was granted on April 24, pending the outcome of a full-fledged court trial. The injunction did not apply to all mining company activity, but placed restrictions on the most delicate areas, such as caribou water crossings and calving grounds, and the most sensitive times of year for the caribou. The government followed up by applying the same restrictions throughout the Keewatin region's caribou migration routes. Justice Mahoney said in support of his action that the Baker Lake Inuit had an exceptionally strong psychological and physiological dependence on the caribou hunt and harvest. The minerals could wait, but the caribou might not.[54]

When the case came to trial in the spring of 1979, the Inuit plaintiffs based their case on aboriginal title to the subject area, verified by occupation from time immemorial. Aboriginal title gave the Inuit land rights that the government could not abrogate unilaterally. Various permits and licenses to prospect and mine for minerals were therefore improper confiscation and should be halted. Corporate and government lawyers mounted an aggressive defense, contending that (1) Inuit had not occupied the Baker Lake area, and (2) Inuit had never had any aboriginal rights. Proof of Inuit presence was then provided by archeologists, who were brought in at considerable expense.

The cost of this litigation became a heavy burden. The hamlet had raised $60,000, but saw their money depleted quickly while obtaining the injunction. Much later, on June 13, 1980, James Wah-Shee, minister of local government in the Northwest Territories, announced that the territorial council would contribute $30,000 to help pay the hamlet's legal bills. Meanwhile, the ITC found itself having to raise $110,000 in costs and fees, and turned for help to the Indian Eskimo Association and its successor, the Canadian Association in Support of Native People.

Aboriginal right was contested by the legal argument that the Inuit were too primitive and disorganized to assert any sort of sovereignty over the land. Justice Mahoney disagreed with this argument, stating that four criteria were essential to establish aboriginal right: the Native peoples must have occupied and used the land, they must have had some level of social organization, they must have been there continuously since European occupation, and they must not permit other groups to take over their territory. The judge acknowledged that the Inuit met all four conditions except in the southwest region, outside the Baker Lake area, where land use by roaming Chipewyan groups overlapped Inuit territory. Justice Mahoney concluded that aboriginal Inuit had an organized society, not with elaborate institutions, but organized to exploit available resources and sustain life.

But what were the implications of Inuit legal rights, if the court chose to recognize them? William Graham, the lawyer for Pan Ocean, contended that such recognition would be tantamount to a loss of jurisdiction by Ottawa. The government's inability to administer the lands of the Northwest Territories would reduce their potential for development; the result would be a wholesale shutting down of resource activity, which depends on access to land. While mining companies acquired subsurface rights under the Canada Mining Regulations, they still needed access to the surface of the land to explore and work the minerals. Aubrey Golden, lawyer for the Inuit, disputed this doomsday scenario. The Inuit, he insisted, were claiming only protection of hunting and fishing rights in this litigation, and were not trying to be intransigent over mining. Lengthy arguments took place on the question of extinguishment of aboriginal rights.

Justice Mahoney finally ruled that what the Inuit possessed were not property rights in the true sense but that their rights had not been extinguished, even inadvertently. He ruled that the rights were vested in common law – guaranteeing the legal right to move about freely and hunt and fish on their traditional land. Whether such rights had been infringed depended on whether the caribou had been adversely affected by mining. Mountains of testimony were advanced by expert hunters detailing the changes they perceived in animal behavior as a result of mining-company activities. Government wildlife experts offered contrary views on the decline of the Kaminuriak and Beverly caribou herds. The judge concluded that there was not enough admissible evidence to prove the point and justify the injunction against exploration. He considered all forms of potential harassment, such as noisy drilling, camps, and low-flying aircraft, and admitted only that there was a special problem at caribou water-crossings.

The case was initially considered to be a mixed victory for the Inuit. A form of usufructuary aboriginal right had been affirmed by Justice Mahoney, but exploration would be allowed to continue, subject to restrictions at areas like water crossings.

Justice Mahoney's ruling, announced on November 15, 1979, was a landmark decision on Native rights in Canada, comparable in importance to the Calder case of the Supreme Court, a split decision which acknowledged the existence of aboriginal rights in 1969.[55] The initial reaction among the Inuit was one of indignation and disillusionment. The ITC expressed its delight that aboriginal title had been recognized, but needed to figure out how this would help the Inuit in general, and residents of Baker Lake in particular. Though *Oilweek* [56] claimed Native rights were now secondary to mining, the ITC believed the door had been left open to other injunctions, if the evidence warranted, so that mining would never be allowed at critical points such as a caribou water-crossing. They understood correctly that the decision would raise concerns about mine financing because clear title was inaccessible. Indeed, mining companies had immediate qualms and pressed Justice Mahoney to clarify his decision by exempting permits and licenses (that is, to declare that aboriginal rights would not prejudice these dispositions). Such an exemption would serve to forestall any protest, regardless of the impact on wildlife. The government lawyers endorsed this petition, but Justice Mahoney refused to act on it and change his decision, as this would invalidate his findings. The questions for exploration now were: How far may a company go before finding itself in violation of aboriginal title? What does it take to cause a "behavioral change" in wildlife? There was now a cloud of uncertainty over northern mining.

As for the Inuit, their right to control events on their lands was indeed fragile, because the legal system gave precedence to third parties, while the Inuit could only resort to after-the-fact injunctions. The resolution of the issue had to come via land-claim negotiations, which were being accelerated (for example, a breakthrough in Inuvialuit negotiations occurred in 1982) but would take some time to conclude. In the meantime, relations between the Inuit and industry would remain strained.[57] The Inuit Tapirisat of Canada began to run advertisements inviting developers throughout Nunavut to negotiate and resolve potential conflicts between their projects and the legal rights of Inuit. The Inuit expected that companies would agree not to erode aboriginal rights or to stake additional claims on Crown lands of possible interest to the Inuit. The companies, however, expressed no desire to negotiate with the ITC, believing there was already too much regulatory red tape to contend with. The companies maintained then and later that it was the government's prerogative and duty to deal with Native concerns.

In the wake of the financially draining Baker Lake case, there appeared to be a new urgency for a comprehensive land-claim settlement wherein aboriginal rights might be ceded to the Crown. A speedy conclusion of the negotiations would put to rest the lingering questions of respective rights in the North. The negotiations were stalled for a full year while the Office of Native Claims sought a new mandate from the federal cabinet. Since a legal course had failed the Inuit, a political settlement was their only resort, although John Amagoalik said the mining lobby could pressure the government to keep the North "dangerously close to colonial status" for an indefinite period.[58] Thomas Suluk, ITC's chief negotiator, said that if the legal system and fair negotiations failed them, then another unpleasant recourse might be necessary. Thus the industry's nightmare – that the Inuit might use pressure tactics and non-cooperation to slow mining down – could become a reality. The new reality of unease in the boardrooms of Bay Street probably strengthened the hand of the Inuit at the negotiating table.

Another point of view about the Justice Mahoney decision, and one close to the heart of some Inuit, was that of Frank Tester, the University of Calgary researcher who was personally involved in the Baker Lake caribou issue and the Polar Gas pipeline assessment.[59] Tester asserted that this was virtually an issue of health: the physical and psychological well-being of a Native population that was at risk, or even endangered. He said that the judge was not concerned in the case about what was happening to the people, but only about what was happening to the caribou. Caribou remained the backbone of the community, especially the elderly, who could not tolerate store-bought sources of protein. The nutrition of children, nursing mothers, and those working outdoors in the Arctic winter climate was also in jeopardy, according to Tester. He contended, "The decision effectively denies the residents of Baker Lake the right of any control over the area in question."[60] The result could be resignation and loss of any spirit of energy and commitment in determining their future.

Tester felt this could be a repeat of the historical dealings with Indian people, their marginalization, and consequent loss of pride and purpose. Neither did he feel that the opportunities the mining companies could offer to young people or the social contact with mine representatives would be good for the community. Social impact was clearly not a consideration for the mining companies, and southern Canadians continued to place others at risk for the sake of economic development.[61] Observers may have seen Frank Tester's comments as an overreaction to events, but his critical powers, seen also in his later writings on Inuit relocation events, should have caught the attention of the mining industry.

The health of Keewatin residents was also the subject of a speech by artist and civic leader Michael Amarook of Baker Lake in 1981, when he was the president of ITC. Disease attributed to the influx of developers from the South, something unknown to his father, had pervaded the population, especially Inuit babies. Bronchial infections and conditions were of epidemic proportions. Even so, not one of the seven Keewatin communities had a resident doctor. The resident nurses were unequal to the task, and there were no Inuit nurses at all in the Northwest Territories. Health and Welfare Canada funded an inquiry into health care in the Keewatin region as a result of his intervention.[62]

To what extent did Inuit set their sights on the mining industry in the years that followed the Justice Mahoney decision? They realized that third-party interests had the upper hand, since their declared aboriginal right did not convey jurisdiction over land. Becoming owners themselves was one sure way to fight back. Another was to negotiate equal rights of management over all lands. The Mahoney court decision bolstered their confidence and helped to establish their vision and position at the bargaining table. They had made an impact on industry, learned to use professional advisors, and asserted Inuit values and beliefs concerning their habitat.

Before examining the tactics inspired by this new confidence, let us look at one case history that illustrates the path of developer action and Inuit reaction.

4. THE NANISIVIK MINE

INTRODUCTION

The best-documented episode of mining history in the Arctic is the discovery and development of the Nanisivik zinc mine at the north end of Baffin Island (73° N, 85°20' W) which operated from 1976 to 2002. Nanisivik provides a unique case history of Inuit in the small isolated community of Arctic Bay living in contact with prospecting, exploration, and development over several decades. Theirs was not a combative relationship like the parallel events in the Western Arctic or the Keewatin (the Banks Island and Inuvik seismic surveys and the Baker Lake uranium exploration); instead, it involved patience and cooperation. The people had no preconceived ideas about the ills of development and welcomed the opportunities for employment. Although the Nanisivik project was nurtured by the federal government and believed to be a model for Arctic Canada, it actually went ahead without meaningful input from the people of nearby Arctic Bay or even from the ITC, the emerging national voice of the Inuit.

During the development process, the people had an earnest desire for consultation as they became aware, by way of rumor, that others were making decisions that would affect them. That this consultation did not happened was a story of omission rather than commission on the part of the developer, who failed to understand the genuine need for meaningful communication. The development of the Nanisivik mine was a success story, and Inuit have used employment there since 1976 to serve their own ends. It was also a huge learning experience that edged them out of isolation into the mainstream of the Native political process.

EARLY HISTORY

The discovery of minerals on Strathcona Sound has been attributed to prospector Albert English, who accompanied Captain J.E. Bernier on his 1910 voyage of the *Arctic*.[1] This government-funded ocean patrol, launched to contact foreign whaling ships in Canadian waters, wintered in a protected harbor off Admiralty Inlet (later named Arctic Bay in honor of the Bernier vessel). During that winter, the crew explored and mapped the general area by dog team, with the assistance of the local Inuit. Numerous mineral occurrences, chiefly copper showings associated with mafic igneous dykes (ribs of black igneous rock), were recorded in Captain Bernier's log of the voyage.

By far the best mineral occurrences were discovered on Strathcona Sound, lying some seventeen miles to the northeast of the anchorage. This area was not notable for its game, but was later described by the Inuit as "the place where people find things." This designation gave the sulphide mineral deposit its name: Nanisivik. The "things" the Inuit referred to were probably the massive pyrite crystals that commonly occur there in outcrop, which were valued as a primitive means of fire making. Diamond Jenness,[2] who visited the northwest Copper Eskimo people during the Canadian Arctic Expedition of 1913–18, observed that "the ordinary method of making fire was by striking together two pieces of iron pyrites, for which there were several sources around Coronation Gulf." Others have also observed this practice in the North.

It seems likely that Inuit dog-team drivers showed Captain Bernier's prospectors the places of interest that were already known to Northerners. Margery Hinds, who taught school in Arctic Bay in the early 1960s, made reference to local guiding of prospectors in her biographical book, *High Arctic Venture*.[3] Amaroali was a venerable elder in the community who had worked as a guide and sled driver with prospectors in his youth. His greatest discovery, however, was not minerals, but an old ship's mast that had washed ashore about fifteen miles from Arctic Bay. Such a bonanza of wood far outvalued minerals in this treeless land. Hence Amaroali reserved it for himself and doled out quantities of wood chips at appropriate times, such as the spring washing of polar bear skins for the Hudson's Bay Company.

The geological reconnaissance of north Baffin Island was carried out by Robert Blackadar and others of the Geological Survey of Canada in the summer of 1954.[4] Blackadar and his assistant Roy Lemon landed on the sea ice at Arctic Bay in early May. For two months, the two men traveled separately with the aid of hired Inuit and their dog teams. They mapped the bedrock of this rugged region, primarily Proterozoic and early Cambrian sedimentary rocks, along routes that totaled nearly a thousand miles. Then they devoted the remainder of the summer season to more detailed examination of the geology on foot and by boat. During this phase of the work, they attempted to find the mineral occurrences reported by Captain Bernier and to appraise the area's economic geology. This task was difficult because the descriptions in the Bernier log were vague or inadequate and because some metal occurrences involved minimal or trace quantities. Disseminations of the copper-bearing mineral chalcopyrite were found within and alongside the intrusions of gabbroic dykes that cut across the sedimentary formations west of the entrance to Arctic Bay. Small veins and pods of magnetite and red hematite were found in similar associations. The ferruginous hematite was known to the local Inuit, who had mined it as a paint rock from deposits located between Arctic Bay and Victor Bay. Blackadar thought that none of these occurrences had any economic potential.

A single occurrence, and by far the most interesting, was located on the south side of Strathcona Sound after some searching within the Society Cliffs formation of dolomite. This was a carbonate rock some nine hundred feet thick overlying the Arctic Bay shale. The showing on Strathcona Sound was described as a zone of weathered pyrite, galena, and sphalerite with a maximum width of five hundred feet and traceable for two miles along the formation. Significant widths of solid pyrite were observed. There was termination or lensing-out of sulphide veins, which suggested to Blackadar that the deposit might not "extend to any great depth." Laboratory spectrographic analysis by the GSC determined that the samples contained major quantities of iron, strong traces of zinc and lead, and a minor trace of silver.[5]

Owing to time constraints and the difficulty of finding the sulphide deposit, Blackadar did not spend very long at the site. He could not find the four veins outcropping in a ravine described by Albert English, the original 1910–11 discoverer. Blackadar did notice several shallow trenches that attested to the work of J.F. Tibbett and J.W. McInnes, who prospected the area in 1937 but never returned. The season's work had been a success, and Blackadar accorded official thanks to his capable Native assistants from Arctic Bay, Attagoocheah (Isaac Attagutsiak), Kakaseak, Levi, and Neuterajuk.

Robert Blackadar was a field geologist noted for his adaptation to the Arctic environment and its Native inhabitants. He learned Inuktitut from contact with Inuit boatsmen during his earlier coastal mapping of the Cape Dorset region of southern Baffin Island. He always accommodated impromptu seal hunting on these expeditions, so that the men he employed did not have to sacrifice the hunting time that was so important to their families.[6]

The north Baffin Island reconnaissance work of Blackadar was published by the Geological Survey of Canada in 1956.[7] It came to the attention of geologists Dick Mollison and Walter Holyk, who were then busy with a zinc–lead discovery at Half Mile Lake in New Brunswick. Their company, Texasgulf Inc., had launched a program of worldwide exploration for sulphur and metal sulphides and was a major participant in the expanding sulphur business of the 1950s. Mollison and Holyk were quick to appreciate the significance of the Strathcona Sound metallic sulphide showing. It was large and unexplored, and it was hosted in carbonate rocks that were reminiscent of the Mississippi lead–zinc field.

Although the showing was eight hundred kilometers north of the Arctic circle, it was favorably located on the coast, in an area frequented by early shipping and by modern patrol vessels. The following summer, Mollison and Holyk flew to Churchill, Manitoba, and then "hitch-hiked their way to Baffin Island with two [un-named] Eskimos, who piloted their whaleboat through heavy seas and floating ice."[8] This was a cold, 2,200-kilometer voyage across ice-choked Hudson Bay, a truly dangerous trip through Fury and Hecla Strait and Prince Regent Inlet to Lancaster Sound, with stops at Arviat, Repulse Bay, and Igloolik. But the journey was worthwhile, as the men staked fifteen claims over the zinc showing before catching a patrol vessel going south. A field program was initiated in 1958, and over the next twelve years Texasgulf explored and diamond-drilled the zinc mineralization along a strike length of 2,800 meters. They also collared an adit at the eastern end of the ore body and completed about seven hundred meters of underground drifting and crosscutting.[9]

The two Texasgulf geologists regrettably left no record of the names of their Inuit crew on that memorable 1957 trip from Churchill. However, the trip was an exemplary display of the skill and character of the Inuit and of the wisdom of employing them for appropriate tasks in their own land. Texasgulf promptly hired local men from the community of Arctic Bay when the summer exploration work began in 1958. Isaac Attagutsiak and others secured yearly employment as carpenters and field assistants. The subsequent owners, Mineral Resources International (MRI), described Attagutsiak upon his death in 1984 as a natural leader among his people and one of the true founders of Nanisivik.[10]

EXPLORATION AND PRE-DEVELOPMENT OF THE MINE

The people of Arctic Bay gave generous support and participated actively in the Nanisivik project. There is a record of their early work history. In 1969, according to a government report,[11] twenty-four whites and ten Inuit were working at the mine site, with an hourly pay rate of $3.50 for white miners and $1.75 for Inuit laborers. An Inuit straw boss – probably Attagutsiak – was also employed to facilitate communications, and he was paid $2.00 per hour. The Inuit and their families were housed at a location of their choice, on a nearby beach on Strathcona Sound, rather than living at the prefab mine camp. Two of the families occupied company plywood houses, while the others lived in tents with plywood floors and oil stoves. Texasgulf provided the Inuit workers lunch and supper at the camp kitchen and supplied them with snowmobile gas, fuel, canvas, and plywood, as well as fresh food for their families. Summer living conditions at the beach were like those at Inuit hunting camps, and therefore acceptable to occupants, but fell short of conditions at established mine camps.

When Baffin regional government administrator Neil Faulkner was informed of the situation, he pointed out the inadequacy of both this kind of housing and the rate of pay, which fell below that paid to service workers in the communities of Arctic Bay and Resolute. Other official complaints centered around the distance to a source of provisions and the fact that children might miss school. In short, Faulkner considered the mine conditions intolerable and unnecessary for mineral exploration employment. He resented the fact that his Branch, charged with the general welfare of Inuit residents, had not been informed by the developers or by resource-oriented branches of DIAND, which were obviously not aware of the implications of development. He contended that "[w]ithout careful planning, the project could easily become another disastrous experience for the Eskimo people, with more serious consequences."[12]

The squatters' beach camp brought no expression of outrage from the Inuit involved, or from the community of Arctic Bay, but such observations by local or regional bureaucrats, made long before there was any certainty of a mining development, drew attention to the possibility of unwitting, injurious social effects upon local people. The spectre of an ad-hoc squatters' village would later prompt government community development officers to make sure that a proper townsite was established at the mine.

Exploration companies operating for a few months in summer apparently did not want to compete for the few skilled or experienced men who held the community service jobs, though they might easily have lured them away by offering higher wages. Nevertheless, both Texasgulf and Panarctic Oil were successful in attracting a few men from Arctic Bay into wage employment during their operating seasons. The average hourly wage paid in the Canadian mining industry was $4.04 in 1971 and rose to $4.82 in 1973.[13]

By 1969, diamond drilling and bulk sampling had established that a substantial deposit of base metals was present at Nanisivik. The ore reserve estimate for the Strathcona Sound claims, calculated to be around six million tons grading a favorable 16 percent lead–zinc equivalent (predominantly zinc), was worthy of serious consideration for development.

During 1970 and 1971, Texasgulf carried out protracted but unsuccessful negotiations with the federal government through DIAND. The principal objective of the company was to secure an export license for sulphide concentrates, so they could bypass Canadian zinc and lead refineries, which had their own sources of supply. DIAND balked at this proposal. Loss of economic benefits to Canadians and concerns for the environment and for the social impact of a large development in Baffin Island on local residents were cited as reasons for rejection. What colored this view was the body of experience within the government on northern developments arising from the Rankin Inlet nickel mine and community rescue mission of 1962, COMINCO's Pine Point lead–zinc mine and government-funded railroad in 1964, and the Anvil lead–zinc mine in the central Yukon, backed by Dynasty Explorations and Cyprus Mines, in 1969.

Texasgulf, realizing that a foreign-owned company with international perspectives would face an onerous task in developing the Strathcona Sound property, began to consider selling out. At this point Mineral Resources International (MRI), a junior Calgary-based resource company, entered the picture. MRI owned an attractive sulphur property with proven reserves adjacent to larger Texasgulf holdings in Mexico, and they could readily appreciate that Texasgulf should be the operator of a development there. Learning that the Strathcona Sound project was stalled, MRI president Frank Agar offered to trade properties. This initiative became all the more appealing after Agar checked the general feasibility of the Strathcona Sound project with consultants Watts, Griffis, and McQuat of Toronto and ascertained from government that an export permit would be available if suitable provision were made for Native employment.[14]

The property exchange was concluded in 1972, on the terms that MRI could acquire a 65 percent interest after development expenditures. These expenditures would include a feasibility study, to which Texasgulf would contribute $300,000. The major company would be entitled to its 35 percent royalty payments from production after MRI had recovered its development costs. Without its own source of development capital or any mining experience, MRI chose to rely on the feasibility study by Watts, Griffis, and McQuat, along with outside expertise provided by Strathcona Mineral Services, to attract loans and government infrastructure investments to finance the project. The buoyant mining economy prompted MRI to attempt this challenge and develop the mine themselves.

SOCIAL ENVIRONMENT

Inuit society was experiencing a very significant change in the early 1970s, reflecting the natural response of the people to a decade or so of community and educational development. On the one hand, the people were being encouraged to assume a larger responsibility for their affairs through settlement councils, but the real power and authority remained in the hands of the government through Northern Service Officers (from 1954), area administrators (from 1959), and settlement managers:

> The people of the far north may be told that the resolution of their social and economic problems will become their own affair, but the degree of autonomy that such a suggestion implies is not at present countenanced by the persons who continue to make the rules.[15]

Hugh Brody, in his study of the often-strained relations between Inuit and Whites, concluded that a new attitude had emerged:

> All Eskimos are awed by the extraordinary range of power and wealth possessed by the government and medical officials, just as they were once impressed by the missionary, trader and police triumvirate. But new attitudes, which contain elements of hostility and resentment, are also growing even in the remotest and least developed communities. Many, perhaps most, Eskimos feel that the Whites have not always treated them honestly, fairly or generously, and many young Eskimos are indignant at such treatment.[16]

As a consequence of this awakening, reaction to the encroachment of developers had already been profound. Oil and gas leasing – and the resultant seismic activity on Banks Island (the site of a prolific trapping industry), the Mackenzie Delta, and Southampton Island – spurred on emerging activism.

It is against this background of ferment that we must understand the reactions of Native participants, including the Arctic Bay settlement, to the Nanisivik project.

The other fundamental problem discussed by Brody – the failure of effective communication between Inuit and Whites even within a single community – also had a bearing on events. If settlement administrators failed in their relations with Native residents and failed to respond to their initiatives, was there much hope for a successful interchange with mining people? This was the era when integration of Native people, as opposed to recognition of their independent rights, was high on the government's agenda. Industry, too, was caught up in this thinking.

Brody was critical of the Nanisivik project and made a strong general case against any frontier development for the reason that wage-economy employment eroded the natural ambitions of the people to sustain their land-based life style. He said that "Eskimos of the Eastern Arctic will be able to live in ways that reflect their own preferences only if the dominant society's intrusion can be minimized."[17] However, he conceded that developers' schemes had found a ready acceptance across the North because of economic need. This weak acquiescence might change when a successful land-claim settlement gave Inuit the political power to control their enclaves and regulate the transition to wage-labor employment by strengthening their renewable resource–based, mixed economy. As the Nanisivik project gathered momentum, was the Inuit viewpoint at Arctic Bay forward-looking and adversarial?

FEASIBILITY STUDY CONTACTS

Upon concluding the deal with Texasgulf in late summer 1972, MRI executives flew to the property with Graham Farquharson, the mining engineer chosen to lead the project's feasibility team. Confirmatory diamond drilling and bulk sampling of the ore body had been completed to establish a working ore reserve estimate (seven million tons averaging 14.1 percent zinc and 1.4 percent lead) and to furnish crushed rock for final metallurgical testing. At the same time, the company met with the settlement council and manager for discussion of their general ideas about the project and the possibility of employing people from Arctic Bay. This early-stage meeting was said to confirm MRI's serious commitment to the Native people. It was promised that the ensuing feasibility study would evaluate human needs as well as economic factors.[18]

The next encounter with Arctic Bay people was a mid-February 1973 visit by seven company and consulting personnel to check potential townsite locations and set up a remote weather station. The party landed on the sea ice at the Arctic Bay settlement and then set out with five local Inuit and their snowmobiles to traverse the twenty-seven kilometers to the mine site. The wind velocity picked up, creating a minus 75 degree windchill during the trip. A serious breakdown of the snow machines occurred, forcing some of the party to proceed on foot. Isaac Attagutsiak and his passengers got through the storm to the ice-cold drill camp, which he promptly fired up before returning to rescue the wandering members of the party. This sobering experience left the White project engineers with an enduring appreciation for the skills of Inuit in coping with conditions on the land.[19] The same visit enabled the party to review their work with the Arctic Bay settlement council.

A number of events during the preparation of the feasibility study demonstrated the insufficient involvement of local residents. This was all the more regrettable because the government had expressed its wish for Inuit participation. Minister Jean Chrétien said in 1972 that the federal government's first priority for the present decade was furthering the interests of the people of the North in a manner consistent with their own preferences and aspirations.[20]

Graham Farquharson, the company's chief consultant, came to Arctic Bay to meet with the settlement council in August 1972, but there is no record of any substantive information being passed to the people during that meeting. Subsequent council meetings in February and August 1973 had both consultants and government officials in attendance. However, neither of those meetings significantly affected the contents of the feasibility study, so the significance of Inuit participation was minimal.[21] On the other hand, various government officers and specialists affected many company decisions. For example, DIAND insisted on a fifteen-year project life rather than the preferred optimum eight years. A longer life was preferred at Nanisivik to avoid a rapid boom-to-bust cycle like that which the people experienced at Rankin Inlet in the previous decade. It would provide more time for Inuit workers to adjust to a wage economy and to undergo the training necessary for more responsible jobs. The employment of Inuit was desirable because of government policy and because of very real labor shortages in mines down south. Both federal and territorial officials advocated a townsite to serve the mine rather than a bunkhouse community.

The people of Arctic Bay were nonparticipants in the consultant-to-government meetings, but they were aware of decisions being made that would affect them directly. In November 1972, they sent an unsolicited letter to the feasibility study consultants (whose role they understood), expressing their need to be heard, in the hopes that they might influence the shape of the development. The rumor that the entire community would be moved to the mine site had come to their attention. Originally written in Inuktitut, the letter was signed by council chairman Levi Kalluk and ninety-eight others (virtually all the adult residents). It began:

> Mr. G. Farquharson
> Watts, Griffis, and McQuat Ltd.
>
> Dear Sir,
>
> We, the Settlement Council of Arctic Bay, are writing our thoughts concerning the mine at Strathcona Sound, and the thoughts the local people have concerning the commencement of activity at the mine. Before activity begins, we are expressing the desires of the people of Arctic Bay and the Settlement Council.
>
> We are aware that the mining consultants have written to Yellowknife and Ottawa to solicit opinions on the mine, and that the governments in both Ottawa and Yellowknife have answered, stating the things that they would like done, but both governments replied without first asking us, the people of Arctic Bay, our opinions on the subject and what we desire here in our own land. And this, despite the fact that we know that the development of the mine is of the utmost importance to us and to our area. Therefore, although we have not been asked to do so, we wish to write to you to advise you of our thoughts.[22]

The letter then proceeded to discuss the important matters of townsite, water supply, road, airstrip, school, and nursing station. Potential employees wanted to live at Arctic Bay and keep their women and children away from the mine site, which was a windy place without a good harbor for small boats. Nor was the drinking water supply considered adequate. A road should be built from Arctic Bay to the mine, and Attagutsiak could provide advice on the route to be followed. Another suggestion was to locate an airstrip designed for jet aircraft near the mine site. Both the school and the nursing station at Arctic Bay were volunteered for the larger good and would possibly need to be enlarged. The letter reiterated what the consultant should understand: the people of Arctic Bay would appreciate being queried whenever the government made decisions affecting their area.

The consultants replied merely to say that the proponents were very concerned about protecting the interests of the community. There was no comment about the contradictory parameters, with regard to the townsite, on which the study was proceeding. Indeed, it was three months before any information was forthcoming or any attempt was made to solicit comments from the people of the settlement. From the Inuit standpoint, the February 1973 meeting and discussion took place in a relative vacuum: they had no prior briefing, nor did they have the time to consider their response to the presentations of the company, consultants, or government officials. The primary purpose of the meeting was information transfer, and because the tenor of response was polite, supportive and optimistic, the government officials concluded there would not be a problem of Native opposition. Nor did they think there was need for further consultation meetings until the feasibility study was completed.

The next encounter was in August 1973, when the company presented the preliminary development plan. No alternative actions were discussed, nor was any attempt made to identify problems that might ensue if the plan were implemented. The lack of critical analysis indicated to Science Council of Canada researcher Robert Gibson that the "meeting was not intended as a consultation exercise so much as a public relations effort."[23] On one point, the timing of the Arctic Bay road construction, the settlement council was asked for its preference, which may have saved the August interchange from being pure tokenism.

Inadequate project funding was a persistent problem for MRI during the year, but by October 1973 executives went to Europe armed with the completed feasibility report to arrange for major financing. That task was completed by the following April, and a complicated financial agreement was worked out. Under its terms, Nanisivik Mines Limited was created to develop the mine. MRI was to hold 59.5 percent of the company, Metallgesellschaft AG of Germany 11.25 percent, Billiton BV of the Netherlands 11.25 percent, and the Government of Canada 18 percent. Eighty percent of the mine's production of zinc concentrate was to be sold to the European partners, and the remainder would go to Texasgulf, which agreed to advance $10 million of overrun capital. In exchange for advancing the $10 million, if the capital had to be used, Texasgulf would acquire 6.5 percent of the Nanisivik Mines shares currently held by MRI.

NEGOTIATING THE AGREEMENT

The final feasibility report by Watts, Griffis, and McQuat was submitted to MRI in September 1973. It covered geological, engineering, economic, and personnel aspects of the project, as well as plans for town and infrastructure development, potential sources of funding, and preliminary environmental issues. It was considered a comprehensive report – and a progressive one too, with a positive attitude towards Native employment that was unprecedented. The feasibility report noted that mining experience at Rankin Inlet, Lynn Lake, and Asbestos Hill had shown Inuit employment to be successful, given adequate assistance in relocation, training, and counseling. At a mining rate of 1500 tonnes per day at the current zinc prices – about 22.8 cents per pound – the project was judged to be profitable and economically viable for at least thirteen years, provided the government was forthcoming with infrastructure investment capital. This last was a crucial ingredient for the success of the venture.

The GNWT and the federal government considered the MRI proposal during the next five months, and during this time Arctic Bay was alerted to prepare for the social impact of the project, but the only serious, face-to-face contact was a survey of the availability of Inuit laborers in five of the regional communities. Finally, DIAND delivered a memorandum to the federal cabinet dealing with many of the project matters, including tailings disposal and social impact, both of which required further study and the development of safeguards. On March 28, 1974, the Cabinet granted its approval in principle, along with a provision for grants and loans to the company of up to $16.7 million. The government, led by DIAND, then prepared a master development agreement, which was signed in June 1974. There was no attempt to involve local people or Native organizations in preparing this agreement.

The people of Arctic Bay and nearby Pond Inlet had available to them certain jobs on Panarctic's oil drilling projects. In 1973, not all of the available jobs were taken, but twelve Arctic Bay men did work for Panarctic on a rotational basis.[24] It was therefore a fact that this community did not suffer severe unemployment, despite the general government attitude to the contrary. Critics like Hugh Brody and Father Guy Mary-Rousselier, editor of the *Eskimo*, saw unemployment in Arctic communities as a bogus issue.[25] This reality accounted for the less-than-aggressive attitude by the local people in pursuing employment.

There was no indication that Arctic Bay failed to support the project in general, although people had misgivings on a few points and were upset about the lack of genuine consultation. An article criticizing the project, published by the Inuit Tapirisat of Canada in *Inuit Monthly*, drew an angry response in a settlement council letter to the ITC in February 1973:

> We, the Arctic Bay Settlement Council, are writing to let you know what our views are towards this development. We encourage it to develop. The same goes for Panarctic Oils Ltd. We want our men employed.
>
> We have no reason to stop the development as it doesn't intend to infringe on our hunting areas and especially where the mine is located there is no wildlife to speak of.
>
> We are co-operating with other people and they are co-operating with us. Exploration groups have been very careful not to disturb the wildlife when they are in hunting areas.
>
> The government has helped us by giving us the tools we need to progress. We are in a period when we are progressing and we don't want to live by the traditional ways. You try hunting without a gun....

The exchange suggested that an uneasy relationship existed between local people and the national Inuit association. The ITC was a newly formed body, with federal funding that enabled it to provide some expertise on problem resolution for Inuit.

A second letter, sent to government agencies during the negotiation phase after March 1974, said: "They have agreed to let the companies in. The people of Arctic Bay do not want to live the old way of life, without a gun, boat, or even a stick." Hugh Brody reported that the letter ended with the following remarkable passage:

> In the present plan the Inuit have nothing. Since we have nothing the Government could start going against us. [*sic*] If we tried to stop them or they can just forget about us. The Government may even start thinking about death or they might split us up. But if our land is going to be worked on, we should like to know the plans before the activity starts so we can help each other. Also, so the companies will have enough workers on their staff.
>
> So we can understand more fully, we would like a letter from the mining company stating the number of workers they want so we can look around for the men.[26]

These passages convey very clearly the dependence of people in the settlement upon government goodwill and largesse. The Inuit were concerned about the inadequacy of this for living the good life with ever-rising costs and scarcer resources. The well-being of the land and its game resources were crucial to Inuit life, but exploiting those resources required arms and equipment suited to the task, and so Arctic Bay needed supplementary jobs and cash income for its men. The letters also conveyed the powerlessness the Arctic Bay people felt and their lack of control over decisions that affected their lives.

It would not be long before all Inuit realized that corrective action lay in the struggle for a land-claim settlement. If land ownership and legal powers were granted them, they might one day secure a viable mixed economy in the North that would be free from overdependence on wage labor jobs or welfare from the whites.[27] At this time, the land-claim movement was in its infancy and little understood. ITC's opinion was not appreciated, nor was ITC invited by the settlement to participate in the Strathcona Sound decision. The organization could understand that Arctic Bay, a comparatively prosperous community with neither police nor any kind of alcohol outlet, was in for some surprises when the mine development went ahead.

The MRI agreement to hire Native workers in substantial numbers – and to strive for 60 percent in three years – was approached enthusiastically by the company. After all, it was a possible answer to the chronic labor shortage facing the mining industry in Canada. The emergence of a corps of skilled, mobile northern workers along the lines desired by government would be a tremendous benefit to nonrenewable resource exploration and development projects.

Housing was needed for both local and relocated Inuit workers, since all labor could not be supplied from Arctic Bay. The rumored threat of moving the entire community to the mine site had prompted the unusual letter of November 1972. In response, consultant Graham Farquharson was conciliatory and noncommittal, and the other respondent, deputy commissioner (and former mining consultant) John Parker, asserted that Arctic Bay people should be able to work at the mine and keep their homes in Arctic Bay, but argued that relocated Inuit should be housed with their families at the mine. A suitable rotational system was also needed. The Panarctic Oils system for Inuit workers had changed from twenty days on, twenty days off in 1971 to a fourteen days on, seven days off rotation to better serve their needs. The company would also accept a replacement worker if more than seven days were required for a hunting trip.

From the start, the GNWT insisted on a new permanent community; later, it rationalized this policy in a position paper refuting some of the Arctic

Bay ideas. Though the company planners originally proposed a bunkhouse community, they adopted the family town model. They developed town planning proposals in the feasibility study, arguing – erroneously – that the settlement council had insisted on it. An all-weather road from the settlement to the mine and airstrip went basically unquestioned, though it would certainly transport social problems from the mine. The Arctic Bay community presented the alternative idea of a much cheaper trail for all-terrain vehicles, but the suggestion was never taken seriously.

TOKEN CONSULTATION AND ITC CONFLICT

The consultative process, encouraged by government policy-makers, engrossed Robert Gibson in his Science Council of Canada background study of the Strathcona Sound decision.[28] He maintained that the affected party, the local people, should have been informed of the proposals in advance and dealt with patiently, in view of the cultural and language barrier. The Inuit should have been defended or supported by some of the decision-makers because they lacked legal power of their own, and also given some technical support. He found the meetings and consultations at Arctic Bay lacking on all accounts. Gibson was himself a specialist in democratic theory and the value of the public participation process, the subject of his doctoral studies at the University of Toronto.

In regard to technical matters, support and advice might have been forthcoming from the ITC, but this new national organization was still struggling to establish its grassroots support. That process was slow because the very idea of a national political organization was foreign to Inuit culture in general. The ITC emerged in the early 1970s from the White-sponsored Indian Eskimo Association, largely through the efforts of the more sophisticated Mackenzie Delta Native representatives.

By 1973–74, ITC had not established a working relationship with the Arctic Bay settlement council – an obvious shortcoming of effective Inuit communication. What chance, then, did the mining industry have of achieving balanced and effective communication, assuming it was wanted at this time? In fact, a critical communication breakdown did occur that had some serious overtones.

At the February 1973 meeting, described earlier as an affront to the Inuit, Graham Farquharson stated that he had received correspondence from the ITC expressing opposition to the Strathcona mine, and he asked the council if the national organization officially represented Arctic Bay in the matter.[29]

The people were annoyed at the idea, since the council itself had just received a request from ITC for information about the project. The ITC had no basis or right to speak on their behalf.

Though the council was alienated from the ITC by this incident, no mutual discussion of the communication with Watts, Griffis and McQuat ensued. In fact, the council wrote a strongly worded letter to Tagak Curley, president of the ITC, asking the ITC to stay out of the Nanisivik debate unless requested to participate. The council affirmed their support for the project. The irony of this event was that the ITC had not written to the consultants any such expression of opinion, but had merely exchanged notes aimed at setting up a meeting that never took place. A meeting had been suggested by DIAND's northern policy and program planning branch, and the exchange with ITC was initiated by the consultants.

Apparently, Arctic Bay people remained unaware of the true involvement of ITC for nearly two years and mistrusted them during the interim. When they discovered the error, the council concluded that the consultants had deliberately misled them. In a March 1975 letter to NWT Commissioner Stuart Hodgson, the settlement council stated that they had been "duped ... into rejecting the potential assistance of Inuit Tapirisat of Canada in consultations." They notified the commissioner that as of January 1975, they had authorized ITC to act on their behalf and to engage in consultation on all matters concerning the project. They had come to realize by that time that local Native concerns and preferences had little impact on decision making. Actually, in December 1974 the ITC flew two Rankin Inlet ex-miners, Donat Anawak and Ollie Ittinuar, to Arctic Bay to discuss their experiences. This event, however, took place six months after the signing of the Strathcona Sound final agreement.[30]

The same letter to the Commissioner tells a sad tale of the misconceptions that beset the settlement, based on a lack of critical information about the construction of a town at Strathcona Sound. The Inuit contended that this was being done without their knowledge and approval, despite their policy letter sent November 1972, which argued for a bunkhouse at the mine. They understood their treatment by the company and government was a shabby case of "tokenism" (their term). They were victims of the company's tactic of "consulting" only on matters of small importance. For example, during the August 1973 meeting they were asked when the road should be built. This restrained the council from making any statement on the real issues at stake.[31] Though completed and submitted to MRI in September 1973, the feasibility study was not submitted to the settlement until February 1975, long after the government had approved the project.

Although some government staff members had concerns about the social impact of the mine on the community, based on a study of the record at the Deception Bay (Ungava) asbestos project and other developments, neither GNWT nor DIAND directed a study of the matter to be undertaken. The only related government effort was a joint tour of five communities in the region to survey labor availability. When asked specific questions about the project, the touring officials were unprepared, evasive, or ill-informed on important matters such as the nature of tailings disposal.[32] The company's plan to discharge mill tailings into the marine waters of Strathcona Sound was vetoed by the government – it was the most serious environmental issue facing the project. The labor survey officer cited above said the waste rock would probably be put back underground when mining was finished! Apparently the people of Pond Inlet refused to deal with these representatives because of the inadequacy of the information provided. In their community newsletter, they questioned why they had not been informed about the mine (240 kilometers away) that would affect their lives for the next thirteen years.[33]

The failure of the government to examine social issues seriously or assume a duty to stand up for the Inuit, or even acquaint them with the real issues, suggests a de facto conspiracy between the company and DIAND. That assumption would have been a fair one on the part of Inuit by this juncture and probably explains their subsequent actions. They turned for help not to the company or government officials, but to Commissioner Hodgson and to the Tapirisat. In spite of this reality, Digby Hunt, assistant deputy minister in charge of DIAND's northern affairs program, made many statements extolling the consultation process and the government's concern for the Inuit. Citing the Nanisivik circumstances as well as the Bankslander case [34] and the Bathurst Island case,[35] Robert Gibson was able to state that "in cases where hunters and trappers have felt threatened by resource exploration activities, the government has consistently sided with the industrial interests."[36]

NATIVE EMPLOYMENT

The Strathcona Sound development agreement set a laudable goal: to employ Northerners for 60 percent of its workforce within three years of the mine's inception. This bold step was taken in the absence of any comprehensive study of Native employment patterns or investigation of the failure of the Anvil mine in the Yukon to achieve an agreed level of 25 percent after five years. The statistical outcome of the regional labor availability study made it necessary for the company to seek men from many northern communities. Panarctic Oil had already tapped the labor pool, successfully recruiting workers from Pond Inlet and Arctic Bay for thirty positions.

The project feasibility study estimated a labor need of 219 workers, of whom 111 would eventually be Inuit and, of those, about twenty-seven would be women. The government labor study did not really ascertain how many women would take full employment – relatively few were already employed in the communities. As for men, in the five North Baffin communities surveyed (Arctic Bay, Pond Inlet, Igloolik, Clyde River, and Hall Beach), around two hundred Native individuals were considered available, more than sufficient for the projected requirement of eighty-four. There was no serious attempt to study work attitudes and aptitudes, but researchers found that the communities looked favorably on the opportunities for wage employment. The exception was Pond Inlet, whose residents took a conservative approach because they were already prosperous, partly thanks to Panarctic Oil.[37] In an interview, Isaac Attagutsiak of Arctic Bay, longtime Inuit foreman for the companies since exploratory work began in the 1950s, said he did not think that more than fifteen local men would go to work at Strathcona Sound.

Whether the government's projections were realistic or not, their researchers did not seriously compare the experience and attitudes of displaced Inuit miners from the Rankin Inlet mine that closed in 1962. Some of those men were relocated to mine jobs at Yellowknife, Lynn Lake, and northern Ungava. Their story was related to Arctic Bay people by Donat Anawak and Ollie Ittinuar in December 1974, as part of ITC's education campaign.[38]

Like the social impact that the mine might cause, the environmental impact was not analyzed before project approval, nor were the Native people ever consulted with respect to their knowledge of the land and their dependence on local resources. The government signed the development agreement not knowing whether satisfactory land-based tailings disposal could be devised, although it properly balked at the company's more economical, original plan for discharge into the sea, which was the practice at the Black Angel, COMINCO's lead–zinc mine in Greenland.

A FUTURE FOR THE INUIT

The government's vision for the far North had come into better focus by this time. The new Arctic development policy of March 1972 was augmented by the comprehensive Strathcona Sound agreement of 1974. It was hailed as a pilot project and a worthy model for future developments in the Arctic. This was the first industrial project in the High Arctic, and others were expected to follow, like the Arvik (Polaris) lead–zinc deposit on Little Cornwallis Island and the Mary River iron deposit on Baffin Island. The Polaris mine did start in 1982: Mary River remained dormant, giving way to the more accessible deposits in Quebec and Labrador. Panarctic Oil was continuing to explore for oil and gas. The new vision was clear: the future well-being of the Inuit and other Northerners should be tied to nonrenewable resource employment. Wage employment would yield training in the trades, skill improvement, and the prospect that Native laborers could hold better jobs. They would be expected to be mobile and willing to move to new communities or commute to the workplace, as both work and rotation schedules would accommodate traditional lifestyles. Nonrenewable resource employment was often short-term or seasonal, so the trained Native workforce would have to be prepared to move from industrial project to project in the future.

This vision had evolved out of necessity, because of the needs and high cost of community living compared to traditional camp living, as eloquently expressed by the Arctic Bay Council. It was now sparked in the 1970s by the vigorous interest expressed by oil companies and prospectors. Developments had penetrated the near North, exploiting Labrador iron and Pine Point lead and zinc, and even beyond to Great Bear uranium and silver. These developments foreshadowed the lead–zinc and gold mines to follow.

In the 1960s, the prospect for northern Native people was recognized as a dismal one. Diamond Jenness, the pioneer ethnic researcher and circumpolar expert, did not see a future in mining for the Inuit, who had neither skills nor capital: mining was considered to be strictly the white man's domain.[39] Jenness recommended, in the face of a rising population and declining game resources, that serious attempts be made to move communities of Inuit down south and train them for new careers. Not many individuals concurred with this idea, or ever would. One might compare the confident vision of Inuit training and employment that Murray Watts expressed.

The government's central idea of promoting wage employment was not really discussed with Inuit themselves, and it drew the ire of northern social critics like Hugh Brody and Peter Usher, as shown below. Unlike mining, traditional occupations like trapping, fishing, and carving, supplemented by hunting – all natural and culturally sound pursuits – were not subsidized or even promoted adequately by government. There could be a flowering of these renewable endeavors, with concerted effort to accommodate the traditional preferences of Inuit, and this would maintain them in their own communities or areas in the long term.

On the other hand, wage employment in nonrenewable projects, emerging as boom-to-bust phenomena, would lure the Inuit into social disaster. A cogent example was the Rankin Inlet mine, which caused social disruption for Inuit and ended up costing the taxpayers heavily for relocation and support. Hugh Brody foresaw a bleak future for migrant workers or casual laborers because government was determined to promote a wage economy and many Inuit were willing to cooperate in taking these jobs:

> The most recent trends are pushing Native people increasingly towards the lowest and least certain rung on the national class ladder: if separated from his own means of production and unable to have a sure relationship to the intruder's means of production, the Eskimo – like many Canadian and American Indians before him – will be turned into a migrant worker, a casual labourer, and – as this *lumpen* proletarian condition develops – prostitute, petty thief and beggar. Abundant signs of this course of events are already visible. The problem will not be cleared away by promises of high pay at the golden frontier: short term booms that are so characteristic of frontier development only worsen the problems that will follow.[40]

FINAL AGREEMENT

How eagerly would the people themselves endorse the Nanisivik plan for their future? Would the land-claim movement have any effect on the outcome? At the final stage of the project evaluation process in 1974, DIAND officials sent a memorandum to Cabinet urging prompt adoption of the plan to avoid possible problems with the Inuit. Their chief reason for urgency was that, although the people of Arctic Bay were thought to be in favor of the project, "future attitudes towards the project could be influenced by the land claims issue."[41] In other words, once the Inuit understood the issues more fully, they might oppose the project. The second reason for haste was that private financing arrangements in Europe would lapse on June 30, 1974 if the crucial government infrastructure assistance were not awarded. All in all, the arguments and proposals were convincing enough for the Cabinet. On March 28, it granted approval in principle for loans and grants totaling $16.7 million.

Also during March 1974, while the DIAND memorandum was being considered by Cabinet, the company met with the Arctic Bay Settlement Council to present a slide show and status report. The company favored a quick decision by the government so that construction equipment and supplies could be sea-lifted during the summer. The meeting was chiefly a public relations effort, as the project decisions had already been made. It was at this meeting that the people of the community, asked to suggest possible names for the mine, proposed *Nanisivik*, "the place where people find things."

The discussions from this point onward served to finalize a development agreement between the company and the government. There was no attempt to involve the local people or Native organizations in the negotiations. The government was far from admitting that the Inuit had a moral or even a legal right to any of the lands involved – too early for that! The complicated document signed on June 18, 1974, was hailed by Minister Jean Chrétien as a major step forward and a model for future Northern developments. The signing took place at Frobisher Bay (Iqaluit). Levi Kalluk and Isaac Attagutsiak attended as guests of DIAND. They went with the mistaken impression that they were party to an agreement between MRI, the federal government, and the people of the Arctic. Their role, however, was simply to witness the signatures on the thirty-one-page agreement, of which only the first page was translated for them.

MINE PERFORMANCE

The Nanisivik project in North Baffin Island now went full steam ahead. The initial communiqué covering Nanisivik, Canada's first Arctic mine, stated that "the federal government would own an 18 percent equity interest in the new lead–zinc mine on Baffin Island designed to give a boost to the economy of the Eastern Arctic and to provide new employment opportunities for Eskimo people in the area."[42] This small mine was expected to employ between 200 and 250 production workers and, as agreed, the company would strive to employ northern residents for 60 percent of its workforce within three years from the start of production.[43]

The known ore deposit at startup was projected to supply 500,000 tons of ore per year for a minimum of twelve years, lasting until 1988. Exploration for ore-body extensions over the years has extended the mine life more than another ten years. Though clouded in doubt at its inception, this "uneconomic" mine has been very successful until its closure in 2002, and the company has been able to boast that it was a low-cost producer. Annual reports over the years documented the company's exploration gains, its repayment of debt, and its success with Native workers, though it never attained the original goal of a 60 percent Native workforce owing to absence from home, alternative employment, and competition from southern workers.

Nanisivik Mines Ltd. began production in 1976 under the guidance and management of Strathcona Mineral Services, the Toronto-based consulting firm founded by Graham Farquharson. Production was achieved in less than five years from the signing of the original option agreement with Texasgulf Inc. Mineral Resources International of Calgary had no previous mine operating experience and had to rely on borrowed capital and grants and the expertise of consultants. Both Graham Farquharson and Jim Marshall continued in managerial capacities at the producing mine. The capital cost expenditure was $51 million. Plant construction was completed by the end of September 1976, with an on-land tailings disposal site 3.7 kilometers away. The first concentrate was shipped from Nanisivik in 1977, on the bulk carrier *Gothic Wasa*. In all, there were five shipments totaling 93,500 tons of zinc concentrate and 12,700 tons of lead concentrate, around 50 percent higher than pre-production estimates based on seven percent zinc. The company's production was under contract to European smelters.

A road was built over the thirty kilometers to Arctic Bay from the mine's airstrip, allowing employees from that community to commute to work. The Company was able to recruit a high-caliber crew with an Inuit segment of approximately 30 percent. Many of the Inuit workers were trained for mine and mill jobs. Some Inuit families lived at the mine site, including Percy Pikuyak, the heavy-equipment operator who was later appointed a director of Nanisivik. The workforce stabilized at 180, including some forty Inuit (22 percent) and forty women employees. The company continued to work toward raising the level of Inuit participation in the workforce by emphasizing on-the-job training programs.

Exploration of the property was advanced by electromagnetic surveys and diamond drilling, which revealed a rare sulphide ore-body system of phreatic tubes with keel zones. This karst-type ore body had, geologically speaking, over six kilometers of strike length. A ten percent increase in ore reserves was made in short order. In 1981, an underground heading was driven through to the eastern end of the main ore body, a distance of three kilometers. By 1982, mining activities had extended to a newly discovered deposit five kilometers from the mill site.

The workforce proved to be exceptionally stable, largely because family housing was available at the townsite. During 1983, fifty employees received five-year service awards. Approximately 25 percent of the 190 employees were of Inuit origin; many were enrolled in a continuing program of trades training.

In 1984, the company reported "the passing of Isaac Attagutsiak, one of the true founders of Nanisivik. A leader amongst the Inuit people of northern Baffin Island for many years, he first became associated with the project in 1954 by assisting the geologists of the Geological Survey of Canada in their initial mapping of the region and subsequently worked with Texasgulf during the period 1957–68 as they explored the area. When MRI entered the project in 1972 he was there to provide guidance and leadership and did so until his retirement in 1978. His leadership and firm belief that the development of Nanisivik was a positive thing for the Inuit people were major factors in the success of Nanisivik and the consequent associated benefits for the Baffin region."[44]

In 1985, the company boasted of innovations and a steady labor force that included 28 percent Inuit workers – all of which made the operation one of the most efficient zinc producers, despite modest ore grades of 10.1 percent zinc and 1.0 percent lead. Buoyant prices allowed the mining of low-grade material running four percent zinc. Cash reserves during the year permitted MRI to purchase the 6.5 percent interest in Nanisivik held by Kidd Creek Mines Ltd., the successor to Texasgulf.

In May 1986, the first northern resident completed the company's four-year journeyman certification, which was indicative of the general upgrading of skills among the staff. The tenth year of the mine's history was also notable for the final purchase of all the remaining outside interests. Of the four original minority shareholders, the Government of Canada was the only one remaining at the start of that year, and this 18 percent interest was purchased in September 1986 for $6 million. In addition, the 35 percent royalty on profits held by Canada Development Corporation was purchased for $11 million, so that MRI through Nanisivik Mines then became the sole beneficiary of cash flow from the mine. The efficiency of the mine and its debt-free position signaled a strong future; thanks to the geological skill of its staff, remaining ore reserves stood at 2.925 million tonnes.

Ten-year awards were presented to thirteen employees from the total work force of 181. Three of these were Inuit workers: John Weetaltuk, chief draftsman; Sakiassie Qaunaq, carpenter; and Percy Pikuyak, equipment operator.

INUIT EMPLOYMENT STUDY

A review of the impact of mine employment was provided in a 1979 study conducted by the Baffin Regional Inuit Association (BRIA) to evaluate the effectiveness of Native employment programs.[45] The study, conducted for the company, the GNWT, and the federal government, was mandated in the Nanisivik development agreement of 1974. The parties agreed to undertake the study after the mine had been operational for several years. BRIA then established its parameters and a suitable budget and appointed Charles Hobart, a sociologist at the University of Alberta, as the principal investigator.

In 1979, the company had a workforce of 216, all of whom were engaged in a work rotation schedule. Sixty-two family accommodation units had been built at the mine site, twenty-seven kilometers by road from Arctic Bay. These units were occupied by supervisory and technical staff, experienced operators, and fifteen Inuit families. The remainder of the current Inuit employees, numbering around thirty-five, lived in bunkhouse accommodation or commuted from Arctic Bay. The Inuit employees were recruited from twenty-three different Arctic communities, but most of them came from Arctic Bay, Frobisher Bay, Igloolik, Clyde River, Pangnirtung, Pond Inlet, Cape Dorset, Gjoa Haven, and Rankin Inlet. Altogether, 297 Inuit with Nanisivik work experience were interviewed in the study.

INUIT WORK RECORD

The personal profiles of these men were striking. They came from a wide range of places, they were young, and a high proportion were single – all indications, it was thought, of a small labor pool and a general dislike of industrial work. A high percentage of the men hired by the mine (26 percent in 1975, 54 percent in 1976, 40 percent in 1977, and 25 percent in 1978) worked for six months or less. The percentage who failed to complete even the first six-week work period was 11 in 1975, 22 in 1976, 30 in 1977, and 23 in 1978. Therefore, during these initial years there was no trend toward an improvement in work duration. As of 1978, family housing for Inuit was provided at the site, but this housing was allocated to the long-term employees. Strathcona Mineral Services computed the annual turnover rate for northern male workers as 106 percent, while that of southern male workers was 63 percent. The long work period, the traditional closeness of Inuit family attachments, and the dependence of Inuit on country food were cited to explain this difference.

By 1978, the right formula had not been found for achieving the same adaptation to the working environment experienced by North Rankin Nickel Mines.[46] In general, the supervisors thought well of their Inuit crews, though they recognized that low educational levels delayed their advancement. Inuit were highly regarded in the mine and shops, where they were judged to be superior to most workers, if properly trained, and more careful with equipment. They had the potential to become good tradesmen and showed no distaste for working underground.

PERSONAL ATTITUDES

The study clearly showed an ambivalent attitude to mine employment among men who needed to hunt for their families, brought about by the long rotation schedule. A small minority also voiced other, minor complaints. Most of the workers interviewed expressed favorable reactions to their Nanisivik employment experience (though their wives, in general, showed much less satisfaction).[47] Although wage income was very significant in some communities – wages to Inuit workers in 1978 totaled $712,000 – interviewees felt there was no detrimental effect on traditional activities. In fact, fur harvesting had increased, in part because wages could be used to purchase better hunting equipment and supplies. Serious social consequences were not observed: indeed, liquor importation declined in some cases as a result of steady employment. The main problems identified were marital adjustment and unsatisfactory socialization of children due to absences of fathers. Some difficulty was experienced by wives wanting to contact their husbands: they regarded the ease of communication by telephone between the company and Inuit communities as unsatisfactory.

RESULTS OF THE STUDY

A case was made in the study for reducing the shift rotation, where economically feasible, to two weeks on, one week off, a schedule similar to that of

Panarctic Oil. Recommendations included a joint community advisory board and the appointment of two Inuit persons to the Nanisivik monitoring committee. The study also urged that monthly meetings be held between mine management and the Arctic Bay hamlet council to achieve prompt discussion of concerns. Supervisors and Inuit employees alike needed to undergo some intercultural orientation programs in view of marked differences in psychology, manners, and communication styles. Grievance procedures needed to be streamlined, family communication links strengthened, and alternative, Inuit-style food served in the dining hall. It would seem that some of these items might have been addressed earlier.

The employment target of 60 percent Inuit workers as set out in the development agreement was not achieved and, in fact, appears to have been unrealistic. The target has never been achieved at Nanisivik even to this day, which suggests that there were compelling reasons for its failure. The BRIA report on the situation after four years of operation tried to suggest ways of achieving an improvement – and some of them may have been tried.

The fact is that the Inuit population base is relatively small, and the major impulse is to remain in the community and work there if possible and, above all, to maintain a strong link with the land and traditional pursuits. It seems that Inuit have exploited the mine opportunity to their own advantage and within the limits of their own numbers. Young men lacking the capital for a good hunting or fishing outfit worked at the mine to acquire it, then returned to their communities, perhaps to exploit some of the skills learned at the mine. The survey asked about the "good" articles bought by workers, and found that overwhelming attention and substantial money were spent on skidoos, boats, hunting equipment, and household appliances. Spending that was classed as "bad" for the communities included alcohol, gambling, and drugs.

For Inuit, the industrial wage economy has merged into their own natural economy, where ever-better skidoos, boats, and other equipment were needed to reach the scarce renewable resources of the land. Thus the participation in mining served Inuit purposes and aims, allaying fears that their future would unfold as Hugh Brody had predicted in 1975, leaving them as impoverished migrant workers. A small percentage of men have pursued whole careers at the mine and taken trades and other training. These have been men who lacked an absorbing interest in land occupations or who found themselves comfortable in the area of Strathcona Sound and able to hunt or fish during their spare time.

Clear preferences emerged from the employment survey.[48] Both workers and wives wanted an adequate job that allowed the men time for hunting and trapping while living in the home community with their families. Most wanted to work for money, but hunt and trap for food. Thus, a full-time job in the home community was a first choice, followed by rotational employment at Nanisivik, where a chance existed for them to keep the family together.

Surveys of country food availability indicated that the mine made no impact on Arctic Bay community figures because the sale of caribou and other game to Nanisivik was not allowed, and many of the men working at the mine were not persistent hunters. As far as trapping goes, Arctic Bay, Pond Inlet, and Igloolik took a higher proportion of territorial fur species after the mine started, but in general, the mine had little or no effect on the traditional activities of the Inuit.

The survey examined the social cost of mine employment, but detected no conclusive trend. Of most concern to Inuit were the effects of the fathers' six-week absences on the socialization of their children. Traditional absences of fathers for hunting trips were much shorter.

ONGOING POLICY

The major recommendation of the BRIA study centered on revising the six-week rotation schedule. A work period of two weeks was urged for nearby communities, on the grounds that this would reduce worker turnover and encourage skill training programs. Since the employer would have to pay return air fares, the economic viability of this scheme depended on the availability of charter plane–load groups (twelve or eighteen men) from each community. Failing this plan, the study recommended a territorial travel subsidy similar to that provided Indian workers from Lac La Martre employed at Terra Mines. For reasons of manpower availability and the sensitive relations with other employees, who worked thirteen weeks at a stretch, these measures were never adopted, and the proportion of Inuit workers at Nanisivik continued to be 20 to 28 percent.

Other recommendations of the BRIA study dealt with communications, cultural orientation, grievances, housing, and food preferences. While the study attempted to explore the Inuit worker's situation on many fronts, the quality of the data and the short experience period were admitted deficiencies.

By 1980, the Polaris lead–zinc mine on Little Cornwallis Island was being prepared for development by Cominco. Both company and government seemed to realize that the Inuit labor pool was small and well distributed. Consequently, the need for the government to set an arbitrary employment target was considered unrealistic. Instead, reliance was put on the company's best efforts, according to the Canadian Arctic Resources Committee.[49] In late 1979, the ITC and CARC were protesting that socioeconomic and environmental studies performed before 1975 were no longer valid and urging the government to hold public hearings on the Polaris project. The hearings never happened.[50]

THE FUTURE

The Nanisivik mine proved to have a life much longer than the twelve years envisioned at the beginning: in fact, it has now reached its end after going twenty-six years. A 1996 article by Graham Farquharson and James Marshall reported that seven years' worth of additional ore reserves had been identified. The company's geological expertise, sharpened on this magnificent ore body, had achieved regular additions and new discoveries. As a result, the 1996 reserves stood at six million tonnes, the same as in 1974, but the ore grade was now 7 percent zinc instead of 13 percent.[51]

Rather than a single-status, bunkhouse operation, Nanisivik was developed as a family-based community, comprising sixty housing units and other structures for its two hundred employees. The payoff for this admittedly more expensive choice has been considerable, according to Farquharson and Marshall, both of whom had worked on designing and operating the mine for the project management firm Strathcona Mineral Services of Toronto. The benefits they cite include a stable workforce; greater Inuit participation in that workforce (20–25 percent); two employees from some families, leading to a record for female employment in mining; lower cost for personnel transportation; and lower total operating cost, which enabled Nanisivik to remain competitive while mining lower-grade ore.

More than 50 percent of the current Nanisivik employees have been there for more than five years, and sixty-four employees have spent more than ten years at this tiny, isolated village in the Arctic. It was recognized that Inuit workers would have to endure long absences from their home communities even though their work period was half that of southern workers. Being able to move to the Nanisivik townsite with their families made the difference, and some of the longest-serving employees were Inuit. Over the years, the percentage of northern residents has ranged from 16 to 28, a better achievement than in any other mine in northern Canada.[52] The ambition for maximum Inuit participation held by the original organizers of this project on Baffin Island has been realized in principle. This is an important achievement with implications for future mining projects in the North. It shows the adaptability of Inuit to the challenges of an industrial economy.

Mining is a skilled enterprise and does not offer endless casual work opportunities. Attracting skilled employees is of primary importance and calls for the provision of special benefits and a suitable work environment. Once attracted, employees are expected to be dedicated and make a personal time commitment to the job, something that relatively few Inuit are able to do.

The Baffin Region Inuit Association, founded in March 1975 as an affiliate of ITC, commented on the Nanisivik socioeconomic study in 1979.[53] BRIA was particularly determined to find out why so many Inuit quit their jobs there. Brian Penny, advisor to BRIA and member of the study's working group (along with Arvin Jelliss of DIAND, Mike Moore of the GNWT, and Jim Marshall of Nanisivik Mines Ltd.) said that there were problems of adjustment on both sides and a lack of Inuit in supervisory jobs. Few of the men were happy with the six-week rotation schedule. Penny also reported, prematurely, that the turnover of Native workers was less than that of white workers. The report released later that year showed clearly that, once corrections were made for summer employees, staff transfers, etc., the turnover matched the company figures. Turnover of male northern workers was higher at 106 percent than that of male southern workers at 63 percent.[54]

How to effect change in this regard was the subject of a 1982 address by Allan Blakeney, former premier of Saskatchewan.[55] He described a multifaceted program aimed at bettering the condition of northern Native people, largely Metis. The Saskatchewan program included the establishment of the Department of Northern Saskatchewan, essentially run by Natives; community improvements; education; and the negotiation of "local employment contracts" with uranium-mine developers. The latter agreements, which were made a condition of approval for leasing of mineral lands, elicited strong Native hiring and contracting programs. Based on a two weeks on, two weeks off rotation schedule, the production workforce at Amok uranium was over 50 percent Native, while the construction workforce at Key Lake was 40 percent Native, including 59 percent of the nonapprenticed trades. Multimillion-dollar contracts were awarded to northern businesses.

It may not be fair to compare these two open-pit uranium mines, with their road (and aircraft) access, to isolated Arctic mines. Nor can it be said that the Indian and Metis situations, with smaller fish and game resources, are analogous to the Inuit condition. The Arctic Native peoples must have come to realize in the 1970s that mining was no employment panacea, and perhaps this explains their growing interest, expressed by Tagak Curley, in equity participation in development ventures. Financial participation of the Inuit Development Corporation was achieved with the Cullaton Lake gold mine in southeast Keewatin district, and similar discussions began on the Polaris mine project on Little Cornwallis Island prior to start-up.[56]

AFTERMATH OF NANISIVIK

The years of the Nanisivik project evaluation and start-up were very formative and dynamic years for the Inuit Tapirisat of Canada. In the decade since its establishment in 1971, the ITC had managed to establish itself as a national organization, develop policies for land claims, Nunavut territory, and cultural preservation – and at the same time elicit the support of individual communities. This was not an easy task, but it was achieved by functioning as an umbrella organization to support local communities and groups in their local struggles, such as the caribou aboriginal rights litigation initiated by Baker Lake and heard by Justice Patrick Mahoney.

It was not part of the will, mandate or capability of ITC to challenge individual developments on its own. However, once the misunderstandings with Arctic Bay were resolved (and BRIA was formed), the Tapirisat got involved in many issues, such as the oil drilling proposed by Norlands Petroleum Ltd. in Lancaster Sound, the sending of Rankin Inlet advisors to Arctic Bay, and assisting that community to monitor tailings disposal arrangements at Nanisivik.

Perhaps the town of Nanisivik will outlive the depleted mine itself and continue as a North Baffin service center. Modern ideas about fuel distribution and rationalization may establish Nanisivik as a major Arctic shipping and resupply center. The present owner Breakwater Resources shipped the last ore concentrates in September 2002. Since that time consideration has been given to decommissioning and reclamation of the mine, disposition of the usable buildings and the possible role of the site with its dock and airstrip in the plans of the Canadian military.

The Nanisivik experience taught what could and could not be accomplished with development. It alerted Inuit to government indifference and developers' single-mindedness, to the crash of events taking place without benefit of their significant input or control. It provided experience with the attendant issues of environmental pollution and employment.

The next chapter will focus on the bigger picture, the united effort of all Inuit in their quest for greater political and economic well-being through land-claim agreements.

5. THE ROAD TO NUNAVUT

A SINGULAR PROPOSAL BY ITC AND ITS AFTERMATH

The decade of the 1970s witnessed great extremes of Native opinion, some verging on the theme of genocide. The most eloquent of these was surely the statement that Frank T'Seleie made to the Berger Inquiry on August 5, 1975. T'Seleie was chief of the Fort Good Hope band (west of Great Bear Lake) and a spokesman for the Dene in opposition to the Mackenzie Valley pipeline. He said they intended to stop the pipeline – even blow it up – because it was a threat to their lands, their future, and their very people. In fact, he linked this development to all the real and potential invasions that threatened his people to the point of extinction: it was a system of genocide. In defense of his children and grandchildren, he confronted Robert Blair of Foothills Gas, one of the proponents of the pipeline, in a verbal life-and-death struggle: "By scheming to torture my land, you are torturing me .... You can destroy my nation, Mr. Blair, or you can be a great help to give us our freedom."[1] The genocide theme was expounded, and rightfully so, by the Inuit of Banks Island and the Mackenzie Delta, in response to seismic exploration, road building, and drilling in the late 1960s. It re-emerged in the form of relentless opposition to uranium exploration and later uranium mining near Baker Lake: the people believed that the Kiggavik mine posed a threat not only to their caribou herds (and hence to themselves), but also to the whole human race, as a result of fallout from the industrial–military uranium chain. The latter idea reflects the broad perspective of the modern Inuit mentality.

Progressive work in pursuit of self-determination also characterized the decade. In September 1979, Thomas Suluk and John Amagoalik presented a discussion paper entitled "Political Development in Nunavut" to the annual meeting of the Inuit Tapirisat of Canada in Igloolik. Drafted as a summation of land-claim discussions to date, the paper (which became known as the Nunavut proposal) was passed by the ITC board. The ITC, representing 20,000 Inuit, called for the political separation of an Eastern Arctic territory of roughly 2.4 million square miles lying north of the tree line. The discussion paper received praise and support at various Native gatherings and meetings. Lobbying by Thomas Suluk and his claims staff at Yellowknife, as well as support within the Inuit caucus, led the territorial council to accept the proposal that Nunavut should be an integral part of land-claim negotiations. Liberal Minister John Munro acknowledged that the federal government would have to make a decision on Nunavut before negotiations could proceed. The 1979 report of special representative C.M. Drury on constitutional development in the Northwest Territories[2] dismissed the idea of territorial division to create Nunavut. Drury understood the frustration in local communities over the lack of power and meaningful political influence and sought to expand the power of local governments through devolution of authority and to boost their finances through block funding.

The growth of concern in the NWT communities led to political reforms recommended by C.M. Drury. The legislative assembly was expanded to twenty-two members and came increasingly under Native influence. At the fall 1979 session of the new Native-controlled legislative assembly, Commissioner John Parker called for unity to make the Northwest Territories strong and advised against division. It was Tagak Curley, member for Keewatin South, who officially called for a new Nunavut Territory north and east of the tree line. Dennis Patterson, member from Iqaluit, was appointed deputy speaker.

Dennis Patterson said that there was universal acceptance for the premise that a separate and divided territory was needed. It was impossible, he claimed, for officials in Yellowknife to administer the Eastern Arctic, parts of which were as much as two thousand miles away: "They are completely out of touch, and we are constantly getting the short end of the stick."[3] A land-claim settlement of this magnitude would enable Inuit to establish the tax base that would make Nunavut viable. Inuit wanted to control the development of the resource-rich Arctic, which would mean capturing the authority to license companies and collect royalties. "This is what any land claim settlement has to be all about," Patterson added. "It will be an important negotiating point over the next 10–15 years." But so far, the federal government had rejected political self-determination as part of land-claim settlements, arguing that this would establish ethnic governments. To Patterson, this was a red herring.

In an early 1980 interview, John Amagoalik, vice-president of ITC, commented on the future government of a new Nunavut Territory, especially its finances: "Well, the way it is right now, the territorial government gets most of its money from the federal government, and the new government of Nunavut will have to operate the same way for a while. But when we are looking at the future, maybe 10 or 20 years from now, this part of the Canadian Arctic is quite rich in minerals, oil, and gas, and it looks as though companies from other parts of Canada are going to come up whether we like it or not; they are going to come up here and they are going to take these resources away, and when they do that, we would like them to pay for it somehow in cash or in royalties."[4] Amagoalik did not think rapid development was acceptable. He said that the Inuit would not stand for it. They would insist on slowing down and doing the appropriate environmental research, since a major oil spill in the North would be devastating. It is important to note that the ITC believed development, albeit controlled, would be vital to the survival of Nunavut, an understandable position for the backers of a proposition that was awaiting government approval. As time went on, the Inuit continued to be ambivalent about commercialization and development in their land. Ironically, the Justice Mahoney decision (a blow to industry) and the Nunavut proposal nearly coincided in time.

As of December 17, 1979, the Baker Lake judicial order restricting uranium exploration was lifted by Justice Patrick Mahoney, and the government was free to issue prospecting and other permits, even if an appeal of the case was launched. Jake Epp, minister of Indian and Northern Affairs in the Joe Clark progressive conservative government of 1979–80, said that land claims were a top priority, but that northern development would proceed as usual during the negotiations. This statement earned him the criticism of Native groups. In a report to the minister by the northern mineral advisory committee, John Bruk of Cyprus Anvil Mining Corp. called for the government to "provide a stable and certain economic and political framework so that the mining industry would then know exactly where it stands, what is expected of it, and what the rules under which it is expected to operate are."[5]

At the Eighth National Northern Development Conference in 1979, John Todd, president of Siniktarvik Limited (a Rankin Inlet multi-ethnic corporation), delivered a sobering address. He spoke on the nature of initiatives in the Eastern Arctic necessary to enable resource development to proceed. The need for development was clearly indicated by the growing young Inuit population and increasing unemployment. He cited the political and legal strength of Inuit organizations and cautioned that overlooking Native concerns could be costly, as in the then current legal battle involving the mining companies and the hamlet of Baker Lake. Todd said, "Inuit are suspicious of development which appears to threaten a traditional life style without assurances of deriving significant benefits. This I believe to be the real issue. It is not only a question of culture and values, but in fact, the credibility and viability of the industrial-corporate future for the North that is questioned by the Inuit." The Rankin

Inlet mine, within the memory of many Keewatin people, was clearly an example of how a development should *not* take place. There had been no socioeconomic impact planning and no consultation with locals, who became dependent on a wage economy and were then thrown out of work after five years. The social cost was high. Now government caution and consultation were in place, "yet the Inuit remain suspicious and unconvinced and perhaps rightly so." Todd urged that industry consider giving equity participation to provide a sense of ownership among Inuit and establish mechanisms that would ensure they derived concrete economic benefits. "I am of the opinion that the industrial-corporate sector has not addressed this crucial issue with sufficient imagination, innovation, energy and resources."[6] Partnerships never materialized in exploration as they did in real estate and other business ventures.

The unpleasant aftermath of the Rankin Inlet nickel experience pervaded the thinking of Keewatin people in many communities where ex-mine workers ended up. Their skeptical attitude, as described by John Todd, no doubt influenced their response to uranium exploration companies, to the Polar Gas pipeline proposal, and also to the rejection of uranium mining near Baker Lake. So the Keewatin people, like those of Banks Island and the Mackenzie Delta, harbored a feeling of opposition to developers. This attitude was pervasive among Inuit leaders during their organizational phase in the 1970s, in response to the hostile feelings encountered in the communities. A notable exception, due to local isolation and exercise of independence, was the early support by Arctic Bay for the Nanisivik mine proposal, which was later reassessed by that community.

The official view of the Inuit Tapirisat of Canada on development, as expressed by president Michael Amarook, a Baker Lake man, in the 1979–80 annual report, was that many actions of government and resource industries threatened the Inuit as a people. He cited offshore drilling in the Western Arctic, the Arvik Mines on Little Cornwallis Island, potential developments in Lancaster Sound, uranium exploration in the Keewatin, and hydro developments in Quebec. The Inuit's major concern was to control the land in order to guarantee continuation of their way of life, but for the most part developments proceeded with little regard for Inuit views. He said that the Baker Lake trial launched by his predecessor, Eric Tagoona, had one positive feature: the recognition of Inuit rights. He was also pleased that the ITC, along with the Labrador Inuit Association, had blocked the Brinex uranium project in Labrador on environmental grounds. Amarook firmly believed that the ITC must press for thorough land-use management schemes to protect the environment, aboriginal rights, and Inuit political aspirations; a Native-controlled

NWT Council might help in this regard. The ITC was an active participant in the Inuit Committee on National Affairs and Inuit Circumpolar Conferences. It held leadership workshops, encouraged Inuktitut television and radio programming, and promoted a Keewatin health study as well as a nonprofit housing corporation. Clearly, the Tapirisat was carrying forward a busy program on many fronts.[7]

The Nunatsiak federal riding – in area the largest in Canada – coincided geographically with the proposed territory of Nunavut. Its sitting NDP member, Peter Ittinuar, who would later become a Liberal, delivered his maiden speech in the House of Commons on October 16, 1979. Ittinuar drew attention to the possibility of economic developments in the extractive industries but deplored haphazard, hurried, or untimely development that was externally imposed on residents, affecting their traditional livelihood. He said this rash of developments benefits the rest of Canada, or possibly West Germany in the case of uranium (a reference to the Urangesellschaft mining project near Baker Lake), but leaves the local residents of the Arctic to deal with the social and environmental consequences.

Citing the push for mining development mentioned in the government's throne speech, Ittinuar criticized the Northern Advisory Committee, which consisted of fifteen members chosen from government and the mining industry. Its recommendations were inadequate, he said, since it had held no discussions with Native organizations or special interest groups, despite its contention that "its achievement must be coupled by an equally strong effort by industry to adapt the particular characteristics and strength of northern people to the needs of industry." The committee had made no mention of the dispute over uranium mining at Baker Lake or the closed-door negotiations between the government and Cominco over the Arvik mine. Ittinuar said such approaches to development can only heighten suspicion and mistrust. He felt that the committee's report was at variance with the truth when it stated that "mining can be a most effective vehicle for increased economic activity and opportunity for employment and training for all people now living in the North or wishing to move there." The record of Native employment at Pine Point, Cyprus Anvil, and Nanisivik was dismal, and the generation of secondary economic activity was minimal. One recommendation in the report read: "Economic development of the North must proceed with full recognition of the legitimate interests of northern people and the protection of the environment." Yet the report urged that mineral rights remain vested in the Crown, regardless of future agreements with Native people, and that the Thelon Game Sanctuary be opened to suit the interests of the uranium explorers. Ittinuar argued that in no case should mining be allowed to interfere with the management of renewable resources or displace the legitimate rights and aspirations of the people. In regard to northern political developments, on which the government was noncommittal, he said the aspirations of the population should be considered. The Inuit earnestly desired the establishment of Nunavut, in keeping with the government's recognition of the boundaries and the uniqueness of Nunatsiak. Ittinuar called on the government to recognize Nunavut, as it had arisen from Inuit aspirations that were not in the least radical.[8]

The effort to intervene in Cominco's proposed Arvik (Polaris) mine, to be located off the end of Lancaster Sound near Resolute Bay, was a notable Inuit failure during the year. The company escaped without an environmental assessment review of the project or any government requirement that Native employment be guaranteed. Cominco refused outright to negotiate such matters with the ITC, arguing that the government had already given authorization to proceed. In 1980, the Canadian Arctic Resources Committee (CARC), a longtime contributor of scientific advice to Berger Inquiry interveners, produced a draft report entitled *Analysis of the Polaris Project Proposal: Arvik Mines Ltd.*[9] This document offered a historical review of the project, including Cominco's accumulated background studies and comments on relevant environmental and socioeconomic issues. It contended that further environmental assessment was warranted.

In regard to Native employment standards it appeared that, because the Nanisivik agreement was unenforceable or unrealistic, the government had given up on targets and requirements and chosen to let Cominco develop its own relations with the communities. Deputy Minister Paul Tellier declared that such a relationship must be developed; otherwise, there would be difficulties in realizing maximum benefits from the project. The authors (Ames, Gibson, Keith and Michalenko) of the CARC report contended that a six-week rotation schedule and lack of family accommodation would largely eliminate Inuit as potential workers. Polaris will "not make a significant or valuable contribution to the employment of Native northerners, at least not in the foreseeable first few years of operation."[10] The Inuit Development Corporation (IDC) held discussions with Cominco on various business contracts, such as catering. The authors said the company was expected to approach this in a businesslike manner, but they considered that, with the project already approved, the IDC was in a weak bargaining position. In their opinion, the local community would obtain no benefit except improvement in air service, and this did not warrant approval of the project. They noted that from an early date, the residents of Resolute Bay, located only a hundred kilometers away from the proposed mine, had opposed the project because they feared that the mine and attendant shipping would damage the environment and the renewable resources of the area. The authors concluded: "In the current absence of a reasonably thorough and comprehensive assessment of the potential environmental impact of the Polaris Project, there is still no basis for assuring the people of Resolute Bay that their fears are not justified."[11]

The apparent intransigence of Cominco as a developer can be seen from ITC's reaction. The ITC had submitted reports documenting various local problems, which "went unanswered to such an extent that this research and presentation became an obvious waste of time."[12] The uncritical approval of the project in February 1980, its development, and Cominco's significant later expansion of mineral prospecting permits all contributed to a hardening of attitude among the Resolute Bay people. Their granite-like resistance during land-claim negotiations (at the land selection stage) led to a significant breakthrough in acquisition of mineral rights for the Inuit. But more of this in Chapter 7.

An editorial note in the *Northern Miner* summed up the mining industry's opinion of the Polaris High Arctic lead–zinc mine. It was considered a shot in the arm for northern development and a boon to the flagging economy. "We were therefore flabbergasted to learn from Ottawa that the Natives of that area, and environmentalists, are opposing this project. The Inuit Tapirisat of Canada, said to represent the 22,000 Inuit in this country, and the Canadian Arctic Resources Committee, a private lobby group, have called for a halt until up-to-date studies determine the impact. This is nonsense. Fact is, this project has already been studied almost to death – no less than 13 have been carried out by the company since 1973 and at a very considerable cost .... Nor have we too much sympathy for the Natives who are apparently rejecting the job opportunities this project would afford them. With few exceptions, these people are living on welfare and government handouts of one kind or another."[13] There didn't seem to be public respect at this time for either due process or the need to obtain local approval of development projects.

Of course, Cominco *had* conducted studies in preparation for the Polaris mine, including at least two on social impact and employment. In 1973, the company drew on its own personnel development services officer, J.E. Barrett, to study the feasibility of Inuit employment at the projected mine. Barrett went north to talk to government and community representatives about this still-distant mining project. His purpose was not to conduct a labor survey, but to discuss the possibility of employment sometime in the future and register a first encounter with the residents at an early stage in the project. Barrett was politely received and educated in one way or another about Inuit aspirations, which led him to quote Tagak Curley (ITC), Zebedee Nungak (Inuit Association of Northern Quebec), and Nellie Cournoyea (COPE) in his report to management. Barrett stressed that those involved in the Polaris project should understand the special needs of the Inuit people. He favored a short-cycle work rotation and the establishment of a mine community that would be welcoming to married workers.[14]

Much later, in 1980, Cominco contracted Outcrop Ltd. of Yellowknife and DPA Consulting Limited of Vancouver to examine the socioeconomic impacts of the Polaris mine on the community of Resolute Bay, the Arctic coastal communities, and the Northwest Territories in general. This was a thoroughgoing study involving many diverse interviews and evaluation of job opportunities and training for northern workers, as well as their social needs. Its recommendations may have played an important role in final company decisions about the project. The consultants recommended that a townsite not be established at the mine, as it would not answer either the extended family needs of Inuit employees or their hunting needs. A townsite might deplete the population of Resolute Bay, which could itself serve a more useful function as a bedroom community for the mine. Work rotation schedules should be monitored and adjusted to minimize the family separation, which was viewed as the major negative factor of mine work. The six weeks on site and four weeks at home might be acceptable and would encourage trades training and advancement to higher positions in the workforce. A shorter rotation schedule would make economic sense for Resolute Bay employees because of their proximity to the mine, but this would induce more Inuit to move there, putting a strain on the community's housing and other services. A subsidized transportation plan, on the other hand, would encourage participation from other, more distant communities. Such a plan was judged to be more worthy of government financial support than the alternative service investment (school, nursing station, and other facilities) in a new townsite at the mine.[15]

During 1973, the minister of Indian and Northern Affairs launched the Lancaster Sound Study in response to three initiatives: the Norlands Petroleum Ltd. application to drill in the Sound, the Arvik (or Polaris) mine, and the Arctic Pilot Project to ship liquefied natural gas by tanker. The final report to the minister, which required public and private input by the end of 1981, was to provide the basis for a regional management plan. The Polaris mine was scheduled to begin production in 1982.

The boom in economic activity from oil-well drilling in the Beaufort Sea caused the Natives around Tuktoyaktuk to call for vocational training centers, so that they could better capitalize on the new jobs, which were mostly going to southerners. Reverend Robert LeMeur, an Oblate missionary who had spent many years in the North, said of the adaptable Inuit people, "Now they are going to have to make the best of both worlds, as they have always done, and use it to their advantage. Eskimos can work for six months and hunt, fish, and trap the rest of the time. It's up to the companies to accept that."[16] He said this is not the worst challenge faced by the people – but they will have to do it for themselves.

Another form of self-help was evolved by the Inuit Development Corporation under president Tagak Curley during 1980. The IDC agreed to lend $25 million to Cullaton Lake Mines Ltd. to develop a gold mine in the southern Keewatin. Curley called it a major breakthrough that enabled the people to participate in mine development on terms beneficial to them, a natural step in implementing the ITC's industrial strategy. The five-year loan at prime plus one percent interest carried a bonus of 100,000 company shares, plus an option to purchase a further 400,000 shares at discount.[17] This financial arrangement achieved the aim of getting a piece of the action, but the risk proved inordinate at a later date. The IDC got out in time, thanks to market equity fund-raising by the company. The Cullaton deal raised the ire of the ITC's chief negotiator, Thomas Suluk, who had not been not made aware of it. He regarded the loan by the ITC subsidiary, without prior consultation, as a threat to unity that Inuit could ill afford. He was quoted by *News/North* (Yellowknife) in February 1981: "We are negotiating a comprehensive package and I, as chief negotiator, will not tolerate any disruption and divisive action."[18] Suluk was well aware of the rocky road being traveled by the ITC in its land-claim program, such as the hurtful dissolving of its land-claim commission and the lack of community consensus a couple of years earlier. Ambivalent aims and practices could no longer be tolerated: traveling along the road to self-government demanded a united effort. It was probably also true that diversity of views would energize the land-claim cause. Consensus building was the Inuit way.

LAND-CLAIM NEGOTIATIONS

The controversial COPE negotiations were reopened in 1981 after being stalled from the time an agreement-in-principle was reached in 1978. That Inuvialuit document included a wildlife park along the Yukon coast, title to land in the Mackenzie River Delta, and a cash settlement of $45 million. Liberal Senator

Davie Steuart, who was now the chief negotiator for the Western Arctic claims, proposed a Native-owned corridor through the park to allow for commercial access and pipeline routing to the Beaufort Sea. There was an immediate outcry at the very idea of allowing Natives the right to bargain with developers on the price to be paid for access to resources. Chris Pearson, leader of the Yukon government, called the proposal "utter appeasement" for 2500 Native persons and "a blatant sellout" for other northern residents. However, Indian Affairs minister John Munro and COPE president Sam Raddi declared that both sides were ready to resume talks, hopefully behind closed doors.[19]

In the early bargaining in 1977, the Inuvialuit found out that the federal government representatives were stingy negotiators. Their position – totally unacceptable to the Inuit – was to cling to the 1975 James Bay agreement, which granted Native ownership of a paltry 1.3 percent of traditional lands (0.32 square miles per person), all of that excluding mineral rights and allowing prospector access. A couple of years later, the agreement-in-principle had leveraged the figures up to 22 percent (14.8 square miles per person) of which about 2 percent included subsurface rights. These were precedent-setting concessions in comprehensive claim terms. Terry Fenge of CARC reported that "Inuit negotiators appreciated as early as 1980 that a land-claim agreement would likely confirm government as the owner of considerable areas of land traditionally used and occupied by Inuit. A key task before Inuit negotiators, then, was to use negotiations to define Inuit rights to participate in managing land and resources in Nunavut that would be owned by the Crown."[20]

In the ITC annual report for 1979–80, Thomas Suluk reviewed the Nunavut land-claim situation as its project director. He said, "A land claim settlement between the Inuit of Nunavut and the federal government should reflect an awareness that the days of trinkets and tiny reserves are over, an awareness that a people cannot retain self-reliance and self-esteem if the power to shape its future is appropriated by others." The era of strong rhetoric had arrived. Suluk added that "an Inuit economic base should be anchored in a fair share of property benefits, particularly in the form of land ownership, harvesting rights and royalties, and a system for regulating development activities that pays some attention to the economic and social needs of Inuit." This required a negotiating process that would show primary concern for the well-being of communities and the environment – that would establish the rights of Inuit to be involved in negotiations between government bodies and development proponents. Their lack of involvement in the past had led to fly-by-the-seat-of-your-pants decision-making by the Inuit and caused open confrontations with government and proponents. Suluk concluded, "A land claim settlement should provide some assurance that Inuit will no longer learn about projects such as Nanisivik and Arvik through the business sections of southern newspapers."[21]

The strategy of including property issues and development policy in land-claim negotiations had great appeal in the communities (it evolved from them). Its potential to delay development made it a strong irritant to the government, and this was likely to accelerate the negotiating process. Suluk described the progress to date as follows: The dissolution of the Inuit Claims Commission in 1979 had caused uncertainty, and the regional associations were not ready to provide direction, although they provided strength to the overall ITC "mini-government." There had been steady growth in the Tapirisat's leadership and staff. The negotiating delay coincided with the political change in Ottawa. In its brief tenure of seven months, Joe Clark's conservative government showed no clear policy direction on Native land claims. The ITC went through a major policy review of its own and came to favor a centralized or national approach that would lead to a unified settlement for all three regions: Kitikmeot, Keewatin, and Baffin. The ITC continued its research and prepared to resume negotiations with the new liberal government in the summer of 1980. In advancing the idea of a new Nunavut Territory, without which no agreement was possible, Suluk said, "We have proven to the rest of the NWT and Canada as a whole that we can develop a political system which contains the unwritten Inuit style of governing, respected for four thousand years, yet conforms to the expectations of the GNWT and the federal government."[22]

EMERGENCE OF THE TUNGAVIK FEDERATION OF NUNAVUT

In 1981, John Amagoalik, a seasoned leader from the days before the land-claim commission, became the president of the ITC for the next four years. He oversaw the 1982 establishment of the Tungavik Federation of Nunavut (TFN), a new entity whose role was to negotiate a land-claim settlement with the federal government. Inuit showed growing maturity and sophistication in evolving this new entity to pursue land-claim negotiations exclusively. The wide representation on its board of directors showed that Tungavik was truly a kind of alliance of Inuit organizations. Board members included Donat Milortuk of ITC; Louis Pilakapsi of Nunasi Development Corporation, formerly IDC; Ole Ittinuar of the Inuit Cultural Institute; Louis Taparjuk of the Co-op Federation of the NWT; Walter Aodla and Kik Shappa of BRIA; Peter Alurut and John Killulark of KIA; and Bob Kadlun and James Eetoolook of the Kitikmeot Inuit Association. Thomas Suluk was the founding project director, and Allen Maghagak was chief negotiator for Nunavut land claims. Bob Mitchell, a Saskatchewan lawyer, was the chief federal negotiator during the first round, which produced an agreement-in-principle on wildlife. Tom Molloy assumed Mitchell's position on behalf of the federal government in 1982, working with Keith Crowe of the Office of Native Claims.

The Tungavik was a major enterprise for the Inuit: it soon had to manage an annual budget of over $2 million, conduct numerous consultation tours, and prepare extensive technical and legal documents and positions. The central Inuit positions were co-management of lands, royalty sharing, outright elimination of private land ownership for anyone except the Inuit themselves, and the inclusion of both surface and subsurface areas in Inuit holdings.[23]

The role and timing of a land freeze were discussed at an early stage, but such a freeze could not be realized, as was the case with the Dene Nation, until an agreement-in-principle had been reached, which would not occur until years later. There was no apparent hostility to mining at this stage owing to the lack of interim protection. In fact, the Nunasi Development Corporation (NDC; formerly IDC), headed by legislative assembly member Tagak Curley, was involved with Borealis Exploration Limited, the holder of iron ore claims on the Melville Peninsula. Curley joined the board of directors of Borealis to gain an inside view of that potentially large development and to protect Inuit interests. He said, "It will be a huge project if it goes ahead, and there doesn't seem to be much involvement of local people [Hall Beach, Igloolik, and Repulse Bay were nearby]. I'm in there because I want to see how all of this is going to come about."[24] The Nunasi board was investigating the Borealis feasibility studies. Curley realized that other Inuit people were not ready for participation, but he wanted to put Nunasi in a good position. After all, land-claim negotiations might take years to complete.

The first phase of negotiations centered on the Inuit need for a system of authoritative boards – such as planning review, impact review, and management authority – to control development on all Nunavut lands, including Crown lands. Federal negotiators resisted any proposal that could be seen to relinquish control by Ottawa and suggested a system of advisory boards instead. The government wanted a separate environmental review of development projects, since it had broader implications, within a scheme that would retain the Federal Environmental Assessment Review Office (FEARO) inquiry process. Although federal negotiators accepted the need for Inuit to negotiate impact and benefit agreements directly with industry, they wanted such agreements to be ratified by the minister.

The two sides were wide apart on these issues. On economic benefits, the Inuit wanted to share royalty income and otherwise get a slice of the action so that Nunavut could eventually achieve economic self-sufficiency. They had no delusions about short-term payoffs, but they believed in the long-term potential. Inuit were not anti-development in principle; rather, they wanted more employment choices for their youth. They also wanted mineral rights on the lands they would acquire outright in exchange for relinquishing aboriginal title, lands which were as yet unspecified. In November 1982, the federal government announced its support and official sanction for the division of the NWT and a new territorial government for the Eastern Arctic, provided an agreement on the boundary could be achieved. The latter condition held a life-and-death power over the process and was not achieved for many years. The declaration followed the April plebiscite across the NWT, which indicated 56 percent support for division, reflecting the aspirations of both Inuit and Dene.

In 1982, the government thought that the Inuit bargaining position was weak, at least for the moment. Clovis Demers, executive director of the Office of Native Claims (ONC), admitted that the government was not pressured to settle the claims because there were no pressing development proposals in the area. In the south, however, development pressure was building up that would eventually have political force. "I think in the next two years we can still resist that pressure and negotiate with some serenity and arrive at something that's agreeable to both parties," he said. "But I think that, say, five years from now, it's going to be different, very difficult."[25] In view of this perception and budget difficulties, the TFN streamlined its operations in the fall of 1982. More of its core funding from the federal government went into priority research work and intensive consultations, which Thomas Suluk saw as necessary. Near the end of the year, frustration and impatience crept in as a result of insufficient funding and federal obstruction. In a special meeting in Eskimo Point with Tom Molloy, the TFN board said they were "disappointed with land claims progress" and wanted a settlement soon. They felt that development in the North was happening too fast, and there were no interim measures to protect their lands while the claim was being negotiated. They requested that Molloy return with a mandate to negotiate the Nunavut lands authority.[26] Molloy said that there would be a better chance of acceptance of that radical idea after some progress on land selection, but Simon Taipana maintained it would be useless to discuss land ownership before Inuit knew what rights they would have on Crown land. Obviously, Inuit were going to need patience for the long haul to get a comprehensive settlement. For the moment, they were completely focused on development abuses and the means of controlling them through administrative structural changes. Of course, their first priority was the crucial issue of renewable resources, which was tackled in the first round, but to their disappointment, the federal ratification of the wildlife agreement was also on hold.

At the last 1982 meeting on administrative structures, held in Winnipeg, there was some agreement that social and economic evaluation of major projects should be linked with environmental impact review, and that the whole process should accommodate Inuit input. Another gain was the acceptance that Inuit and developers should negotiate benefit agreements applicable to developments anywhere in Nunavut. There would be a role for a Nunavut impact review board (NIRB) and an environmental assessment review panel (EARP) on projects of national concern. The federal negotiators opposed the NIRB powers proposed by the Inuit, in the form of a joint management board. This stiffened the resolve of Inuit to become more than advisors – to negotiate a fully shared management over the use and development of lands and resources. If that point could be finalized soon, Suluk believed that a comprehensive agreement could be completed in two years. This was not to be.

SOME FEEDBACK TO DEVELOPERS

Tagak Curley addressed the 1982 Northern Development Conference in Edmonton as president of Nunasi Corp. He was one of several Native speakers to the predominantly business and development attendees interested in the conference theme: Partners in Progress.[27] Curley's point was that it was time for a new approach to economic cooperation between industry and Native people. He sketched some of the history of Native obstructionism, which had led to the myth that Natives were militantly opposed to all development. He said the James Bay hydro issue and the Baker Lake court case had helped to shape this myth. The result was the launching of administrative reforms and land-claim negotiations, but progress on the latter was painfully slow. Northern Native people could not afford to wait for implemented settlements to become involved in economic development. Increasing population and higher expectations were making crushing unemployment all the more burdensome. He said that although Native people were slow to support the concept of business development, the time was right for a major policy shift among northern Native people. Now there were economic organizations (like Nunasi) in place for partnership ventures, and Inuit were rapidly gaining financial and managerial know-how. Capital for development corporations stemmed from advances pending land-claim settlements – expected to reach $1 billion in the North. After settlement, in the case of Nunasi, all the corporation's shares would be in the hands of Inuit.

Curley described many avenues for investment, such as Nunasi Air or the recent $25 million loan to Cullaton Lake Mines Limited. He urged the initiation of partnerships or equity participation early in a project to take advantage of Inuit skill and experience, to say nothing of their influence before the National Energy Board and other agencies. Inuit would gain by building their expertise in management and subcontracting. At present the economy was government-supported – and the inevitable budget cutbacks would gradually erode the present lifestyle. "In the North, we must welcome development. We cannot afford to do otherwise. Secondly, industry must also change its attitude. We should be welcomed as the serious investors we are, the serious businessmen we are, and industry should be ready to share the profits of development with us."[28] He ended with a plea for cooperation.

At about this juncture, the Sogepet Limited seismic testing program off Southampton Island in Hudson Bay was at the approval stage, requiring the oil and gas lands administration to lift the 1971 moratorium on restricted areas. In the meantime, Sogepet president Frank Joubin met with the Inuit land-claim staff and held discussions with the Nunasi Development Corporation. The last was invited to be a participating partner in the Sogepet venture for 25 percent. The NDC considered this a genuine proposal that was worthy of consideration, not just a token relationship.[29] Clearly there was a proactive side to Inuit thinking on resources, which was often expressed by Tagak Curley, one of the more impatient Inuit leaders. The Inuit also felt the need to secure more than casual and unskilled jobs from resource development: otherwise they would never escape the casual labor trap decried by Hugh Brody and described in Chapter 4.

The same Partners in Progress conference heard an address by the Honourable Allen Blakeney entitled "How To Effect Change," in which the premier described northern Saskatchewan initiatives, which came about because one-half of the northern Native population was on welfare in 1971, and these people were strongly impelled to move south. The Department of Northern Saskatchewan was therefore established at La Ronge to administer many reforms that responded to Native demands. In one crucial reform, development approvals were linked to local employment contracts: approval was not granted unless the developer committed to hiring Native workers. The experience with local workers was very good. AMOK, a French uranium consortium, filled over 50 percent of its workforce from Native residents and awarded over $10 million in contracts to northern businessmen. At the Key Lake Mine then under construction, 59 percent of nonapprenticed trades were Native Northerners and service contracts worth $50 million were issued to Northerners. Steady employment was achieved without disrupting communities, as employees worked twelve-hour days on a one week on, one week off schedule. The mines effectively retained their Native workforce by the fly-in, single-accommodation method.[30] The Province of Saskatchewan became an advocate for Native employment through the sheer mounting of Native numbers. However, the seedbed for these progressive ideas had been planted earlier by Metis prospectors and political activists Malcolm Norris and James Brady.[31]

John Parker, Commissioner of the NWT, admitted at the same 1982 conference that constitutional and aboriginal matters were having some effect on resource development projects. He said that the lack of progress with the claims process (for example, the COPE agreement) had influenced the timing and nature of federal cabinet decisions in this area. He suspected that it acted as a brake on corporate plans as well.

> The Dene Nation has stated clearly that it is not opposed to development per se, but wishes to have its claims settled before further land alienation takes place. The Dene have not seriously opposed mining activities. I am confident that upon settlement of claims, they will seek out joint ventures and other means of themselves participating strongly in a broad range of resource development. The Inuit people also have accepted mining activities, except for the people of Baker Lake, who fear uranium exploration for several reasons. The Inuit also oppose tanker traffic, as evidenced by their testimony at the National Energy Board hearings on the Arctic Pilot Project.... The feelings of the Inuit and Dene people about the animals, fish, and birds of the land and sea must be understood. They depended on the land and its resources for their existence from time immemorial, and their relatively short experience with the white man is that his economy is up and down, and not seen as dependable. The land and animals are part of their culture, their heritage, and they want to safeguard their heritage.[32]

Interestingly, John Parker, the author of this perceptive statement, was educated in geology and drawn to the Northwest Territories as a participant in mineral exploration.

David Searle, former speaker of the NWT Legislature, interpreted the "Partners in Progress" conference theme as referring to southern and northern Canadians working together. His view of the processes taking place in the NWT behind closed doors was a jaundiced one. He described territorial division as balkanization, which would widen the understanding gap between north and south, and claimed it had little support from northern communities. The name of this game was to grab power from the federal government and establish a homeland for the Inuit and the Dene. The Dene Nation in particular was virtually holding out for a "homeland" and dismissed land-claim settlements like those made in Alaska and James Bay out of hand as "mere beads and trinkets." On the matter of resource development, Searle said Native groups were unanimous that none should take place until a settlement had been concluded. He cited actions of the current legislative assembly and the Berger Inquiry that supported this view. The "no development" stance had been softened, according to Commissioner Parker, to a "no further alienation" stance, indicating that anything interpreted as "prior alienation" might still go on stream. Fortunately for developers, said Searle, the federal government was still in control and could call the shots. He thought the government would render settlements subject to prior alienation because of the enormous cost of appropriating third-party lands. The big challenge for developers was how to respond to the social and economic issues of the day, for no project would proceed without addressing them. He said developers had to be sensitive to northern issues and "go behind the public rhetoric" – that is, get past the politicians and make a deal. Early-stage liaison by the developer would inform people at the community level: the people who would be affected, and those who were unemployed and wanted to blend a traditional way of life with a regular income.[33]

Professor Walter Kupsch, the Saskatchewan geologist who headed the Churchill River Study, was also a member of the Carrothers Commission, appointed in 1965 to investigate the development of government in the Northwest Territories. He described to the Northern Development Conference the modern expectations for the role of governments in development. He pointed out that southern interests had to be met, and that the federal government must represent the interests of all Canadians. He deplored the closed-door nature of land-claim negotiations, which made it difficult for other Canadians to know whether the federal government was adequately representing their legitimate interests. The speed with which the COPE agreement-in-principle was reached had caused alarm, but its delay in ratification seemed justifiable in the interest of the Yukon Territory and all Canadians.[34]

The Northern Development Conference of 1982 produced a number of resolutions calling on government to speed up the land-claim process.[35] It urged that developers offer participation, employment, and business opportunities to Natives and minimize impact on local tradition and culture. Orderly development would require stable regulation, tax, and marketing policies. The conference resolutions on revenue sharing with Native peoples and open-door negotiating were defeated in open session, the delegates deferring to government on these matters.

Political analysts were also studying nonrenewable resource questions. In 1983, Simon McInnes examined the consequences of policy and northern developments such as oil drilling in the Beaufort Sea, lead–zinc mining on Baffin Island, and uranium exploration in the Keewatin. He concluded that the direct consequence of government policy was the advance of development to serve Canada's energy and mineral needs, while not adequately meeting local and environmental concerns. The unintended consequences of development were twofold: the political awakening of aboriginal Northerners, and the desire for responsible government. Coming on the heels of the NWT plebiscite on division, McInnes' remarks suggested that the price of a Nunavut Territory might be the federal government's insistence that it retain control and management of nonrenewable resources.[36]

THE WESTERN ARCTIC CLAIM: THE INUVIALUIT FINAL AGREEMENT

The COPE negotiators took over five years to achieve the Inuvialuit Final Agreement – from October 1978 to June 1984 – owing to changes in the federal government and lengthy debates on constitutional and Aboriginal rights issues. In Section 7 of that agreement, which was ratified in 1985, the Inuvialuit were granted title to land and resources above and below ground, so as to provide the basis for cultural protection and possible income from resource development. The 7(1)a lands comprised 4,200 square miles in fee simple, including minerals, oil and gas, selected in blocks of seven hundred square miles, one near each of the six communities. Added to this was another block of eight hundred square miles in fee simple, including subsurface rights, on Cape Bathurst, where a moratorium on exploration and development was in effect. 7(1)b lands comprised 30,000 square miles in fee simple, less oil and gas, coal, native sulphur, and minerals.

In total, the Inuvialuit received title to 35,000 square miles, including the beds of lakes and rivers, although the Crown retained ownership of water per se.[37] The lands would be held by Native corporations with limited conveyance rights, so Inuvialuit were protected from the loss of land that occurred in Alaska. Another concrete benefit of the Inuvialuit Final Agreement was the payment of $45 million (1977 dollars), as well as special loan privileges and grants for social development. The settlement extinguished any further claims that the Inuvialuit might have regarding land rights, but this did not concern COPE members at the time of signing; in fact, they reiterated that all the rights the Inuvialuit wanted – political, cultural, civil, and educational – were in the agreement.

To summarize, the Inuvialuit got twelve square miles of land surface and two square miles of land with subsurface rights per person. The total allotments correspond to about 19.4 percent and 3.2 percent of the traditional area of 154,400 square miles, including land and sea. In his summary of Canadian comprehensive claims, Keith Crowe of the Office of Native Claims stated, "The Inuvialuit agreement was the first in Canada to include land ownership, both surface and subsurface in fee simple absolute title, including the beds of water bodies. The land however can only be conveyed within the Inuvialuit community or to the government of Canada. The quantities of land per capita are greater than in any other claim and represent title to about 45 percent of the lands traditionally occupied by Inuvialuit, still a record among comprehensive claims."[38] (In fact, Crowe underestimated the lands traditionally used by the Inuvialuit. The correct figures from the Inuvialuit Nunangat were 100,900 square miles of land and 53,500 square miles of sea, for a total of 154,400 square miles. Thus, the percentage gained was 34.6 percent of the traditional land or 22.6 percent of the traditional land and water.)

During the period when the Inuvialuit were building their case for substantial land ownership, say from 1974 to 1978, Canada was undergoing the OPEC energy crisis. The domestic oil industry was experiencing a burst of growth, and the government was laboring to encourage it. In 1978, for example, when the COPE agreement-in-principle was signed, the national oil company Petro-Canada was on a growth track. It negotiated the purchase of Pacific Petroleum that year under government sponsorship.[39] Industry investment in land rights and drilling was at record levels. The startup of the Syncrude tar sands plant in Alberta offered a promising new domestic source of petroleum.[40] Minister Hugh Faulkner estimated a petroleum industry investment exceeding $300 million during the year. This was the third year of Beaufort Sea drilling, and the approval for a Davis Strait test well was given (and later deferred). Other notable events of 1978 were the first submarine gas well completion and the first environmental assessment review of offshore drilling in the Eastern Arctic. Faulkner reported the signing of the COPE agreement at Sachs Harbour in October, an item that went unnoticed in the oilpatch media, despite its precedent-setting assignment of land rights.[41] The South tends to ignore Arctic affairs even when its vested interests dictate that it should pay attention.

In the new year, 1979, the government announced its interim federal land regulations, whereby Petro-Canada could option 25 percent of lands put up for sale, and could "back-in" on any applicant with less than 33 percent Canadian ownership. The industry did not always appreciate this nationalistic oil policy. In the Arctic, Dome Petroleum secured a seasonal drilling extension beyond September 15 to work on its Beaufort Sea gas discovery. Panarctic proved up gas reserves in the Arctic Islands, and Polar Gas, after spending $60 million on studies, released its pipeline route proposal to bring natural gas to southern customers by 1985.[42]

It can be safely said that the treatment of northern Native issues during the period – for example, the appointment of the Berger Inquiry – was influenced mightily by the energy crisis and the perceived need to promote northern petroleum resources. The federal acquiescence to Inuvialuit demands at the bargaining table in 1978 was not surprising in the light of such pressures.

LAND-CLAIM NEGOTIATIONS ARE STEPPED UP

The strategy of Inuit negotiators, who knew from other agreements that they would only secure ownership of limited areas of land, was first to determine what influence and control they could establish over Crown lands in Nunavut. They concentrated first on hunting rights and game management, and second on the management, control, and benefits from nonrenewable resource developments. Preliminary discussions about Inuit-owned lands elicited a rigid federal position: Inuit land selections would be for hunting and fishing purposes only, consistent with Aboriginal rights, and areas with third-party interests would be off-limits. The ground rules of comprehensive land-claim negotiations were set by the federal government, of course. The process began with the original receipt of the claim. A clarification stage followed, when both sides sat down to discuss precise meanings, policies, and mandates. For example, government policy was not to acknowledge Aboriginal title, but to recognize traditional use and occupancy of the land. Following clarification, the process was one of extended negotiation leading to a package of all the elements that might form an agreement-in-principle.[43]

Significant changes in federal policy emerged, and by April 1983 the TFN had successfully negotiated a broader understanding. An agreement was reached to provide guidelines for Inuit land ownership in a final settlement, though it did not specify the process of selection or the amounts of land – both were to be matters for resolution later. Several important principles regarding the kinds of lands available were enshrined in the agreement. Chief negotiator Allen Maghagak reported as follows:

> In one of the most important achievements, the federal government now agrees that Inuit land ownership can include subsurface areas where mineral deposits have been identified or other land areas that could be used for mineral development. We emphasized that Inuit land should do more than just allow people to continue hunting, particularly since our wildlife agreement-in-principle already gives hunting rights to Inuit. We had to convince the government that Inuit should own land not only for the purpose of wildlife hunting, but also for its mineral value. This agreement was difficult to reach because the federal government had said that aboriginal rights do not include subsurface rights. Inuit land ownership, as seen by the federal government, covered primarily hunting and wildlife. The federal government also wanted to protect the subsurface rights held by developers. Now, the federal government agrees that Inuit of Nunavut want to work towards economic self-sufficiency and, to that end, we will need subsurface rights to mineral resources.[44]

The 1983 guidelines for land identification set out numerous categories; land value would be recognized for renewable resources, nonrenewable resources, and archeological, historical, or cultural importance. Lands in which third parties had an interest, such as leased lands, were not excluded from the negotiation, but the rights of those parties would have to be treated fairly: leases, licenses and permits would not be canceled arbitrarily. The offshore was excluded from this agreement altogether, and land selections would be subject to hundred-foot corridors along certain shorelines of major water bodies.

By 1984, the TFN had a fourteen-member negotiating team under chairman Donat Milortuk (from Repulse Bay). The team signed three side agreements on municipal land, hiring policy, and priorities and goals of land-use planning. There were tours of communities for consultation and negotiations with Inuvialuit and Dene people regarding the overlapping use of land. Training was arranged through Employment and Immigration for ten Inuit to become community liaison officers on land-claim matters. Other issues, such as national, circumpolar, constitutional, and women's issues, occupied the attention of ITC leaders. One affiliate, the Keewatin Inuit Association (KIA), was a participant in the Hudson Bay oil and gas committee, whose purpose was to ensure that only safe, controlled, and locally beneficial exploration occurred in the waters of Hudson Bay.

The government continued to make arbitrary decisions advancing the interests of exploration and development. In July 1986, the new DIAND minister, Bill McKnight, approved an oil and gas exploration agreement in Lancaster Sound, soon to be the site of a land-use planning study. The agreement enabled Consolidated Magnorth Oakwood to undertake geophysical surveys immediately, without waiting for the land-use study. This was tantamount to a development commitment. Immediately the ITC, TFN, and BRIA protested on the grounds that this decision prejudiced land-claim negotiations and exposed environmentally sensitive waters. They were ably supported by the Canadian Arctic Resources Committee (CARC), an Ottawa-based public interest environmental group. The disappointment experienced over this issue evoked a renewed effort to seek some legal recourse or press for a change of policy that would enable Inuit to share the government's authority to manage the offshore.[45]

Tungavik Federation of Nunavut spokesmen like Donat Milortuk were encountering more and more questions in the communities, although they found acceptance that negotiation was a slow and difficult process. Bob Kadlun, chief negotiator, made an effort to describe the track record and pointed to the real progress they had made to date. After four years, they were still discussing the creation of a new system to screen and review the impact of projects. Inuit wanted a co-management role (i.e., not just an advisory role, but one recognizing traditional knowledge and allowing meaningful input for decision-making) and tried to achieve it by establishing a Nunavut impact review board (NIRB). They would soon start on a similar process designed for the offshore, on the grounds that Inuit were a truly coastal people. Inuit had been consistent in their demands for a decade: an ITC requirement, stated to the Berger Inquiry, was that they must have substantial and effective influence on the planning and regulation of development activity.[46] Soon to be placed on the agenda was a process for identifying what should constitute Inuit lands, as community land-use and occupation surveys were progressing to establish where the camps, hunting areas, archeological sites, mines, and potential parks were located. The land identification process would be a cooperative one involving all communities.

Kadlun reported that, up to this point, Inuit had reached agreement in seventeen different areas, a very good achievement in the light of the government's restrictive 1981 policy (described in the white paper *In all Fairness*).[47] He was confident that a final settlement would be the key to eventual self-sufficiency. Inuit would have access to capital and would be "able to participate in deciding on issues that directly affect them rather than reacting to what government and developers decide upon." Political change would be achieved through the creation of Nunavut, and that achievement would give everyone improved democratic government.[48]

Paul Quassa, regional negotiator, reported on the TFN's wildlife agreement-in-principle, which had set up the interregional wildlife management board. Though the agreement had been initialed five years earlier, it had still not been ratified by the federal government because the Department of the Environment opposed it. The TFN was therefore withdrawing support for any new national parks in the Nunavut settlement area, such as Bylot Island, Wager Bay, and Blue Nose Lake.

The federal policy of devolving powers upon the territorial government, in the absence of a final land-claim agreement, caused consternation among Inuit. Devolution would throw revenue sharing and development matters into doubt, and it would add a third party, namely the GNWT, to the negotiations. Inuit leaders met with David Crombie, the Progressive Conservative DIAND minister, who assured them that devolution of powers was part of the same process of territorial division and land-claim settlement. Nick Sibbeston, the NWT government leader, supported division and offered to cooperate with Inuit by suspending these resource management talks with the federal government.[49]

In January 1986, the impact and benefit agreement was initialed, and its provisions were vigorously debated in the Legislative Assembly. The debate provoked an open letter on the subject from the TFN. While the negotiations themselves were confidential, all agreements-in-principle were released to the public. The object of the agreement was to ensure that Inuit received economic benefit from resource developments (defined as projects requiring over $35 million in investment), not to prevent or discourage them. The Nanisivik and Polaris mines were cited as examples of developments where the number of Inuit working was well below the percentage promised by the companies and the federal government. The statement continued, "We conclude from this that the federal government poorly represents our interests in negotiations with developers and that we should conduct these negotiations ourselves."[50] This intention applied throughout Nunavut, on both Crown and Inuit-owned land, because otherwise land selection could become a hopelessly complicated process that was biased by geology. "Through land claim negotiations, we are trying to put in place institutions and processes to plan for and to manage the development and conservation of land, water, wildlife and the offshore; and to ensure that Inuit benefit from resource development. We are not and never have been interested in a cash and land settlement. We are seeking a more imaginative settlement of our land claim that will give Inuit a major role in making decisions about development."[51]

THE COOLICAN REPORT AND MORE NEGOTIATIONS

The Federal government task force headed by Murray Coolican reviewed land-claim policy over many months and presented its report to Minister David Crombie on March 19,1986.[52] The report, entitled *Living Treaties, Lasting Agreements*, dealt with the framework, substance, and process of negotiation. While policy had to be consistent with the national interest, one of the report's stated objectives included establishing "a framework of certainty concerning land and resources that accommodates the interests of aboriginal peoples and other Canadians." Another objective was to "enable aboriginal societies to develop self governing institutions and to participate effectively in decisions that affect their interests." Some of the substantive points of the report included eliminating blanket extinguishment of Aboriginal rights; granting secure rights to lands and resources that encouraged investment and development; empowering groups to negotiate decision-making participation in the management and use of land and resources, that is, boards and commissions that were more than advisory and consultative; and enabling negotiated ways for aboriginal peoples to share in revenues from resource development. Where traditional activity involved the use of ocean resources, the offshore should be treated as if it were land. In regard to the negotiating process, the report said that aboriginal rights should be acknowledged and protected during negotiations, which should be resolved in a reasonable time period. To ensure a fair and expeditious process, a commissioner was recommended as a "keeper of the process," but most aboriginal groups objected, preferring to deal directly with a negotiator who had more authority. The TFN supported the major policy recommendations of the Coolican report and expressed the hope that the Cabinet would see them adopted.[53]

Nevertheless, the federal government continued to pass legislation reflecting co-operation with developers, which served to complicate the Nunavut process. The Canada Laws Offshore Navigation Act, by redefining the ocean boundaries of the NWT, raised the possibility that legitimate Inuit hunters could become entangled in provincial game laws and imposed restrictions that were at odds with the wildlife agreement.[54] Despite representations by Inuit before the legislative committee examining the Canada Petroleum Resources Bill to regulate frontier lands, Bill C-5 was debated, given third reading, and passed on October 14, 1986.

None of the concerns expressed were acted upon, and only the NDP members defended the Inuit position. This event added to the disillusionment with government resource policy and swelled the support for an outright land freeze. TFN president Donat Milortuk wrote to the Prime Minister prior to the bill's approval, but to no avail. He said, "Nunavut is our land, therefore we protest in the strongest possible terms the enactment of any federal statute that facilitates the alienation of our land and resources from under our feet." Pointing out that the federal government had maintained at the negotiating table that the rights of third parties must be protected, he insisted that every time land is handed over to an oil company, the Inuit claim is prejudiced. Biologically rich areas such as Lancaster Sound were not protected by the bill, nor did it entrench an enforceable benefits agreement. The bill's environmental studies management board did not include Inuit representatives. "It is particularly insensitive to exclude Inuit from the board when hydrocarbon exploration is occurring in their homeland over which they assert aboriginal title." In the Commons debate, MP Ian Waddell said of a failed amendment that it was "a clear example of the priorities of the Government. It is a Government by and for the big oil companies whom they represent. Indeed, the big oil companies wrote the policy of the Government." It was also disconcerting that Energy Mines and Resources (EMR) Minister Marcel Masse, in wrapping up the debates, noted that "[t]he entire process has hinged on consultation. We have been sensitive to the concerns raised by all parties, especially representatives of the Native people."[55] Policy indeed had not changed: for the government, it was business as usual during the land-claim process.

John Amagoalik defended Inuit policy in a rebuttal to Gurston Dacks' article entitled "The Case Against Dividing the Northwest Territories," which appeared in the journal *Canadian Public Policy*.[56] Amagoalik said, "The search for minerals, hydrocarbons and the means of transporting these through sea ice, across rock and muskeg to southern industrial consumers has also put under threat the very lands and waters from which our people live and have always lived. This too is sponsored and closely overseen by the Canadian Government." Nunavut, which was proposed as a collective solution for these problems, had its first formal expression in 1976. "Nunavut ... is not a local fantasy or a single idea, but an expression of a changing north and of the emergence of our Inuit people from colonial-style domination by others."

Amagoalik maintained Inuit were a coastal and Arctic people, whereas the rest of the NWT is dominated by inland and Subarctic interests. "Our approaches to economic development, cooperation and collective control in economic management, conservation and sustainable development differ from the western NWT," he said. He had praise for Canadian values in a world torn by race relation issues, and hailed Nunavut as a clear assertion of Canadian integrity and social progress.[57] It is interesting and significant to note that, once again, development issues received credit as a motivating force or primal cause of the efforts to create Nunavut, whereas community issues went unmentioned. Nor did Amagoalik comment on the splitting of the territory's economic base, to the detriment of eventual provincehood, which was Dacks' point.

In June 1986, after five years of negotiations, the Nunavut wildlife agreement was re-initialed. The TFN got almost everything it wanted, including a Nunavut wildlife management board with equal representation by government and Inuit. Signatory to the agreement Paul Quassa said, "Now we've got a Nunavut Wildlife Management Board that will have decision-making powers. And we've got a disallowance clause that will help to guarantee that government cannot automatically override decisions relating to the Inuit share of wildlife. This is what we have spent five years fighting for."[58]

A major understanding was also achieved (though not ratified) on a boundary agreement with the Dene/Metis people that honored mutual hunting and fishing rights in overlap areas and proposed a joint management system. It was agreed that revenues and royalties from nonrenewable resources would flow exclusively to the party on whose side of the boundary the resource occurred.[59] One issue that remained to be settled was the status of the Inuvialuit settlement region in regard to the Nunavut political region.

Another piece of the interlocking system of managerial boards was actually put into place in January 1986, with the initialing of the Inuit Impact and Benefits Agreement (IIBA). This made the fifteenth formal accomplishment at the table.

So far, the Nunavut impact review board (NIRB) had not been successfully negotiated. In concept, this board would review all proposed projects. If the board gave its approval, and if the project involved more than $35 million capitalization and offered two hundred person-years of employment, the developer would have to negotiate an IIBA with the Inuit of the region. Thus it was proposed that mineral development would require impact benefit agreements. The benefits of IIBAs were to include training, hiring, housing, business opportunities, and the like attendant to a large development project, but such an agreement would not prejudice the rights of other Nunavut residents (possibly Whites) to obtain benefits also.[60] The control system being fought for so strenuously in negotiations would also include a land-use planning commission, a water board, and a wildlife management board, which would mesh together in a comprehensive co-management scheme.

In December 1986, chief negotiator Bob Kadlun spoke to the Yellowknife Geo-Science Forum to explain the land claim to the mineral industry. This meeting of mineral exploration professionals and prospectors convenes annually to hear the latest geological reports by government and industry spokespersons. Kadlun said the TFN, formed in 1982 for the purpose, was about halfway through the land-claim process. Eighteen subagreements had been initialed to date, and approval had been secured for the eventual establishment of a separate Nunavut Territory under Canadian law. He said the policy of government at the start was to give Native claimants a little land and a little money and to extinguish aboriginal title. This was totally unacceptable. He said it was far more important that "the Inuit seek to establish co-management structures to deal with land and renewable and nonrenewable resource development that is occurring in our own back yard. I believe it is only fair that Inuit participate with government in determining how development is conducted and ensuring that it is orderly."[61] At the moment, the negotiating plan was to extend the same coverage to offshore areas as to land. Kadlun said Inuit must gain confidence in management systems (which remained to be finalized) before dealing with land ownership. The land base would help economically, but Inuit really wanted to achieve self-sufficiency by participating in economic development, when appropriate, through access to capital. Inuit wanted to work and pay their way to economic self-sufficiency and also enhance their democratic rights through Nunavut public government.[62]

The Inuit objective of regulating offshore activity, which the federal negotiator had no mandate to deal with at that time, was predicated on their position as fundamentally a coastal, not an inland, people. Peter Ernerk, TFN executive committee member, had expressed a cautionary attitude about seismic activity, drilling rigs and oil tankers. His view, which echoed the genocide theme, arose from the failure of government and industry to consult the Inuit, "I think that these drilling rigs and these oil tankers are for Inuit what the train tracks and the locomotives were for Plains Indians – the end to a way of life. But we cannot stand idly by and watch as 'progress' risks destroying our means of survival through the inevitable oil spill or other industrial accident." In 1982, the Inuit sought government funding so they could establish a Hudson Bay oil and gas committee to present views on environmental issues. Funding came slowly and was cut off when the exploration activity in Hudson Bay ended in 1986, but the lesson was slammed home that it was necessary to have a marine management system as part of the overall claims settlement. The long-awaited review of federal policy, announced by DIAND Minister Bill McKnight on December 18, gave hope that marine harvesting and management might finally be on the table.[63]

The claims policy recommended by the federal task force (the Coolican Report) was a long time in being approved by cabinet. Meanwhile, the Inuit speculated that the land-claim process could be wrapped up in two years if the federal chief negotiator were suitably mandated to proceed. There especially would have to be movement on co-management boards. "We are not prepared to settle for just advisory bodies, as has been the case in other land-claim settlements. We are already up to our ears in advisory bodies," said Bob Kadlun and TFN vice president Jack Kupeuna when appearing before the standing committee on Aboriginal affairs in February.[64] What was needed was a single management system for both renewable and nonrenewable resources found either on land or offshore. Third parties were getting first-refusal rights on lands. It made Inuit indignant that third-party interests were so highlighted in government policies, including the newly announced northern mineral policy, while Inuit interests were said to be protected by existing mechanisms such as land-use and water permits. The inadequacy of the existing mechanisms – for example, the failure of land-use regulations during the 1970s uranium crisis – was at the very core of the Inuit policy of establishing new structures of management. Kadlun concluded, "We have to have a lot of confidence in any management structure before we are prepared to give the baby away for adoption. Otherwise, by the time we negotiate our entire claim, it will be the third parties that will be breaking open the champagne and we'll be left with the beer."[65]

Inuit were particularly proud of the Nunavut impact review board (NIRB). Intensively negotiated over a long period of time, it was able to control the pace of development while involving Inuit in the process. The primary task of securing decision-making powers throughout Nunavut had now been accomplished. The last piece of the puzzle was in place, and they could turn their attention to land selection and other benefits.

The northern mineral policy referred to above, which was meant to reassure investors, declared that development would not be put on hold while land claims were being settled. Mining companies were to be issued registered mineral claims, which would remain valid even if land ownership were transferred.

There also appeared to be a chance that conservation areas might be reduced (despite the conservation areas agreement initialed in 1983) as the international biological program sites were clarified and initiated. Kadlun said, "If it's not a public process, then we are back to the days when DIAND [acted] unilaterally and arbitrarily to impose its decisions on the North."[66]

Overall, the government was responding to Inuit positions, although it set the parameters for negotiation. Terry Fenge describes the fundamental point: "In this regard, the changes to the land-claim policy in 1986 were crucial. Without these changes, and the concomitant broadening of rights and benefits that could be included in the agreement, negotiations would likely have broken down."[67]

PUTTING THE LAND IN LAND-CLAIM NEGOTIATIONS

The land identification process formally got under way in the fall of 1986, when fieldworkers Luke Suluk and Keith Hay joined the team of Rick Riewe (leader), Michael Haqpi, and Jaypaitee Akeearok. These fieldworkers gathered critical information from each community and plotted the data on large maps. Suluk concentrated on Keewatin and Haqpi on Kitikmeot, while Hay and Akeearok covered Baffin and the High Arctic Islands. Moe Kilabuk and Andre Uttak later joined the team. Coordinator Riewe had forty-one map sheets in preparation to start the Nunavut Atlas, using 1983 guidelines. The community-by-community work of delineating parcels of land wanted by Inuit would take an estimated two years to complete.[68]

The delicate question of land ownership was now being approached, and carefully. While the ultimate end was the acquisition of title to some lands, Inuit knew this process meant "giving back" land they were already occupying and using, and what they gave back was bound to exceed by far what they received. Bob Kadlun said the TFN position was that every one of the twenty-six communities would have to be involved in the process of selecting Inuit settlement lands. Local residents had stated firmly that they must be able to participate directly in any negotiations dealing with land title.[69]

In 1987, the TFN began sending negotiators, board members, and staff members on community visits to present land-use maps to the people and seek their advice on land ownership negotiations. The issue of land relinquishment was sensitive and controversial. Previous attempts through the courts had not enabled Inuit to assert ownership, though the court had agreed on vaguely defined aboriginal rights. Despite these "rights," the TFN observed, government action to promote development continued unabated, though there remained the possibility that Inuit could act to stop it. Negotiation toward a land-claim settlement was indeed the best solution – and one that would give the government certainty of title.

The manner of deciding the location and amount of land required a consensus of local people. The TFN preferred that each community compile its own land position; the aggregate of these would then comprise the target quantum for the whole of Nunavut. The federal government preferred to negotiate a quantum figure at the main table, the land to be distributed among the regions and communities later on. The success of the whole claim package, on which all Inuit would eventually have to vote, would hinge on the success of land selection. To this end, Bob Kadlun said, "We think it would be a good idea if each community established a small committee composed of hunters, trappers, council members and elders."[70] Local people showed great interest in the land ownership meetings and supported the committee concept. They understood the trade-off of gaining decision-making powers, some legal title, and the hope of economic self-sufficiency in return for "giving up" the majority of the land. They wanted to be involved in negotiations, and they also wanted more information on general progress. Published information, like the newsletter *Nunavut*, had to be supplemented by more personal contact and communication from negotiators.

On August 12, the main table negotiations produced an initialed agreement on entry and access to and across Inuit lands. This agreement dealt with public, government, and third-party access. The last had relevance for developers who might hold subsurface or adjoining rights on Crown land. In some cases, existing third-party rights would be grandfathered or entrenched in the final agreement, even if Inuit became the landlords and collected the rents and royalties. General access would be regulated with the help of a surface-rights tribunal, whose concern would be fees and compensation for damage. This agreement was believed to give Inuit protection and a role in determining who could enter their lands and for what reason.[71] From the prospector's standpoint, this measure would impose a limitation on the free-entry system, insofar as Inuit surface rights would overlie Crown mineral lands.

At the end of 1987, a full year after the Murray Coolican Report, the government spelled out the federal negotiation mandate. There were significant changes from the fifteen-year-old policy voiced in the 1981 white paper *In All Fairness*. A briefing by DIAND Minister Bill McKnight and Assistant Deputy Minister Rick Van Loon included some bad news for the TFN negotiating team and its president, Donat Milortuk: The government considered some of the initialed agreements "too detailed" for the new mandate. "A claim agreement can guarantee Inuit membership on a management board, but its structure and operating details must be designed through public government," said McKnight. Significant changes would be sought for the proposed NIRB, which would reduce the effectiveness of development screening and separate the entire management system from the claim process. The government wanted to limit the application of impact and benefit agreements and set aside the discussion paper on Inuit equity participation. It appeared that the agenda for future negotiation was going to be reduced – and this had not been the expected outcome of the report. Inuit were greatly disappointed that the new policy narrowed the scope of negotiations rather than broadening them, as had been discussed at the table over many months. Now, the sought-after completion time of fourteen months would be impossible to achieve.[72]

A pattern of deliberate delay seemed to be emerging.

The new policy definitely denied some specifics, like economic guarantees, equity participation in developments, and hunter income support, but a significant broadening of federal policy in 1986 was the inclusion of royalty income sharing, clarified decision-making, offshore application of management boards, and toned-down aboriginal rights extinguishment.[73]

On January 25, 1988, provisions of an agreement-in-principle on marine areas were initialed by Tom Molloy, the chief federal negotiator, Ross McKinnon, representing the GNWT, and Bob Kadlun, chief negotiator for the TFN. It was gratifying to Inuit as a coastal people to have documented rights to the ocean and its resources and a guarantee that Inuit and federal government would share decision-making powers over the ocean. These were the first Aboriginal claimants in Canada to achieve protection of their rights to the offshore in a land-claim agreement. But after all, twenty-eight of the twenty-nine Arctic communities were on ocean shoreline. As Paul Quassa pointed out, "The most important thing about this agreement is that now all our successfully negotiated agreements relating to the management of Nunavut's land will relate to the management of Nunavut's offshore as well."[74]

A major hurdle was also overcome on February 23, 1988 with the initialing of a paper called "Inuit Land Identification," which met the concerns of both sides. The TFN would set up land identification and negotiating teams (CLINTs, for short) in every community. Each team would be made up of the mayor or council member, a representative from the hunters and trappers' association, one from the regional Inuit association, a community elder, and a delegate from the TFN. Each group would shoulder the task of setting down on detailed maps the amount and location of land that the community wanted to own (its area of interest). The total of such areas, computed at a major workshop, would then become the negotiating position or quantum of land for the TFN. One year was set as the time frame for the TFN and the government to come up with an overall quantum. Actual land ownership negotiations at the respective communities would follow. The government, to allay fears that Inuit would be saddled with less desirable lands, guaranteed that the Inuit would end up owning at least 75 percent of the quantum from lands identified earlier as their areas of interest. Thus the CLINTs were impelled to identify more land than they expected to get in the end. To expedite the selection process, the TFN hired three regional coordinators and organized land identification workshops.[75]

The land-use knowledge of community elders was indispensable during the preparation of land inventories and area calculations. Though familiar with maps, the elders often lingered at this task, savoring reminiscences of travels and camp life before getting down to the work of selection. By way of contrast, according to Bill Erasmus, Dene negotiators during this same period had to plead for cooperation at tribal meetings.[76] John Bayly reported observing four old Chipewyan men at Snowdrift community hall poring over a map with much delight, but clearly hesitant to mark it: "The map was like an icon which represented to these elders that which has been good and worthwhile in their lives." Native elders treat a map with respect, even reverence, and sometimes recoil at the thought of marking it. Maps are pictures of something sacred – like a stained glass window. "Illuminated by the sunshine of their memories and the shared experience reflected in their weathered faces, I began to see that there is something ... yes, something holy about maps."[77]

The year 1988 witnessed a lot of progress despite the negative feelings expressed among Inuit about the federal negotiator's mandate earlier in the year. Sound understandings were reached as well as some initializations on the ratification process, which would be triggered upon signing of an overall agreement-in-principle. The parties agreed that the Inuit would have access to carving stone, and the government to sand and gravel. In this discussion, the Inuit acquired a clearer understanding of surface versus subsurface ownership under Canadian law. Implementation provisions were agreed to, including the roles of various working groups and the already started Wildlife Management Agreement. The parties also agreed to arbitration provisions and processes, bringing the total number of land-claim subagreements to twenty-four.

Participants at the TFN annual general meeting were in an optimistic mood and spoke of a possible completion by March 1989. Rosie Oolooyuk of Rankin Inlet said, "Because our negotiators have been more aggressive than most other claim negotiators, and because we have not been in a rush like other claimants to settle our claim, we have been able to get a better deal from the government. I believe we are also in a better position to push to try to claim more land than other claimants."[78] John Magsagak of Cambridge Bay complained about weak communications, but he called the Iqaluit land-identification workshop a highlight and said he was looking forward to regional land quantum talks. David Aglukark of Eskimo Point thought the number and variety of agreements was exciting. Regarding land selection, he said, "Local people have been working hard on maps of their regions and drawing lines showing their areas of use and occupancy. When the public gets involved in our work, it's a good sign." Though many would want to keep all the land they had selected, the portions allowed them by the government, including resources, were still reckoned to be a good deal, in Aglukark's opinion.[79]

Surrender or extinguishment of aboriginal rights was a thorny question that Aboriginal people had opposed for sixteen years. The TFN board arrived at a decision in March 1989 that led to a position paper entitled "Cession and Surrender of the Aboriginal Title to Lands and Waters in the Settlement Area."[80] In essence, this was an offer to give up title – not rights – to the bulk of land in return for a quantum of land over which Inuit would have legal title and, of course, a system of management boards giving them real control throughout Nunavut. It was also established through negotiation that Inuit lands would be immediately vested upon final ratification, that boundaries for the most part would not require surveying, and that these lands could never be alienated or sold (as opposed to rented or leased) to third parties – they would remain Inuit lands in perpetuity.

Progress was being made with land identification. For example, the Baffin Region advanced its previous quantum designation of 33,000 square miles by adding a further 230 because of evidence of occupancy and use in the Fosheim Peninsula, on the western side of Ellesmere Island. Baffin people also gained recognition of the fact that they hunt at the floe edge of the landfast ice, well beyond the twelve-mile limit that is generally defined in offshore agreements.[81]

Detailed agreement was finally reached that would establish the Nunavut impact review board (NIRB), one of the most important bodies of all, for the purpose of screening and reviewing development proposals and applications. The board would comprise nine persons: four Inuit, two from the federal government, two from the territorial government, and one chairperson. The purpose of this board was to protect the ecosystem integrity of Nunavut and the well-being of residents and communities. Municipal developments and those of a minor nature defined in a schedule would not be subject to screening or monitoring. Decisions sent to the minister by the board could include the necessity for a review. If such a review was environmental in nature, a federal environmental assessment review office (FEARO) panel might take over, but one-quarter of the panel members would be nominated by Inuit and the department's guidelines for impact statements would be honored.

A NIRB report, with or without public hearings, could be rejected by the minister for three reasons: if it were not in the national or regional interest, if the terms would not mitigate adverse impacts, or if the terms were so onerous as to undermine the viability of a project deemed to be in the national or regional interest. To reject or alter a NIRB report, the minister would have to present sound reasons in writing. In any case, the board could modify and resubmit its report. The two processes were intended to dovetail, so that the regional interest would always be properly appraised.[82] This twenty-fifth agreement was the last one needed to complete an integrated management system covering all aspects, and it enabled TFN to proceed confidently with the final step of clarifying title. It was a profound achievement by Native people in Canada in pursuit of co-management rights covering land and environment.

It had been the goal of Inuit to establish an integrated system of institutions on which they would be equally represented. To summarize, they formulated the following system of five entities: a Nunavut wildlife management board (NWMB), a Nunavut impact review board (NIRB), a Nunavut planning commission (NPC), a Nunavut water board (NWB), and a surface rights tribunal (SRT). The decision-making procedures were spelled out for all except the SRT. These entities would become vehicles of public government upon the ratification of the final agreement, or within two years after ratification. They would operate to screen development proposals: "Oil, gas, mineral and other proposed development projects are to be considered in the context of approved land-use plans. Projects that conform to land-use plans are to be subject to a vigorous environmental and social impact assessment by NIRB or through the federal environmental assessment and review process (EARP). If proposed projects successfully navigate screening and/or assessment by NIRB or, if a water development project, by NWB, a project certificate is issued to outline terms and conditions under which the project may proceed."[83] New authority for the management of Crown subsurface resources could not be wrested from government, so the hopes lay in eventual devolution of such powers from Ottawa to the territories. However, Inuit surface ownership of significant acreage over Crown mineral land and the operation of the relevant boards in the case of development would likely influence the scale, pace, timing, and location of nonrenewable resource activities in Nunavut.[84] The skill with which Inuit handle these powers should influence the federal devolution of Crown land resources to the Territory.

The long-standing debate with the Dene–Metis alliance over the boundary issue was still unresolved because Manitoba and Saskatchewan Indians wanted to get out of the 1986 overlap agreement. Both sides advocated the use of an arbitrator in an attempt to resolve the problem once and for all. John Parker, former commissioner of the NWT, was eventually selected for this task.[85] His results played a role in actual land selections as described in Chapter 7.

AT LAST, AN AGREEMENT-IN-PRINCIPLE

In late April 1990, amid celebrations at Igloolik, the Agreement-in-Principle (AIP) was signed between the Inuit of the Nunavut Settlement Area and Her Majesty in Right of Canada. Signing for the government were Thomas Siddon, minister; W.T. Molloy, chief negotiator; and F.R. Drummie, associate deputy minister. Signing for the GNWT were Dennis Patterson, government leader, and Titus Allooloo, associate minister for aboriginal rights and constitutional development. Signing on behalf of the Tungavik Federation of Nunavut were Paul Quassa, president; Bob Kadlun, vice president; Mark Evaluardjuk, secretary-treasurer; Louis Pilakapsi, second vice president; and Pauloosie Keyootak, Charlie Lyall, Pudloo Mingeriak, James Eetoolook, and David Tukturdjuk, Board members.[86] The Inuit committed themselves to working toward a final ratification of the agreement within two years. That would require two steps: concluding the land ownership negotiations, and deciding upon the final text.

Certain provisions of the Agreement-in-Principle are relevant to the subject of land and mineral rights. The agreement says, "The primary purpose of Inuit Settlement Lands shall be to provide Inuit with the rights in land that promote economic self-sufficiency of Inuit through time, in a manner consistent with Inuit social and cultural needs and aspirations" (Section 17.1.1). Lands are expected to be valued for their renewable resource potential, commercial value, and areas of archeological, historical, or cultural importance, but may also include "areas of value principally for reasons related to the development of non-renewable resources, including areas of known or potential mineral deposits, areas of value for various operations and facilities associated with the development of non-renewable resources" (Section 17.1.2b i and ii). A number of principles apply to the identification of lands, including the provision that "identification may take place in areas subject to third-party interests; any rights or interests of third parties shall be dealt with equitably; the identification may be made on a case-by-case basis" (Section 18.1.1a). The Nunavut Settlement Area and its regions were identified and defined in the Agreement-in-Principle. The High Arctic islands of the Sverdrup Basin were excluded from land available for selection because Inuit did not occupy or use that land.[87]

The land quantum totals arrived at by negotiation were specified in the AIP as follows:[88]

Negotiated Land Quantum Totals specified in the Agreement-in-Principle signed in April 1990

Region	Total Area (sq. mi.)	Quantum (sq. mi.)	% of total area
North Baffin	192 519	33 230	17.26
South Baffin	112 675	25 500	22.63
Keewatin	204 943	36 890	18.00
Kitikmeot East	81 577	14 275	17.49
Kitikmeot West	149 971	25 495	16.99
Sanikiluaq	1317	to be negotiated	
Nunavut Total:	743 002	135 390	18.22

The negotiated quantum was only a fraction of the land used by Inuit, but according to Bruce Gillies, the land section coordinator on the TFN team, the negotiators recognized that the communities would be able to acquire ownership of the essential lands, of the quality lands they desired, by a process set out in the AIP.

Moreover, the subsurface land quantum was set at 14,000 square miles for the whole of Nunavut, which is 1.88 percent of the aggregate area or 10.3 percent of the surface quantum.[89] The allocation of these lands was to be the subject of negotiations with community land-identification negotiating teams (CLINTs) and regional representatives.

The rules of procedure and the timetable for the identification of Inuit settlement lands were spelled out. The government agreed that the final negotiated lands would comprise at least 75 percent of the lands originally identified by the communities as areas of interest. Procedurally, the government asked the Inuit to identify more land as a starting point than they would gain title to, that is, more than the agreed quantum areas (Article 19).

Interim protection measures under the AIP were to be put into force within five weeks of the completion of the areas of interest on identification maps. Blocks of land up to 20 percent of the agreed quantum would be withdrawn from further disposition (or "frozen") under the Territorial Land Act, subject to existing interests, benefits, and privileges. (In this provision, prospecting permits already in existence could be explored and converted to mineral claims under the Canada Mining Regulations).[90]

Also of interest is the resource royalty–sharing provision of the AIP, whereby Inuit would be paid each year 50 percent of the first $2 million of royalty revenue plus 5 percent of any additional royalty revenue received by the government (Section 27.2.1). The details of payment, applicability to all Nunavut lands, and capping if the amount yielded an equivalent average Inuit income exceeding the Canadian average per capita income, were also provided. Resources under this section included coal, petroleum, precious and base metals and uranium, but not stone, sand, and gravel, which are legally part of the land surface, not the subsurface. The royalty sharing benefit would be paid to the Nunavut Trust (Article 27). Prior to the Final agreement the TFN was required by Article 34 to establish a Nunavut Trust to receive capital transfer payments and royalties.

Paul Quassa observed in his April 30 speech at Igloolik, on the occasion of the AIP signing, that this was a remarkable achievement of ten years of work by the Inuit and the government, notably federal negotiator Tom Molloy. He said there was still much work to be done, but the AIP was a step in achieving the Inuit's own goals and a commitment to Canada. "The key to a better future for Inuit," he said, "lies in us taking control of more and more of the political and economic levers that affect our lives. We must determine our own future."[91] Quassa discussed the many features of the agreement, as well as the commitment to proceed with negotiations on the final agreement, implementation, and land ownership. He said political development discussions, initially with the territorial government, would also start within weeks; their purpose would be to frame enabling legislation for the establishment of Nunavut. The final package, which should be completed within two years, would then be put to the people of Nunavut for their discussion and ratification.

One dissenting opinion was voiced right away by broadcaster Saali Peter from Iqaluit. He contended that the AIP was a sellout to the federal government, which would continue to dominate Nunavut as a territory, since recent constitutional accords had tightened the rules for creating future provinces.

When about 80 percent of the land reverts to federal property and Inuit fundamental rights are extinguished, then the political future is lessened. The AIP was already outdated and should be rejected, he said, so that pressure could be maintained toward constitutional guarantees of self-government: "The TFN's Agreement-in-Principle threatens to shrink our past, present and future onto a dotted line. If the land we depend on turns into federal property, then the political influence we've enjoyed over the past 17 years risks being vaporized. Shouldn't we examine alternatives before accepting such a drastic measure?"[92] The self-government route that Peter was advocating would be a long and arduous one, whereas the Nunavut agreement was near at hand. It would enrich the Inuit with enormous powers and responsibilities, despite a dependence on Ottawa and dominance of Crown land (on which the Inuit had already negotiated hunting and fishing rights, carving stone priority, royalty sharing, and co-management privileges). Inuit were ultimately going to vote on the issue as part of the whole land-claim package. What remained was the important task of selecting which particular lands would comprise the 18 percent reverting to Inuit ownership.

At this point, we should explore the non-Inuit side of the coin. Exploration and development appeared to proceed rapidly during this era of Native activism and did not undergo change, other than a more acute awareness of land-access issues. Examples of pre-emptive agreements between government and Inuit to establish northern national parks were beginning to appear – and in such a way that these would not prejudice land-claim settlements. The Auyuittuq national park preserve had been established on Baffin Island in the early 1970s.[93] Such parklands offered substantial benefits to Inuit, both by giving them some control as joint managers and by taking out of circulation lands that might otherwise have been acquired for development. We might also remind ourselves that oil industry representatives in Alaska had pressured the government to expedite the Native claims settlement there in 1971, but John Bayly, in presenting the COPE–ITC final submission to the Berger Inquiry, expressed the view that Canadian pipeline proponents had washed their hands of the fate of Native claims.[94] Inuit were given some cause to harden their opinion of developers as they came into the final phase of the negotiating process: selecting the all-important settlement lands. They also became sufficiently empowered as negotiations evolved to take a strong stand on development issues.

6. EXPLORATION PRESSES ON

CALMING THE WATERS

The news of Justice Patrick Mahoney's decision on aboriginal title was headlined "Inuit Given Title at Baker Lake, But Mining to Continue." The court affirmed aboriginal rights over the 130,000 square kilometers under contention in the matter of user rights – the right to hunt and fish and possibly to claim for compensation. Justice Mahoney did not find in this case that the mining companies were significantly responsible for the decline in the caribou herds, as alleged. His sixty-five-page judgment stated that the Inuit do not own surface rights as such, and he added that on December 17, 1979, he would lift the Court order that had restricted exploration in the Baker Lake area since 1977. Luther Chambers, lawyer for the government, considered that the department would then be free to deal with the land and to issue mining permits. However, the question of certainty of tenure was a puzzling one for the mining industry, since Native residents would now be on more solid ground for claiming compensation. Marc Denhez, lawyer for the Tapirisat, said they were cautiously optimistic, since the federal court had given recognition to the "principles of aboriginal rights," which "can be invoked in the future to protect the Inuit from careless development." They were not sure just how the judgment could be used to further restrict development and help the people of Baker Lake. There was talk of the fund-raising that would be necessary to start an appeal of the case (but this never happened, as it turned out).[1]

About the same time, the Northern Mineral Advisory Committee reported to Minister Jake Epp through its chairman, John Bruk, president of Cyprus Anvil Mining Corp. Bruk stressed how important mining had been to the North, how it had persisted despite the attendant risks, and called on the government to "provide a stable and certain economic and political framework so that the mining industry would then know exactly where it stands, what is expected of it, and what the rules under which it is expected to operate are." The minister had already said in recent Arctic tours that land-claim negotiations were a top government priority. He reassured the mining and petroleum industries that northern development must proceed while those negotiations were going on, although this policy had drawn criticism from Native groups. On the official side, the Yukon had been particularly vigorous in its agitation for more development to broaden and strengthen its economic base.[2]

A BUOYANT BUT CHANGING INDUSTRY

Northern mineral exploration was already quite diverse. Typically, 30,000 claims were being staked each year, and 100,000 claims were in good standing in the NWT in 1977. Base-metal activity had declined during the year, setting a trend that was expected to continue. The Polaris lead–zinc deposit on Little Cornwallis Island was judged to be a significant discovery, but it was economically problematic pending the resolution of water supply, tailings, and shipping arrangements. Uranium exploration, however, was upbeat all over the world, since the radioactive metal was selling for a record $40 per pound. NWT permits for uranium prospecting were issued in large numbers, especially to three companies, Uranerz Exploration & Mining Ltd., Gulf Minerals Canada Limited, and Urangesellschaft Canada Ltd., the first of which was working in northern Baffin Island.

Pan Ocean Oil Ltd., one of the active companies and a defendant in the Mahoney trial, carried out work in the Yukon, Saskatchewan, the NWT, and overseas. In the Baker Lake uranium play, Pan Ocean had large land holdings in the Kazan Falls area, which were currently being explored under option by Cominco. In addition, the company held 838,000 acres of permits and claims in the Angikuni–Yathkyed Lakes area, where drilling for radioactive minerals was planned for 1978. Activities by Pan Ocean, Cominco, and Urangesellschaft were hampered by the land freeze intended to protect the Kaminuriak caribou herd until the trial.[3] Iron ore properties, subjected to the harsh realities of the global market, were in a holding pattern, with claims being maintained on the basis of assessment work already performed. Elsewhere, exploration was busy: Panarctic Oils had drilled seven exploratory wells in 1977 and was budgeting for ten more wells in 1978 to establish Arctic natural gas reserves.[4]

In 1980, all major metals were down in price, with some consequent effect on new exploration activity. Continuing projects were active, such as Cominco's fine discovery at Pine Point, where 6 million tons averaging 10 to 12 percent lead–zinc had been blocked out. Some clouds overhung the Canadian uranium industry because of the Bates Royal Commission on uranium mining, which was followed by British Columbia's outright ban on exploration for seven years.

That decision was brought down by the government of William Bennett, Jr., because of concern about the orchards of the Okanagan valley, where exploration attention was focused. Editorially, the *Northern Miner* argued that this ill-conceived political move would harm the whole B.C. economy: "For what mineral is next, it might well be asked? By bowing to a vociferous and fanatical minority group of environmentalists and activists, flush and ecstatic with their victory, they have only encouraged the fanatics to make further – and possibly even more damaging – demands."[5] Rhetoric of this kind did not inspire public confidence. Canadian attitudes toward mining (and uranium mining in particular) were approaching a low point, and professional organizations such as the Mining Association of Canada (MAC), the Canadian Institute of Mining and Metallurgy (CIMM), and the Prospectors and Developers Association of Canada (PDAC) were launching programs designed to improve their image.

Bill Padgham, DIAND's resident geologist for the Northwest Territories, was optimistic in his 1980 report on the mineral industry. Mineral and petroleum production had set a record in 1979, at $450 million. Exploration was increasing as well, and the government was encouraging industry to do more:

> Exploration also set new records because of the continued expansion in the search for uranium. Claim holdings of 3.65 million hectares, January 1, 1980, are at an all time high and more prospecting permits (288) are held than ever before, with the issuing of 98 on February 1, 1980 and 100 in 1979. Thus, the mineral rights of nearly 8.35 million hectares (31,234 square miles) are held under mineral exploration permits of one type or another. This, however, is only 2 percent of the area of the NWT.[6]

What was causing all this excitement and desire for acquisition? Prospecting expenditure was large, possibly as much as $30 million, as some twenty-four companies submitted applications for prospecting permits to DIAND. Most of the interest was concentrated in the Baker Lake area, well known for its uranium potential. Strong possibilities existed for the discovery of deposits near the basal contacts (unconformities) of the Dubawnt and Amer Groups of Precambrian sedimentary rocks. Similar geological models emanating from Saskatchewan were motivating uranium exploration in many other parts of Canada as well. Urangesellschaft undertook major programs in the Thelon region, including diamond drilling on the important high-grade Sissons Lake discovery west of Baker Lake. This occurrence, which was to become known as the Lone Gull or Kiggavik deposit, convinced other companies that economic concentrations were likely to be found. Indeed, the Baker Lake area was experiencing a uranium rush of large proportions despite the opposition of local people.[7] Nothing motivates prospectors more than a good discovery.

Padgham also reported vigorous activity due to strong precious-metal prices. Echo Bay Mines was going underground to test the Lupin gold property at Contwoyto Lake, and rumors of a production decision by O'Brien Energy at Cullaton Lake were considered meaningful. Both the Yellowknife and the Kaminak volcanic belts were attracting the attention of gold prospectors.

The Northwest Territories, like the Yukon, continued to be the domain of private industry. For this reason, it may have siphoned some free-enterprise activity away from mixed-economy sectors like Saskatchewan, Manitoba, and Quebec, provinces whose governments participated directly in the exploration business, despite its well-known risks. The mineral industry gained a certain amount of comfort from a new study by the Centre for Resource Studies at Kingston, entitled *The Impact of Policy Change on Decisions in the Mineral Industry*.[8] The authors, Owen and Kops, said that where the government is both rule-maker and participant, this can lead to a slowdown of private-sector development. A major cutback was noted in Manitoba under the NDP initiatives, which committed government to active mineral exploration. Private sector managers were also put under more pressure in a mixed economy to recognize the public interest by hiring and contracting local people. Clearly, they would rather not be distracted from their preoccupation with geology and minerals. The political pressures in some jurisdictions turned the spotlight on the more appealing frontier regions of the territories.

SOME INUIT RESPONSES

In early 1981, proposed federal legislation to control frontier petroleum exploration and development in so-called Canada Lands came up for discussion in the Commons committee on national resources. One of its key articles would allow Petro-Canada, the national oil company, to claim a 25 percent interest in oil and gas discoveries not yet in production. The ITC, representing 23,500 Inuit, jumped at the chance to exploit this legislation, proposing in a brief to the government that Petro-Canada be allowed to take more than the proposed 25 percent, so that Ottawa could stream the revenue into the Eastern Arctic. The ITC saw this potential revenue as a source of support for a postulated Nunavut Territory, which would otherwise lack funding. The same argument had been made earlier by the territorial governments, who thought this legislation could shatter territorial dreams of provincehood by preventing them from getting control over resources.[9]

This legislative event happened less than a year after Graham Rowley proclaimed, in a 1980 speech to the Royal Society, that the Inuit had been spoiled by massive government spending on the part of overgenerous bureaucrats. He added that the legacy of such spending "makes it difficult for the Inuit to successfully make the transition to wage employment and could cause problems for an Inuit government if the goal of a separate territory of Nunavut is achieved."[10] On the contrary, the "spoiled Inuit" of the new generation not only possessed the motivation for change, but exhibited through their leaders their having come both to understand and to exploit the Canadian political system.

In 1980, the Ottawa-based Inuit Development Corporation (IDC) agreed to provide loans of up to $25 million to bring the Cullaton Lake gold mine into production. The president of IDC, Tagak Curley, called the gold mine project "a natural step in implementing our industrial strategy." He was thinking of their actual participation, in the form of 100,000 bonus shares and options on 400,000 shares at $10, in Cullaton Gold Mines Ltd. The company had already spent $7.5 million in exploration dollars and estimated ore reserves to be 315,000 tons, averaging 0.74 ounces to the ton on a "cut and diluted" basis.[11] The Cullaton deposit looked very promising, with apparent reserves of over 200,000 ounces in place and gold prices running over $500. Buoyant gold prices have always caused mining excitement and activity; the IDC did not, however, appreciate the geological risks attached to such a venture.

Cominco representative H.M. Giegerich addressed the National Northern Development Conference held in Edmonton, Alberta, in October 1982. This was the ninth such conference, and its particular theme, *Partners in Progress*, was primarily an attempt to reconcile Canadian Native and developer views. Indigenous people made presentations to the conference and participated in the discussions. Commissioner John Parker also addressed the group on territorial division and on Northwest Territories affairs such as Native self-expression: Native groups were not pressing for a cessation of development per se, but merely requesting a delay until claim settlements were achieved. Parker also said that Inuit, who were against tanker traffic as proposed in the Arctic Pilot Project (which never materialized) and feared uranium exploration around Baker Lake for several reasons, were calling for impact studies.[12] Giegerich, however, stressed Cominco's previous partnerships in the North, calling them "partnerships between private enterprise, the public sector or government, and the peoples of the North."[13] He called on government to help defray the cost of the social infrastructure required by mines to advance their good work in the North: "Such a commitment should include upgrading of transportation, power and other infrastructure, but it also means creating a climate favorable to growth. The existing royalty regime should be maintained, the status quo on taxes should be maintained, and a stable approval and regulatory system should be described and fixed for at least five years. All these factors would assist the private and public sectors in the North to be true partners in progress."[14] Notwithstanding these remarks, the first resolutions of the conference recognized that the land-claim impasse was the major problem standing in the way of orderly development. Delegates urged the government toward a cooperative, equitable, and speedy settlement.[15]

URANIUM DEVELOPMENT PROJECTS MAKE NEWS

Following the extended hearings in 1980 of the Key Lake Inquiry Board, commissioned by the Saskatchewan government, a production lease was signed in August 1981 with the Key Lake Mining Corporation, allowing for the province's fourth uranium mine, which by 1984 would become the largest in Canada.[16] The major concerns at the inquiry were project-related effects on the environment and health: contamination of the water supply, tailings containment, radioactive exposure of workers, and local drainage disruption. The five-member inquiry board was chaired by Robert Mitchell, the Saskatoon labor lawyer who subsequently assumed the job of federal negotiator for Inuit land claims and later became a provincial politician.

The momentum of uranium exploration was generally maintained in 1981, with its chief focus on Saskatchewan. Exploration moved forward despite the downward trend in prices of the previous year, when overall uranium expenditure for the country had been about $130 million. In the Baker Lake area, encouraging results helped to sustain additional drilling efforts. For example, Pan Ocean Oil, a company that had been drilling since 1975, encountered near-surface mineralization west of Bissett Lake. The occurrence had some similarity to the unconformity-related deposits that were the principal targets in the Athabasca basin of Saskatchewan. More importantly, Urangesellschaft had success drilling east of the main zone of its Lone Gull property, situated about 80 kilometers west of Baker Lake. The main zone, first delineated in 1978, was reported to contain an estimated 7,700 tonnes of uranium metal.[17] The German-owned company was actively looking for Canadian partners to participate in property development.

The year 1980 had brought bad news for the uranium industry, demonstrating once again the complex emotional overtones of uranium mining, already well appreciated in Saskatchewan since the contentious Cluff Lake Inquiry of 1977–78, which preceded the opening of the French-owned Amok Ltd. mine. In Labrador, Brinex Limited and Commonwealth Edison deferred development of the Kitts–Michelin project, pending improvement of market conditions and further consideration of the environmental implications. (Commonwealth Edison had recently bought out Urangesellschaft's 40 percent interest.)

A significant factor in this case was Native opposition by the Montagnais-Nakapi (Innu) people. They convinced the Newfoundland government to accept the recommendation of the environmental assessment board and withhold a development license until the companies could demonstrate safe disposal of radioactive waste. In British Columbia, the announcement of a seven-year moratorium on uranium exploration and development stopped all activity. The B.C. government took this action, which preceded the receipt on October 31, 1980 of the final report of the Bates Royal Commission, because it believed that such activities would do irreversible harm to the environment or, at any rate, it considered the environmental costs to outweigh the potential benefits.[18] The major impact of the B.C. decision fell upon the Rexspar and Blizzard Mountain properties; the latter had a promising mineral deposit close to the fruit-growing Okanagan Valley.

Arctic mineral development was going ahead smoothly at Cominco's Polaris mine near Resolute Bay. The company announced its plan to train and employ Native mine workers, because of the value of having a local workforce, but proceeded with a single-status camp and refused to construct a townsite for families like the one at the Nanisivik mine.

When Northern Affairs chief geologist Bill Padgham released his usual report on Northwest Territories mineral activity for 1982–83,[19] he noted the severe impact of the general mining and exploration recession beginning in 1981, including the closure of the Pine Point mine for six months in 1983. Exploration expenditure had dropped from the boom level of 1979–80, when it had approached $50 million, to approximately $25 million in 1982–83. He said that uranium had been the major Northwest Territories target from about 1977 to 1980, largely replacing base-metal exploration, which had rated number one throughout the previous fifteen years. Gold then surpassed uranium as the major target commodity in late 1980. This transition was gradual, because many costly drilling programs to test the prospects of earlier uranium discoveries continued during the review period, whereas gold exploration was carried out by junior companies that were slow to raise funds during that inflationary period. Padgham said that 1983 mineral exploration was characterized by a balance of effort, which included uranium, base metals, silver, tungsten, coal, and, increasingly, gold. The partial statistics on land acquisition reported by Padgham[20] indicated these trends:

Claims Staked (in thousands of hectares)	1976	1977	1978	1979	1980	1981	1982	1983
Keewatin	214.0	108.0	620.0	438.8	518.1	323.2	65.0	52.6
Arctic Islands	0.9	1.4	82.1	16.0	9.7	3.0	22.8	4.0
Number of Prospecting Permits Issued								
Churchill Province	53	25	50	72	57	54	17	1
Arctic Islands	0	0	10	0	10	2	53	87

The reversal in 1982 of the downward trend in prospecting-permit and mineral-claim disposition for the Arctic Islands was chiefly due to Petro-Canada's new program of coal exploration on Banks and Ellesmere Islands. Interest was also accelerating in other areas: these included the Milne Inlet area of Baffin Island, near the Nanisivik mine; the Melville Peninsula, for its iron-formation potential (Borealis Exploration); and the Minto Arch on Victoria Island, for copper (Panarctic Oils). Padgham noted that "the absence of new prospecting permit applications for the areas of uranium potential in the Baker–Thelon and Hornby basin underscores the shift of exploration to other commodities."[21] As of January 1984, there were 40,952 mineral claims in good standing throughout the Northwest Territories, covering 3,114,606 hectares, or about 12,000 square miles.

The outlook for uranium brightened considerably during 1985, especially in Saskatchewan. Eldorado Resources commenced the mining of the Collins Bay ore body with concomitant environmental improvements to its mill, since the Rabbit Lake ore body had become mined out in 1984. Cluff Mining's phase-two development included underground recovery from the Dominique–Peter deposit. Key Lake Mining Corporation became the world's largest producer. In exploration, the budgets of the ten largest corporations accounted for 95 percent of the $35 million expended nationwide, but chiefly in Saskatchewan. Included in this group of ten were Urangesellschaft Canada Ltd. and PNC Exploration (Canada) Co. Ltd., both of which had large investments in the Baker Lake area.

The federal government was reviewing its policy on foreign ownership of uranium mines and also on upgraded processing of uranium (to hexafluoride form) before export. The new policies were designed to encourage investment and thereby contribute to economic growth and job opportunities.[22] DIAND issued a discussion paper and began consultations leading to the adoption of a changed northern mineral policy in 1986. This change incorporated specific planning policies for the Lancaster Sound and Mackenzie–Beaufort areas.[23]

In 1987 came a proposed free-trade agreement with the United States, which had significant unfulfilled uranium demand, and a new policy on foreign ownership. The latter specified that Canadians must own 51 percent of an individual uranium property when production commences; however, a reduction of that percentage might be permitted if the project was Canadian-controlled. More flexibility in financing such projects was therefore made possible. The three projects next in line were Cigar Lake Mining Corporation and Denison Mines Ltd. at Midwest Lake, both in Saskatchewan, and Urangesellschaft's project at Kiggavik (formerly known as Lone Gull) in the Keewatin. A 1986 "pre-feasibility" study of the Kiggavik project recommended a production level of over 1400 tonnes of uranium per year, attainable in the mid 1990s, for an investment of $200 million.[24]

THE EFFECT OF LAND CLAIMS ON EXPLORATION

In 1988, there were thirty-five prospecting permits issued in the district of Keewatin, compared with five in 1987 and twenty-six in 1986. Most of the interest was centered on the Archean Kaminak volcanic belt (also known as the Rankin–Ennadai belt), and the target was primarily gold occurring in iron formations, quartz veins, and shear zones. Several uranium projects were also underway following the success of Urangesellschaft in exploring the Thelon sandstone unconformity near Baker Lake.

A feasibility study was now in progress to examine the Kiggavik situation thoroughly. The whole area had problems not found elsewhere, it was noted, because large tracts of favorable geology lay within the Caribou Protection Area. According to reporter James Borland, "[t]hat means air access to exploration camps can be halted within two days notice anytime from mid-May to mid-July if government biologists determine that caribou herds are too close …. In 1988, Borealis, Inco, and Noble Peak Resources all suffered shutdowns lasting about six days. There is also the potential for land-claim settlements to have some impact on the area, but there is no indication at the moment that either land claims or the caribou herds will have any significant effect on mineral exploration or development (if development becomes warranted) in the area."[25] It was business as usual, or so said Borland, basing his conclusion on comments by J.L. Griep, resident geologist for the Keewatin district, in the year's exploration review. Borland's article did not disclose that explorationists were paying attention to Dene and Inuit land-claim activity and staking more mineral claims that year.

The stock market funds raised by mining companies had contained a strong component of "flow-through" (or tax write-off) shares since 1983. The year 1988 witnessed a decline from the enormous volume of flow-through funding (estimated at nearly $1.2 billion) in 1987, when Canadian exploration spending was 80 percent devoted to gold. The value of such financing would draw public comment in 1989, as between 1983 and the end of 1988, flow-through shares had raised some $3 billion for mineral exploration in Canada. This was an enormously productive period: the decade had more gold discoveries than ever before, and most of the new gold discoveries after 1983 were made with the assistance of flow-through funding.[26] The pace of exploration activity in the North therefore reflected the ready availability of cash from investors.

The annual DIAND exploration overview, authored by Bill Padgham and other staff geologists, reported weakness during 1991. Claim-staking was down again, for the third year in a row, but the NWT was energized by the copper exploration carried out by Noranda on Victoria Island and the diamond prospecting by Dia Met and BHP-Utah near Lac de Gras. The trend of activity was reflected in new stakings for the territory:

Mineral Claims Recorded, in Thousands of Hectares

1984	1985	1986	1987	1988	1989	1990	1991
240.9	294.8	360.3	552.3	739.9	321.6	281.9	182.6

The peak year of staking, 1988, saw the total area of NWT mineral claims in good standing rise to 3.115 million hectares. This area would fall to 2.413 million hectares in 1991 through a combination of declining rates of staking and systematic lapsing. Prospecting permits also experienced a peak in 1988 and then another peak in 1990, followed by a decline in 1991. Padgham reported an increase in the number of projects being worked in the Keewatin, with this explanation: "Some of the increase in exploration in the Keewatin may have been triggered by the expected TFN land-claim settlement, which could have a significant effect on the ease of restaking mineral lands if overlying claims lapse after the settlement has been made. Existing holdings (or maintained claims and leases) will not be significantly affected by the settlement."[27] Some of this activity consisted of diamond drilling, of which the drilling around Kiggavik continued to be an important element.

The Inuit Agreement-in-Principle was signed in 1990, and the final land selections and ratification were under way. The Inuit had agreed to respect third-party rights, including those conveyed by the Canada Mining Regulations. So, if mineral claims were kept in good standing, they were destined to be protected: the essential terms would remain unchanged, although the royalty recipient would change. If the claims lapsed, however, Inuit ownership would come into effect, with consequent changes in rules and royalty rates. The land-claim awareness that Padgham referred to was interpreted by industry, and companies acted upon this information in terms of their own self-interest. The record of exploration interest gave the Inuit reason to believe that mine developments would engulf the North and the need for planning, monitoring, and regulation would be large.

PROJECTS UNDER NATIVE AND ENVIRONMENTAL SCRUTINY

Things were happening at Baker Lake in 1988. Urangesellschaft, according to Michael (Mick) Stuart, its executive vice-president, was growing impatient that the federal election would slow down development of its Kiggavik project. In fact, the federal environmental assessment review office (FEARO) had delayed naming the members of its Baker Lake project review panel, but manager Paul Scott hoped the government review process, likely to take a year, could start with hearings in November. The panel was to be made up of local people and uranium mining experts. They would be working with the company's environmental impact study of the mining project, which Beak Consultants Limited of Toronto had prepared for Urangesellschaft.

Already concerns were being voiced, by Dan Prima of the environmental group Nuclear Free North, because the study would deal with the impact only on the local site, and not on the surrounding communities, and it would not examine what the uranium was to be used for.[28] The residents of Baker Lake (and Inuit in general) were going to get an opportunity to participate in the approval of the project, a decade after their 1978 rejection of the Polar Gas pipeline project.

Concurrently, a small controversy was brewing at Cominco's Polaris mine. Lead and zinc concentrations had risen in the surface waters of Lake Garrow, which was being used for subaqueous discharge of tailings, contrary to the provisions of the Water Board license. This practice created a risk of contaminating the nearby ocean, posing a threat to the marine environment. The causes of the problem were accidental spills and other tailing management practices that did not carry the solids slurry below twenty-six meters into the depths of the lake. Some government agencies, including the Department of Fisheries and Oceans, were opposing Cominco's application before the NWT Water Board to raise the limits of lead and zinc; these agencies proposed that the company undertake stringent surface-water treatment. Furtive sampling of the tailings lake by an independent researcher had touched off the row, causing Cominco to assert its sole rights over the lake, which was properly designated for tailings and leased for that purpose.[29] The Water Resources Division excluded the whistle-blower from the tailings pond after the incident, but the affair contributed an element of bad public relations to the company image.

Near Yellowknife, a tailings pond application by Treminco Resources Ltd happened to cover lands that interested the Dene and the Metis. Their specific land selections had not been concluded, and the process was in abeyance pending a final land-claim settlement. The negotiating secretariat under Ted Blondin made a private agreement with the company, thus avoiding a confrontation and ushering in a new era of cooperation between Natives and industry.[30] This agreement manifested what David Searle, then speaker of the NWT legislature, had meant when he spoke to the Northern Development Conference in 1982: "Putting it simply, the developer must go behind the public rhetoric of 'no development until land claims are settled' and make a deal.... If you ... make your deal with the people at the community level, then you're playing their game from a position of strength and almost anything becomes possible."[31] Another clear example of the new cooperation was the agreement on social and economic benefits between the Dogrib Tribal Council and Neptune Resources Corporation. The benefits included a guarantee to provide on-the-job training and hire 25 percent of employees at the Colomac gold mine (140 miles northwest of Yellowknife) from the nearby communities. The Colomac agreement paved the way for a $3 million loan guarantee by the GNWT.[32]

In the Keewatin, community concern over the matter of land-use permits was evident as never before, because exploration activity continued apace. Mark Kalluak, mayor of Arviat, criticized the DIAND permit system and demanded that more Inuit be represented in the issuance process. Dennis Trudeau, speaking for DIAND, said that thirty land-use permits had been issued to mining companies in the Keewatin. He added that the department seriously sought community input from various groups when applications were received, and he hoped that the use of fax machines would help in the future. But Joe Niego, mayor of Baker Lake, said he had never known ahead of time when a land-use permit was going to be issued; his hamlet received the applications from the government only *after* the land-use permits had been issued.[33]

The Baker Lake uranium inquiry and the sentiments of the local population came to the attention of political candidates of the day, including Peter Kusugak, the Nunatsiaq federal candidate for the NDP. He reported that Baker Lake people asked "a lot of hard questions about the risks from ionizing radiation and radioactive wastes to the environment, their own health and the health of generations of Inuit to come." He said, "It will be crucial that the FEARO review of the proposal provide the people of Baker Lake and the rest of the Keewatin with all the information they need to decide whether the expected economic benefits are worth the risks. Hopefully there will also be a forum to discuss the aspects of uranium mining which the FEARO panel will not be allowed to discuss. For example, the legislative assembly has declared the NWT a nuclear weapons–free zone. I don't believe that Inuit want the raw materials for nuclear weapons to be taken from our land."[34]

THE KIGGAVIK PROJECT REVIEW HAS ITS DETRACTORS

Early in 1989, the NWT Chamber of Mines conducted a presentation at Rankin Inlet that was intended to give the residents an overview of the mining industry. The residents pressed the visitors to arrange a more detailed discussion of the proposed Kiggavik mine, 75 kilometers west of Baker Lake, which was very much on their minds. What transpired was a debate that was started by Louis Pilakapsi, the newly elected president of the Keewatin Inuit Association, who claimed that uranium mining raised the incidence of cancer, caused long-term pollution of the land and animals, and yielded a product that was used for potentially dangerous purposes. Mick Stuart, president of Urangesellschaft (UG), a company that had explored the Keewatin district for thirteen years, said the mine was not in an area that would affect the migratory patterns of caribou. He said studies indicated less environmental impact was anticipated at Kiggavik than was occurring in northern Saskatchewan. He explained that Canada's uranium nonproliferation treaty would ensure only peaceful end-purposes; in fact, there was now a need in the world market for nuclear-power fuel. He claimed residents had nothing to fear, as the company would take adequate precautions and the upcoming FEARO investigating panel would enable them to voice their concerns and provide input.[35]

As correspondent Jim Shirley reported in the *Nunatsiaq News*, that was just the beginning of the debate. Jack Hicks of the Keewatin Regional Council (KRC) said he had been closely monitoring the FEARO process on behalf of mayors and municipalities, and there were many problems. To date, there had been virtually no input from Keewatin residents in establishing the terms of reference for the federal government's review. There was a dearth of information supplied so far, and participants were not going to be allowed to discuss the end use and destination of the uranium.

The discussion that followed included vigorous challenges from Tagak Curley, Peter Ernerk, Donat Milortuk, and many others, who expressed concern about the long-term storage of nuclear waste and the moral issues involved. Ernerk, as current territorial MLA, said that he was receiving many letters about the mine and termed this "the single most important issue of 1989."[36] He questioned the large scope of the mine and also the environmental effect of ancillary facilities such as roads, docks on Baker Lake, and the shipping corridor through Chesterfield Inlet. Milortuk cynically sensed that the federal government had already made a deal, and he compared the current situation to the influence that European interests had once had in fur-trade policy. Urangesellschaft, of course, was a German-owned company, and its partner CEGB (Central Electric Generating Board) was British. Jose Kusugak of the CBC tried to put the project into perspective by saying that "while the mine will only be in production for ten years, Keewatin residents will be faced with the prospect of living with a contaminated environment for several thousand years."[37]

Last to speak, with the aid of slides, was John Moelaert, an anti-uranium activist from B.C. whom the Keewatin Regional Council had hired to prepare its case against the mine. He said uranium mining is totally different from other mining, and those who benefit from it are rarely those who face the environmental risks. He described genetic effects from radiation occurring in U-mines, for which federal exposure limits were dangerously inadequate. Tailings containing radium and radon would continue to emit dangerous radiation for thousands of years. He noted that air pollution is one of the ways that radioactive contamination can spread from a mine, citing the contamination of Saami reindeer herds from the Chernobyl disaster. Plant life, water, and game animals can be affected by air pollution. Speaking for UG, Mick Stuart said that no final plan had been devised for the disposal of radioactive tailings, but one was under preparation.[38]

Clearly, the battle lines were being drawn even before the Kiggavik environmental assessment panel began its work, and such was the intensity of feeling over the issue that vigorous statements were already beginning to divide the Inuit community. A uranium intervention coordinating committee of Keewatin regional groups was formed after the Chamber of Mines meeting, with Tagak Curley serving as its spokesperson. In this capacity, Curley wrote an open letter to Environment Minister Lucien Bouchard, calling for more time, more intervener funding, and more changes in the FEARO terms of reference that would enable the committee to make a meaningful contribution to the process. He listed the groups that comprised the committee and stated, "We strongly oppose the mine and have serious reservations" about the FEARO review. The list included the Keewatin Regional Council, Inuit Association, Health Board, and Wildlife Federation, as well as the TFN and Peter Ernerk, who represented the riding of Aivillik.[39]

Ernerk publically questioned the intent – and especially the extreme stand – of the committee as expressed in Curley's letter, of which he claimed to be no part. He denied that Keewatin residents, including himself, had already made up their minds on the issue: on the contrary, they looked forward to the public presentation of facts that would enable them to draw sensible conclusions. Nor did he believe that some of the organizations participating in the intervention committee, such as the Health Board, the TFN, and some of the mayors on the Regional Council, had already reached a final decision on the Kiggavik project. Ernerk reiterated some of the concerns expressed by Curley and pledged to call on the Legislative Assembly to debate the issues of funding, duration, and timetable of the hearings. Ernerk and Curley were political foes, and Ernerk chose to quote some of the Curley's more questionable statements to the Assembly.[40]

Dennis Patterson, the territorial government leader, and the GNWT itself came under fire when Donat Milortuk, president of TFN, labeled their silence on the uranium-mine issue unwarranted and disturbing: "Opposition to the project by Inuit in Keewatin is clearly growing based on moral, environmental, social, cultural, economic and land claim grounds," he wrote. "Yet the government of the Northwest Territories has said remarkably little about this proposed development, except to support a public review of it by a panel."[41] Milortuk lamented that the review would be limited in scope because the terms of reference dealt only with the direct environmental and social impacts. He thought this wait-and-see attitude was unbecoming to a government that intended through devolution of powers to assume Ottawa's role of planning and managing mineral development in the North.

Commentator Jim Shirley observed that the Kiggavik case was bringing to the surface the many conflicts that would be on the minds of residents. There was the conflict between the traditional lifestyle and the modern wage-labor economy, each with its own vision of how the land should be used. The land quantum agreements were nearing completion, and the land-claim settlement – which would place Inuit in a position to direct the pace of growth – was in sight. They would be wondering how much influence they really would have in this debate, and how well their strong leaders, such as Tagak Curley and Peter Ernerk, would represent public interests, not just their own political ones. The press had a responsibility to focus on the issues, according to Shirley, rather than on personal rivalries and ambitions that might arise.[42]

Keewatin people had a reputation for their ability to stand up for their own interests and to produce strong leaders who, in turn, influenced them. While personal ambition was always present in public affairs, there was no question about the dedication and determination of leaders like Tagak Curley, who marshaled a very powerful case in opposition to the mine. The usually pro-development Curley was strongly critical of UG. As already noted, the company took no overt action to draw in the Inuit or involve them at an early stage. Polar Gas, behaving in a similar vein, had evoked the wrath of local citizens over its pipeline proposal. Curley, who was among them, noted that the arrogant company had encountered "Eskimo red tape" as a result.

DIAND information sessions were presented in Baker Lake during March in response to the requests from KIA and TFN to learn more about the mine. Pierre Laporte organized the meetings for the department, using government and industry material. The meetings were denounced as a "snow job" by local people, who were represented by Jack Hicks of the intervention committee and Jim Harding, a Saskatchewan university professor who was brought in to speak for the committee. The Inuit side took this meeting to be a debate, a contention that Mick Stuart of UG vigorously denied. They resented the use of "ten-year-old" American material and pro-development scientific reports and objected to the company's new tailing disposal plan – and noted that nothing at all had been translated into Inuktitut. Harding found the meetings paternalistic and insensitive.[43] Another attendee, Chris O'Brien of Ecology North, found the meetings inadequate (mine abandonment and restoration were not even dealt with), one-sided, unannounced, and incommunicative, he wrote a letter to this effect to Bill Stephen, DIAND director-general in Yellowknife.[44]

In response to the DIAND fiasco, public workshops were held in Rankin Inlet and Baker Lake under the sponsorship of the intervention committee, now called the Northern Anti-Uranium Coalition (NAUC). Organizer Jack Hicks, executive director of the Keewatin Regional Council, said five of the best possible experts were brought North to provide an overview of the nuclear cycle. This was the first formal opportunity for those who opposed the mine to present their views.

Tagak Curley, the official spokesperson for the NAUC, chaired the April 14 workshop, which focused on uranium industry moral issues, at Rankin Inlet. Robert Del Tredici, an artist and writer, presented a slide show on the horrors of Hiroshima, which included the picture of the melted statue of Buddha. The purpose of his presentation was to establish the link between the processing of uranium and the production of nuclear weapons, especially the fact that the waste products from reactors served as military raw materials. (He didn't mention that Japan relied on nuclear-generated power.)

Gordon Edwards, of the Canadian Coalition for Nuclear Responsibility (CCNR), had spoken to the NWT Legislative Assembly in 1981. He discussed radioactivity and the food chain, alleging that nuclear contamination from American atmospheric testing was still present in the Keewatin twenty years later. Every dose of radiation was potentially harmful. He said that UG wanted to mine in the Keewatin because of the Canadian government's lack of policy on the disposal of hazardous wastes. The government also turned a blind eye to end uses because known by-products of atomic reactors were used to make bombs. Edwards contended that the common man should wrest control over decision-making from the politicians.

The next speaker was Rosalie Bertell, author and president of the International Institute of Concern for Public Health (IICPH). She cited numerous examples of the health threats posed by the nuclear industry throughout the world – most of them on lands owned by Native people. She called on everybody to focus on the future, on their children, and take a good look at what the rest of the world was doing before going down the same road.

Speaker Mike Simmons, a London journalist with a background in nuclear issues, confirmed Curley's statement that the British Central Electric Generating Board (CEGB) was a 20 percent owner in the proposed Kiggavik mine. He said that nuclear power in Britain was found to be 10 percent more expensive than coal-generated power, but that didn't matter: nuclear power was important because it secured the country's production of nuclear weapons. Britain was also a significant exporter of waste product plutonium to the USA. "The nuclear industry is expensive, dangerous and unnecessary," he concluded; "It's only there because of the bomb."[45]

The last and most powerful speaker at Rankin Inlet was Nette Wiebe, a farm wife from Warman, Saskatchewan. Wiebe had been instrumental in organizing her home community in its fight to stop the construction of a uranium processing plant there. High prices were offered for land, ostensibly to be used for an industrial park, in this pacifist-minded Mennonite community. When the truth emerged, the voices of opposition became welded and the community came together in the belief that "uranium companies lie." She saw a parallel situation at Baker Lake and urged concerned Keewatin citizens not to let themselves be bullied by the public-hearing process. Their strength lay in their own expertise concerning the land they lived on, and they should not be fazed by technicalities. She said, "When people collect together, persevere and make a single-minded effort, they can succeed."[46]

SCOPING WORKSHOPS AND AFTERMATH

The FEARO panel decided to go ahead with "scoping workshops" in Rankin Inlet, Baker Lake, and Yellowknife at the end of April, even though Keewatin organizations had serious concerns about the haste of the process. Paul Scott, executive secretary for the panel, said that delaying the meetings into the busy summer months would reduce participation. The workshops could be used by participants to voice any and all concerns, including those not included in the terms of reference, like end-uses for uranium, or energy and land-claim policy. Co-chairmen of the six-man panel were Bob MacQuarrie, former Baker Lake school principal and Yellowknife MLA, and David Marshall, from FEARO in Western Canada. Other members were Thomas Kudloo of Baker Lake; David Malani, an environmental consultant for Keewatin Environmental Services who had formerly lived in Rankin Inlet; Rod Douglas, mining engineer and ex-Cominco executive, formerly of Yellowknife; and Dr. Ernest Mastromatteo, an expert on environmental health and safety. The panel's mandate was to examine what environmental, social, and economic impact the Kiggavik mine would have on the community of Baker Lake and the Keewatin region as a whole. The panel was to set guidelines for the environmental impact statement (EIS) to be prepared by the proponent, Urangesellschaft Canada Limited, and then approve or disapprove that EIS after a public review of it.[47]

Joan Scottie, a young Baker Lake woman with education and experience as an outfitter and land-use officer, was concerned that Inuit leaders were preoccupied with matters other than the environment, and that little information about Kiggavik was getting to the people. Using local radio and a petition, she secured the views of some two hundred community residents opposed to the project. She teamed up with another concerned citizen, Samson Jorah, to form the Concerned Citizens Committee (CCC). Its primary purpose was to get information, as many people thought they were being kept in the dark. Scottie had prepared a list of questions for the DIAND consultants who visited Baker Lake in March, but there is no record that the consultants responded to them. The CCC eventually joined the Northern Anti-Uranium Coalition (NAUC) of Rankin Inlet and received a small financial grant from them. Scottie intended the committee to play an important role in the panel hearings.[48]

At the initial scoping workshop at Baker Lake, Scottie began by challenging the objectivity of two panel members, but her comments were dismissed on the grounds that she was unable to document her claims. The meeting was charged with emotion. David Simailak, town mayor, who was thought to support the mine, said he was "scared to death" about the radioactive tailings. The caribou-management board formally announced it opposed the mine out of concern about the effect of radioactive contaminants on the migrating caribou, on which so many depended for food. Scottie said the meeting was well attended and everybody spoke, indicating general concern and opposition to the mine.

The Rankin Inlet scoping workshop was also well attended, and contributions came from citizens, students, and teachers. Peter Ernerk spoke in the afternoon, referring to the community's past mining experience and urging a strong program of training for Keewatin residents if the mine went ahead. Ollie Ittinuar, a veteran of North Rankin Nickel Mines, wanted to hear some of the positive things about the new mine, which he considered unstoppable, but he wondered if the land would be ruined by it. Mayor Lavinia Brown and NAUC representative Jack Hicks outlined the health concerns that they wanted incorporated into the company's impact statement. They recommended a baseline health study that would include a comprehensive review of the current health status of Keewatin residents and of the impact of nuclear contamination on country food, breast feeding, cancer rates, congenital disease, malformation, school days lost, and other indicators. Cathy Towtongie urged the panel to consider intelligent use of resources: tourism would be more useful to the economy that uranium mining. Rhoda Karetak said that local people were not properly educated to participate in the wage-labor economy. She saw a strong and continuing dependency on caribou as a food source. Mary-Rose Anguhadluk asked if their opinions meant anything – would the mine be stopped? Lizzie Ittinuar recalled again the experience of the nickel mine. People were happy working at the mine, but when the company shut down the mine, it left them with nothing. In this case, were they going to leave their children contaminated land? Dave Nutter of the Chamber of Mines said the industry had a serious concern about pollution. He said that if there were any danger of long-term damage, the mine should not proceed. Chris O'Brien of Ecology North suggested a regional referendum on Kiggavik, adding that the panel should consider whether the results of such a vote should be binding on UG or the federal government. The FEARO panel now had a month to review submissions and prepare the draft guidelines for the company and future hearings.[49]

The uranium issue was getting wide publicity in the North. School children were studying about the issue and expressing their concern for the environment and the availability of future jobs. Distant readers were expressing their views. Brian Pearson, a white businessman and politician from Iqaluit, wrote a blunt letter to the editor about the depressed economy, the dreadful housing shortage, and the plight of half-educated youth in the Keewatin district. He urged that elders should not "sit and rage at these meetings," but give the youth some hope – and prevent the epidemic of suicides. He claimed that nuclear energy was clean and that many of the residents' concerns about the mine were based on ignorance. He said Japan had reasons to be afraid of nuclear energy, yet they produced most of their power with it.[50]

The FEARO inquiry now entered a new phase upon release of the terms of reference. Intervener groups like the Baker Lake Committee applied for more funding so they might generate strong comment on the environmental impact study being prepared by Beak Consultants. Joan Scottie said the CCC would keep its office open to answer people's questions. The company opened its own drop-in center in Baker Lake as part of a public relations and educational campaign, and it distributed newsletters and fact sheets. Scottie regretted the loss to the cause of Jack Hicks, who resigned as executive director of the Keewatin Regional Council in June, the day of his performance review. She said he had greatly helped the residents with his own knowledge and by bringing in experts to speak to them.[51]

The Hicks resignation raised a storm of protest from concerned citizens like Peter Williamson of Rankin Inlet, who saw this as an act of political interference by Gordon Wray, MLA for Baker Lake and minister of economic development. Apparently, a southern business journalist named Ray Silver, who had not attended the recent meetings in the Keewatin, had had an unsatisfactory telephone conversation with Hicks. He subsequently wrote him a fiery letter, sending copies to NWT politicians, including Wray. The letter openly criticized Hicks, saying that the "consultants" he brought north were not credible persons regarding the proposed uranium mine, and furthermore, that he was patronizing toward the Inuit. This expression sounded reminiscent of industry criticism over transplanted southerners or radicals stirring up the otherwise amiable Natives on environmental issues. Industry never hesitated to garner the best legal and technical advisors for its own purposes, however. Northerners' confidence in Jack Hicks has been sustained; he is a valued government administrator to this day.

Gordon Wray repeated the criticism and accused Hicks of manipulating the Keewatin mayors to focus on the Kiggavik issue to the detriment of other municipal matters, such as the new military runway at the Rankin Inlet airport. Baker Lake mayor David Simailak was also critical of Hicks' overstepping his authority.[52] Williamson denied these charges from firsthand information and pointed out that the people of the region were "losing one of their most resourceful and energetic individuals and in the process might lose the fight to stop the proposed Kiggavik Uranium Mine near Baker Lake." He reminded Wray that every recognized Inuit organization and their boards of directors had put the proposed uranium mine on the priority list.[53] Jack Hicks was retained by NAUC and continued his involvement in the uranium issue.

Urangesellschaft president Mick Stuart told a public awareness committee in Baker Lake that the company would give preference to Northerners seeking employment or contracts. Baker Lake residents would have first priority, and training and educational programs would be introduced in conjunction with the GNWT Department of Education, Canada Employment and Immigration, and Arctic College. The company was working with the committee to survey local households on existing job skills and desire for upgrading. Furthermore, the company pledged to operate fly-in, fly-out schedules that would enable Inuit to maintain family and community relationships, and to allow seasonal leaves of absence so that workers could attend traditional activities.[54]

The company hired Ed Schiller, one-time resident geologist for the NWT who had been particularly interested in Inuit prospecting near Bathurst Inlet, to present its case in Baker Lake. Schiller distributed bulletins on uranium mining safety issues that described problems elsewhere, but he insisted that Kiggavik would be a clean mine. The tailings would be impounded and covered with water, which would serve to block the radiation. Airborne dust and the associated radiation created by open-pit mining would dissipate quickly, and the effect on the environment would be insignificant. He said lung cancer was associated with underground mining, where ventilation with fresh air was not adequate. Today, the most serious environmental problems were depletion of the ozone layer, the greenhouse effect, and acid rain, and these flowed from the use of fossil fuels, not from uranium by-products.[55]

There was no doubt that the mine issue was getting wide attention. Nunatsiaq member of parliament Jack Anawak, who grew up in Repulse Bay, succeeded in arranging a Liberal party caucus meeting in Iqaluit, so that his colleagues could experience the North. He wanted the caucus to discuss three important issues: Arctic sovereignty, economic development, and the environment. He knew these issues would affect the expansion of southern interests northward, but he wanted Aboriginal people to play a role in determining the direction of that expansion and share in its benefits. He said people wanted to be involved in the planning. He cited Kiggavik as a case in point, in which Inuit people to date had not played a part, and he rhetorically asked them: "Do we want to stand by and allow people to come in and do whatever they want to do, or do we want to have a part in what they do?" Anawak himself had not yet taken a stand on Kiggavik, as he considered it a complex issue. He was well known for his business attitude and his support of Native development corporations.[56] The government did not visibly enter the debate.

Significant changes in strategy of the anti-uranium forces occurred over the summer after the initial experience of the scoping workshops in May had served as a turning point for the consolidation of views. Ordinary people had demonstrated serious concerns about the mine, if not downright opposition to it. On the other hand, the impression came through that governments already favored the mine and could exercise their power over organizations such as the Keewatin Regional Council, as in the case of Jack Hicks. Consequently, more "people groups" were thought to be needed, in addition to the protesting organizations. In Rankin Inlet, Janet Onalik and others founded a concerned-citizens committee, got on local radio and circulated an anti-nuclear petition, which by August had garnered 531 names. "We wanted to demonstrate our support for NAUC," she said. "We wanted to show there was community support, not just organizational support." Now there was a plan to present the petition signatures to federal and territorial politicians and encourage them to take a public position parallel to that of the communities.

Chesterfield Inlet citizens, as well as those of Baker Lake, had also raised petitions. Joan Scottie returned to Baker Lake after her summer of work as a guide at Ferguson Lake Lodge, determined to throw her energies into the cause. She was struggling to get more funding to open the committee office and make use of the films, books, and other materials they had acquired. She was saddened by the loss of Hicks as an organizer and technical consultant, and more determined than ever to see that NAUC and other anti-Kiggavik forces had someone at their disposal who could act as a consultant and advisor on technical matters.[57]

As the petition campaign gained momentum, it came to the attention of both the Legislative Assembly in Yellowknife and the House of Commons.

The petitions presented in Yellowknife bore 970 signatures, including 531 from Rankin Inlet (thirty-six percent of total population). The people's voice reached the House of Commons on October 8, 1989, when MP Jack Anawak tabled petitions entitled "No Uranium Mines Here" bearing 1083 signatures from Rankin Inlet, Baker Lake, Arviat, and Whale Cove (twenty-six percent of total population). He urged the federal government to recognize and consider the views of the petitioners. The petitions said, "No economic gain could balance the damage that would be done [by the Kiggavik project]. We believe a uranium mine anywhere in the Keewatin is unhealthy for both the land and the people, today and in the future."[58] In a covering letter, petitioners asked Gordon Wray and Peter Ernerk, as well as Anawak, to speak up on this issue on behalf of the people. During public appearances and media interviews, Anawak stuck to his view that the response was too emotionally based to influence the government. Residents would have to come up with more concrete reasons why the mine should not proceed. He challenged them to suggest alternative ways of creating the kinds of job opportunities the mine might provide. Joan Scottie replied that the position of her Baker Lake group was based on practical concerns, not emotional ones. To her knowledge, no tailings disposal system had ever been created that would protect the environment from radioactive contamination.[59]

Scottie may have been thinking of the reports that had followed the previous summer's visit to Saskatchewan uranium mines at Key Lake and Rabbit Lake. The FEARO panel members and six Inuit observers were taken on an orientation tour that happened to follow on the heels of a tailing spill at Key Lake. The mine was not yet in full operation, and officials were reluctant to talk about the incident. The tour afforded the opportunity to visit three Native communities in northern Saskatchewan, Fond du Lac, Black Lake, and Wollaston Lake, where residents voiced frustration over the mining activities. Thomas Tiktak, an observer from Rankin Inlet, said that a public meeting at Wollaston Lake had revealed a lot of anger over the Key Lake accident and company hiring practices. Despite early promises to hire residents with a grade ten education, the company had soon released those candidates and raised the educational requirement to grade twelve. The Rabbit Lake mine now provided only a few menial jobs for local residents. Barnabas Piryuaq of Baker Lake told the residents of Wollaston Lake that the Kiggavik mine would be located beside the Thelon Game Preserve, where Inuit had been forbidden to hunt for years to protect the animals. The Indians of Wollaston Lake said they could not do anything now that the mine was on their doorstep. The mine personnel even refused to buy their fish catches because, it was alleged, they knew the fish were contaminated. All in all, the Saskatchewan briefings seemed to confirm the original concerns of the Inuit observers.[60]

During this time, Urangesellschaft (UG) continued its program of uranium exploration in the Baker Lake area. The discovery of two new mineralized zones, announced by UG president Werner Sposs on CBC radio, was confirmed by public relations consultant Ed Schiller, who said the company was continually trying to enlarge ore reserves. Sposs used the term "ore body" in his interview, as he suggested that the new finds, a short distance from Kiggavik, might extend the potential life of the mine by another ten years. As reported by Jim Shirley, the new finds seemed to confirm the contention of some anti-mine spokespersons that "the proposed mine (with a ten year life) was only a prelude for much more extensive uranium mining activity in the future." Shirley added, "This point was made during the scoping workshops held last spring and during the public information sessions held in the communities of Rankin Inlet and Baker Lake prior to the workshops."[61] The size of the mine seemed to be one of the uncertainties that haunted the population. Inuit unfamiliarity with land ownership and multiple claim holdings led to suspicion that the mining company had a devious plan. Similar suspicions arose over the Polaris mine because Cominco had numerous claim holdings distant from the site of the proposed zinc mine.

Another subject of apprehension was the ongoing exploration north of Schultz Lake, in the vicinity of the Beverly–Kaminuriak caribou herd protection zone, or calving area. The vulnerability of caribou to radioactive fallout was well known, though the contribution of exploration to this fallout (through drilling and rock trenching, which create local dust) was conjecture. The former high levels of cesium found in Keewatin residents were attributed to their dietary dependency on caribou affected by atmospheric testing of nuclear weapons. However, Department of Health and Welfare sampling of hunters during the year at Baker Lake indicated that cesium-137 levels had dropped more than 50 percent since 1964.[62]

The FEARO panel required Urangesellschaft to present its environmental impact statement (EIS) by November 1, 1989, as specified in the twenty-nine-page guideline statement issued in June. Parallel supporting documents were also requested of federal and territorial government departments. All the documents would be made available to the public and used to structure a series of hearings at which informed and productive discussions were anticipated. The June document described the guidelines as follows:

These guidelines are organized by groups of issues that are of concern to local communities and interested government and non-government organizations. The sections on potential impacts are broken down into baseline data, potential agents of chance and potential consequences. The EIS should make clear distinctions between these types of information. Each section begins with a statement on the focus of concern and the spatial and temporal boundaries. Mitigation and monitoring and the regulatory framework form separate sections but clearly apply to all potential impacts. Rather than repeating requests for mitigation plans and regulatory requirements, the request is made once in each of the appropriate sections.

The issues addressed in this document are organized into the following categories: ecosystem impacts, including plant life, wildlife, physical environment, surface and groundwater, and atmospheric environment; socio-economic impacts; human health and safety, including worker and public health; risk management; tailings management and decommissioning; and mitigation and monitoring.[63]

In the fall of 1989, UG held public information forums, using slides, maps and models to raise company credibility in the communities of Baker Lake, Rankin Inlet, and Chesterfield Inlet. The last-mentioned community naturally was interested in marine travel through the Inlet waters and barge transport of uranium concentrates and supplies. Vice-president Mick Stuart stressed the company's interest in environmental integrity and its desire to address the concerns of Keewatin residents. He presented plans for the containment of an estimated four million tonnes of radioactive tailings. He pointed out that processed uranium concentrate (or "yellow-cake") was a relatively safe product and would be barged out to deep water. The atmospheric release of radioactivity at the mine would not be much higher than what is presently released at the site naturally. He repeated that, on balance, the use of fossil fuels produced a greater environmental impact, which was being reflected in the "green-house effect"; uranium was regarded as an attractive alternative fuel. Stuart said there would be 370 jobs on construction and 287 during production at the project.

Residents challenged Stuart's contention that the mine would be safe. Louis Pilakapsi expressed disbelief that the product would not be used to make bombs and that Inuit would get many jobs at the mine. Stuart denied any involvement on the part of UG in the sale of uranium for weapons. An offended Tagak Curley demanded that Stuart apologize for earlier comments by UG President Werner Sposs, who had expressed disbelief in the sincerity of

Inuit comments at the scoping workshops in the spring. Stuart said that Sposs's comments were "misinterpreted" – that Sposs "has great respect for Inuit, as I do myself."[64] Sposs had observed that a real phenomenon was taking place at Baker Lake: "What surprises me is that people seem to be extremely well organized. Obviously, it was a case of an overkill performance on the negative side. Everybody was 100 percent against uranium mining, without knowing anything about uranium."[65] Perhaps Sposs was all the more surprised at the local response because UG had been exploring in the Baker Lake area for fourteen years. Certainly the local people must have been exposed to considerable information about uranium during that time.

The existing $200,000 of intergovernmental funding for FEARO intervener groups was raised by another $62,500 in the fall. The territorial minister of energy, mines and petroleum resources, Nellie Cournoyea, reported that the remaining $187,500 would be distributed in November by an independent committee. The funding was intended to stimulate a more thorough public review. Northern groups and individuals needed the money for research to develop positions that might not otherwise be raised at the hearings.

The amount of intervener support drew fire from John Murray of Nuclear Free North at Yellowknife. Both his organization and Ecology North had made presentations at the scoping workshops but were now being funded at one-third of their requests to prepare for the hearings in 1990. Moreover, FEARO had specified how the money must be spent, that is, on workshops in Yellowknife. Murray wanted funding to do research or bring in experts to counter the "steady stream of AEC [Atomic Energy Canada] and Urangesellschaft propaganda on the so-called benefits of the mine. The critics of the mine have very little in the way of funds for fighting million-dollar industries, and what little funding they get is tied to ridiculous restrictions on its use. To me it is obvious with whom the government is siding."[66]

The issue of mine tailings at Rankin Inlet stirred up a controversy around this time. Community anger was directed not at North Rankin Nickel, the company that had left the tailings exposed to saltwater wave action twenty-seven years earlier, but toward the hamlet council, which had kept a consultant's report quiet for several months while exploring possible remedies. The study by Thurber Consultants had revealed tidal invasion of tailings beds rich in arsenic and nickel and therefore the likelihood of heavy-metal pollution of the waters of Hudson Bay. The affected harbor area was much used by small craft, and the adjacent beach for camping and recreation. The study, though not comprehensive, had been commissioned by the Department of Municipal and Community Affairs of the GNWT to test the area's suitability for building construction under a new community plan. Some citizens were outraged that, once the area had been identified as a health threat, their council had not warned them immediately of the danger. However, most citizens recognized that they had already lived with the situation created by the breach in the tailing dam for years. Following the report, considerable effort went into the search for reclamation funding and a more thorough evaluation of the environmental impact. The whole matter revealed an active public concern for the purity of the environment and a healthy community, a sentiment which pervaded the whole uranium issue.[67] During the year, concern was expressed over PCB levels at Broughton Island, on the northern shore of the Cumberland Peninsula. It was reported also that cesium-137 levels in Keewatin residents had declined over the past twenty years because nuclear testing had been curtailed and Inuit were eating less caribou.

GATHERING OPPOSITION

At 1989 year-end, the territorial government spokespersons reiterated their official position of waiting for the FEARO process before reacting to the Kiggavik issue, although they were co-funding the intervener grants with the federal government. However, Gordon Wray said he was nervous about the mine: "It hasn't been proven to me how it will help Baker Lake." Wray admitted this sentiment was a holdover from his antagonistic involvement in 1978 as an advisor to the hamlet council, when he and David Simailak had organized and contributed to the court challenge against the uranium exploration companies. Now, as minister of economic development and tourism, he was overseeing UG presentations to the GNWT, and he encouraged the Baker Lake chamber of commerce to identify opportunities for local residents if the mine should go ahead.[68] It seems appalling that the minister, representing Baker Lake, would not have been apprised by UG of the employment and contract benefits the mine would bring to the community. This says something about the ineptitude of the company's public relations efforts.

John Bayly, Yellowknife lawyer and former (1967) resident of Rankin Inlet, commented that "the jury is still out on Kiggavik." He said that, given mining's influence on the early history of Rankin Inlet, nobody should have been surprised at what many residents had said about their own mining experiences during the Kiggavik meeting he had chaired the previous spring. Peter Ernerk, for example, had reminded EARP panel members that the local people were mindful of the short-term benefits of mining, having had to bear the brunt of two mine closures, at the Rankin Inlet nickel mine and the gold mine at Cullaton Lake. The best jobs went to people from the south, and after the closure it was the government and the local people who had to deal with the resulting social problems. Donat Anawak had been lured from Coral Harbor to work in the nickel mine. He said the mine had polluted the harbor and the inlet; where once there had been plenty of seals, there were now only a few. Recalling his life as a driller and driver, he said the work was a lot of fun; they worked hard, learned a lot, and sometimes earned a bonus in addition to their pay. Alvin Kamuk had worked at many mines, with the result that he was now almost completely deaf. He had received no compensation for this disability. Despite their misgivings, many residents saw that mining could provide job opportunities, and they knew that wage-earning jobs were going to be important for the young people, but the people were still not prepared to support a uranium mine in their region. Many spoke with passionate concern for the health of living things and the preservation of a clean environment. Their experiences had "taught them to be skeptical about the assurances of would-be developers in their land."[69]

EIS FUELS MORE CONTROVERSY

The Kiggavik environmental impact statement was completed a couple of months late because, according to UG vice-president Mick Stuart, the FEARO guidelines were so extensive: "Because of this, putting the assessment together took longer than anticipated." In its seven hundred pages and supporting documents, the EIS covered the environmental and socioeconomic impacts of the proposed mine. A summary in Inuktitut was being prepared.

Joan Scottie said the people of Baker Lake were concerned about the content of the report for a special reason: "There was nobody from the community hired to help develop the study. We thought that some of the local hunters might have had some input since they are very familiar with the local environment and the wildlife. But they weren't." In addition, Scottie felt the consultants plainly displayed a bias: they had come to town and told people that they had nothing to fear because there would be no environmental damage.[70]

Shortly after this initial response to the EIS, Keewatin groups resumed their activity. Jack Hicks returned to Rankin Inlet to direct the efforts of the Northern Anti-Uranium Coalition (NAUC), after discussions with the Inuit Circumpolar Conference (ICC), the Tungavik Federation of Nunavut (TFN), the Canadian Arctic Resources Committee (CARC), Ecology North, and its own member organizations. The Concerned Citizens Committee held a public meeting on January 31, with about fifty people in attendance. All the groups had been following mine-closure news from Elliot Lake, Ontario, as well as Native reactions to uranium mining in Saskatchewan.[71]

At about this time, a departmental report by Bill Stephen, director-general of DIAND, was leaked to the press, causing considerable embarrassment. The report, written on the basis of candid conversations with territorial deputy ministers, revealed that the GNWT had offered its support to the project. It said, "After very thorough environmental screening and extensive public hearings and political debate, the territorial government will approve the Kiggavik Uranium mine," adding that "construction will start about 1995." The report came to the attention of Peter Ernerk, a member of the legislative assembly, and Louis Pilakapsi, president of the Keewatin Inuit Association. The KIA issued a press release demanding that government leader Dennis Patterson explain "why the territorial government, in public, is sitting on the fence about Kiggavik, and why his deputy ministers tell the federal government, in private, that they believe the project will go ahead?" Patterson replied that neither he nor his ministers had offered any go-ahead for the Kiggavik uranium mine project, despite federal documents stating otherwise.[72] Meanwhile, Ernerk said he would call for a motion on the government's position in the next few weeks, in the hope that policy would be decided by consensus, not by cabinet. His motion would echo the position taken by the Inuit Circumpolar Conference held the previous summer in Greenland, that until all the research had been completed, the ICC would not support the project. He stubbornly maintained that he thought the territorial government "had placed a lot of hardships on the people of Baker Lake and the people of the NWT by not telling the people of the Keewatin that we're at least going to say 'no' to uranium mining until we know the facts."[73]

In a letter to the *Nunatsiaq News*, Louis Pilakapsi clarified the stand of the July Inuit Circumpolar Conference in Greenland. The pertinent resolution said that the ICC should "support Inuit of the Keewatin region in opposition to the proposed Kiggavik uranium mine" and "collect information on the environmental, health, ethical and political questions pertaining to uranium mining, and make this information available to communities, organizations and the FEARO panel." Another resolution advocated the prohibition of uranium mining or processing throughout the Arctic region and called on all jurisdictions to declare themselves uranium mining–free zones. The Greenland people themselves had outlawed uranium mining shortly after attaining home rule in 1979.[74]

Paul Scott, executive secretary of the FEARO panel, referred to the leaked Stephen report as a "planning document." Its release would not affect the future plans for the review. Both governments had originally requested the environmental review, and neither one had made a decision on the project. He said that the panel would meet with four technical consultants to review the EIS before any meetings were held to hear the public's views, which are "all-important."[75]

At the same time, Peter Hubachek of Lucky Eagle Mines announced an important discovery of gold mineralization at Meadowbank Lake, 100 kilometers northwest of Baker Lake. A two-year diamond drilling program was announced by the operator (the mine was jointly owned by Hecla Mining Company and Agnico-Eagle Mines Limited) on property optioned from Comaplex Resources International and Cumberland Resources Ltd. Gold mineralization was described as exceptionally rich for the Keewatin, comparable to that of the Lupin mine at Contwoyto Lake. Hubachek said he expected that the project would come to depend more on Baker Lake and less on Yellowknife for logistical support. The company might hire some local people to assist in its summer program.[76]

Dave Nutter explained that the NWT Chamber of Mines, a group representing mine operators and explorers, supported responsible containment of tailings, now possible through adherence to modern standards. He sad that certain mines, like Pine Point, had natural containment, but some old mines, such as the Discovery Mine north of Yellowknife, were potential problems. The tailings deposited twenty-seven years earlier by North Rankin Nickel represented a problem for both municipal and Crown authorities, according to Nutter, and DIAND would probably take the lead in remedying the situation.[77]

The KIA executive invited representatives of UG and Beak Consultants to address their annual general meeting in Rankin Inlet. Don Lush and Ed Ferris spoke after presenting a promotional video. Lush equated concerns over Kiggavik with fear of the unknown: "It is very important for you to know what is going on so you can make knowledgeable decisions. Most people fear things they don't understand." He asserted that the environmental impacts of the mine would not create a hazard for the residents or for the wildlife. Lush apparently also said that Inuit were impressionable – and for this, Ruby Arnaan'aaq, the secretary-treasurer, demanded an apology for any suggestion that they were being misled, as Inuit know more about their environment than anyone.

Tagak Curley also found the presentation "condescending." He said, "We have the freedom to express our conscience …. The integrity of Inuit culture should be respected at all times." Mike Kusugak then pointed out that very little of the video was shot at the actual site of the proposed mine. Donat Milortuk, president of the TFN, questioned the choice of an open-pit mine, which is subject to winds, rather than an underground mine. Several resolutions were passed at the meeting: they condemned the mine, demanded that the GNWT take a public stand, and insisted that the entire EIS be translated into Inuktitut.[78]

The subject of uranium mining continued to dominate the discussions at the meeting. KIA president Louis Pilakapsi questioned the NWT government leader's position on the project and restated his opinion on resource development in general: "We are in favor of mining development in our region. I am delighted to hear about promising gold discoveries close to Baker Lake .... But we oppose uranium mining, and we have the facts to support our reasons for doing so." He said that the leaked Stephen report confirmed the worst fears of many residents that a deal had already been cut. Tagak Curley commented, "I am beginning to wonder if it's really worth going through the FEARO review .... Maybe we're just playing with ourselves if the decision has already been made." Paul Scott of FEARO also made some remarks about the report and the process, insisting that no decision had been made. Jack Hicks, coordinator for NAUC, claimed that public opposition had never been stronger, and discussed some approaches to a coordinated opposition. Hicks criticized the EIS document, saying it was "late, shoddy and inadequately translated." He added, "The Inuit were not consulted during the study and there was only one paragraph on the effects of radiation released at the mine site on caribou."[79] The sources of irritation remained the same – lack of involvement and respect, language, impact on caribou – but whether these caused "great anxiety," as Ernerk claimed, was questionable.

The "one paragraph" mentioned by Hicks may refer to part of section 2.5, "Potential Off-Site Radiological Effects," contained in UG's Summary Report dated January 1990. That section discusses potential radionuclide contamination of two groups of people as a result of releases from the project of dust, radon, and fluid that might accumulate in area lake sediments, soils, plants, and animals:

> The first group is assumed to live in a hunter's cabin for 4 months of the year on the northeast shore of Judge Sissons Lake, 15 km (9 miles) southeast of the ore bodies. It was assumed that the group consumes a large quantity of caribou from the Kiggavik study area, fish from Judge Sissons Lake, and various other country foods from the local area (e.g., ptarmigan, goose). The modeled additional radiation dose received by

this group is 1.4 mSv [millisieverts] per year, or 28% of the annual radiological dose limit of 5 mSv set by the Atomic Energy Control Board (AECB). This means that the additional risk of getting cancer as a result of this radiation exposure is approximately one in 36,000 for a four-month exposure period.[80]

The other group, permanent residents at Baker Lake, would receive less than 2 percent of the AECB allowable dose limit.

This scanty treatment of such a crucial issue and the technical jargon in the report, as translated into Inuktitut, caused understandable disappointment among the residents, who could easily see themselves traveling on the land closer to the mine than the 15 km specified.

Political pressure had escalated with Peter Ernerk's efforts to get a territorial resolution on both Kiggavik and uranium mining in general. The Hamlet of Baker Lake took a new direction with the election of a council headed by mayor Garry Smith. While the previous council had maintained a neutral position, the new representatives decided that the community should take a stand in the form of a plebiscite, to be held on March 26. The Concerned Citizens Committee had already secured half of the residents' support in a petition, but expressed their enthusiasm for a plebiscite through chairman Martin Kreelak, who said, "It's important that Inuit be able to make choices. The plebiscite will answer questions." A new consensus was emerging on the issue that included teachers and government employees, who were usually white residents, but resident Fred Ford, who had attended all the local meetings on the matter, still had the impression that opposition groups were getting nowhere.

Ford was especially impressed with the KIA efforts, which had recently secured a delay of the panel's hearings until all of the EIS could be translated into Inuktitut. He criticized the way the UG research was conducted and the final document submitted, saying that Inuit resented the fact that there were only two paragraphs describing the impact on caribou, and that the people had not been consulted despite the wishes of the panel in that regard. Inuit felt left out. Ford expected a pro-mining group to emerge from the Keewatin business community, and feared that people might be swayed by the argument that the mine was the surest way – perhaps the only way – to secure job opportunities. The Concerned Citizens Committee would have to clear up a lot of misconceptions, in his opinion.[81]

Peter Ernerk's unequivocal motion in the territorial legislative assembly was stalled by a motion of the justice minister, Mike Ballantyne, which sent it to the Committee of the Whole for later discussion. Ernerk said, "The motion was proposed on behalf of the thousands of Inuit who live in 'day-to-day fear' of the proposed mine. The government's tactics are clear. They [the Cabinet] don't see relieving the great anxiety of the people as a priority."

Ernerk vowed to keep up the fight for his people. By this point, Keewatin residents had clearly demonstrated their opposition by way of petitions with 1700 signatures from five communities. The Inuit Circumpolar Conference had offered its full support to the Inuit, and the mine had been publically opposed by the Keewatin Inuit Association, the Keewatin Regional Council, the Keewatin Wildlife Federation, the Beverly–Kaminuriak Caribou Management Board, the Inuit Tapirisat of Canada, the Tungavik Federation of Nunavut, concerned citizen's committees, the NWT Federation of Labour, Ecology North, and Nuclear Free North.[82]

Ernerk's motion brought an immediate and angry response. The NWT Chamber of Mines, in the *Yellowknifer*, charged Ernerk with turning this decision into an emotional issue and usurping the government's power by not waiting for the FEARO report. Ernerk insisted the Inuit had good reasons for opposing uranium mining, which would put their lifestyle in jeopardy, and pointed out that the GNWT had taken official positions on the Mackenzie Valley pipeline and new pulp and paper mills in northern Alberta. Inuit were against both uranium mining and exploration – and Ernerk included both activities in his legislative assembly motion. The citizens of Baker Lake had opposed exploration twelve years earlier in the courts. The Chamber of Mines said that the motion "doesn't show good faith in the mineral exploration industry." On this point, Ernerk said, "The people of Rankin Inlet had expressed concern about tailings from the abandoned nickel mine to the government and industry officials for years. Only now are we beginning to learn the truth. The Cree and Chipewyan who live near the uranium mines in northern Saskatchewan still aren't being told the truth about the impact on the environment there. The mining companies and the federal and provincial governments refuse to fund research projects."[83]

Jack Anawak, Inuit member of parliament, was also under pressure to abandon his neutrality on the issue of Kiggavik. However, he challenged some leaders and organizations who believed that any resource development is good, so long as it is not uranium. The fragile northern ecology demanded the strictest standards for all developments, whatever their form. In this case, Anawak advocated a negotiated strategy, whereby Inuit would negotiate with the mining company before the FEARO hearings were completed, so as to maximize their economic benefits if the mine were to go ahead. He commented, "I don't think that people have enough information on hand to reach the right conclusion about the mine. We also have to realize that there are people who are dead set against the mine and no amount of information will sway them." Jack Anawak at this time was notably concerned about literacy issues, Native communications, federal budget cuts, and the responsible handling of the land-claim dollars.[84]

BAKER LAKE REJECTS THE MINE

On March 26, 1990, as predicted, the people of Baker Lake voted in overwhelming opposition to the mine. The plebiscite elicited a 72 percent turnout of eligible voters, who recorded forty-three votes for the project, and 397 – more than 90 percent – opposed. The hamlet council was now able to support the popular opposition to the project. Fred Ford said the opposition was not an emotional appeal: people had given it a lot of considered thought after thorough discussion of the issue.[85]

The outcome of the Baker Lake plebiscite prompted mayor Garry Smith to write to UG on behalf of the hamlet council, requesting that the company abandon its plans to proceed with the mine. Smith reminded the company of its statements, made during public presentations, that it would abandon its plans if the residents of the community indicated they did not want the mine. Smith directed another letter to MLA Gordon Wray, seeking his support in opposition to the project and indicating that the hamlet would be supporting other anti-Kiggavik forces in the future.[86] In turn, the hamlet council received a letter from Power Reactor and Nuclear Fuel Development Corporation (PNC) of Japan, indicating that the firm would be carrying out exploration activities in the area the following summer. The letter said that PNC had no connection with UG and would be interested in meeting the Hamlet officials and others. The PNC uranium property was beyond Kiggavik, near Schultz Lake. Despite Peter Ernerk's contention, the community residents were willing to accept early exploration, provided the wildlife was safeguarded.[87] PNC took the view that the company had a legal right to explore its Crown mineral claims, but might establish a closer liaison with the community. UG was under no compulsion to abandon its plans, as Stuart had impulsively suggested. The company had legal rights, as well as obligations, under FEARO.

FEARO policy director Paul Wolf commented on the process and the panel's decision to delay the hearings until UG had completed more work on deficiencies in the EIS. He said the company's omissions had been substantial, given the FEARO guidelines and the accepted standards for similar work, for example those submitted by Canadian oil companies. The weakest part of the EIS, according to Wolf, was the section on the assessment of the social impact of the project, where the costs can really show up. He pointed out that, contrary to what people think, "the final decisions of the panel are rarely determined by technical matters." So the public need not feel that the proceedings were over their heads or beyond their ability to contribute. He admitted, however, that there was a lot of cynicism, especially in the initial stages, when people thought this was just a drill.

This seems to have been the case with Kiggavik project. The Canadian process allows for a high degree of public participation and is aimed at protecting the aspirations of the Canadian people. Looking to the future of environmental assessment in Nunavut, Paul Wolf observed that the Nunavut advisory boards created in land-claim negotiations would ensure a democratic process: "The Inuit will be in a strong position to control their future and protect their environment once land claims are in place."[88]

UG SUSPENDS THE KIGGAVIK PROJECT

UG's decision to suspend the Kiggavik mining project indefinitely was announced by territorial government leader Dennis Patterson, who said the decision reaffirmed the GNWT's original call for a full public review via the FEARO process. Dr. Werner Sposs, president of UG, had communicated that "because of concerns raised by the FEARO Panel, the GNWT, and residents of the Keewatin, plans for the mine have been halted indefinitely. The Company does not wish to proceed with the project unless the concerns raised by the FEARO Panel can be satisfactorily addressed and the concerns of the federal, territorial, regional and hamlet authorities can be resolved. They must be addressed if the project is to proceed." Nellie Cournoyea, territorial minister of energy, mines and petroleum resources, said that the EIS "left too many unanswered questions about the project, including the short and long term effects on the land, wildlife, people and the economy. We need more information and a greater level of public understanding." The economic development minister, Gordon Wray, was pleased that the company had responded positively to the people's concerns. He said, "The TFN land claim will give Inuit a major role in determining how future developments such as this will take place."[89]

The Kiggavik project near Baker Lake never went ahead. It is possible that, because of changing market conditions and the superior ore grades in neighboring Saskatchewan, 1989–90 had provided a window of opportunity for its development that was never to be repeated.

As NAUC reported to the panel during the scoping workshops, Urangesellschaft had created a misunderstanding in the communities from the beginning:

> Stuart has repeatedly stated that the proposed project would only proceed if it had the support of Baker Lake and the other Keewatin communities. At the Rankin Inlet scoping workshop, [UG president Werner] Sposs modified Stuart's ambiguous assurance to read: "UG seeks the support and cooperation of the people of Baker Lake and the other Keewatin communities to utilize their experience and their know-how of living in the North. That does not mean support in a political sense of the word." Then, in the Rankin Inlet school on the morning of 5 May 1989, Sposs expanded on his statement of the previous evening: "I think in the beginning there was a misunderstanding, in a way, that somebody thought that the community has to decide whether there is a mine or not. That is obviously wrong, without any doubt. The decision is made by the federal government.... It could be, for example, that the federal government has an interest which is in the interest of all Canada to develop a mine, and that this interest of the whole of Canada is against the interest of a special community in Canada. In such a case the government has to decide what is the higher level of interest," Stuart agreed, and added that he felt it was "highly unlikely" that the communities would not eventually support the proposed project.[90]

There does not appear to be any indication that UG attempted to negotiate a separate deal with the Inuit, as advocated by Jack Anawak, along the lines of the impact and benefit provisions intended to promote Inuit goals on Crown lands that were initialed on December 13, 1988 and described in the Agreement-in-Principle. Section 28.2.1 of that agreement reads: "No Major Development Project may commence until an Inuit Impact and Benefit Agreement (IIBA) is finalized in accordance with this article." In fact, there was no obvious awareness of Inuit interests and strategies during the long years of negotiation to reach this point, including the struggle for a Nunavut Impact Review Board (NIRB) with screening and development review functions, which were described in detail in the Winter 1989 issue of *Nunavut*. In September 1987, in the same newsletter, TFN president Donat Milortuk had stated that, "Negotiations on the Inuit right to co-manage Nunavut's natural resources, with government, have been going on for four years and are now almost complete.... Our main goal in the negotiations on natural resource management has been to obtain as much control as possible over what happens to the land, fresh water, ocean and animals in Nunavut. We've been using our claim to conserve and protect all of the land in Nunavut, whether we end up owning it or not." Included among their achievements were a wildlife management board, a water board, a land-use planning commission and an impact review board. The nine-member Nunavut Impact Review Board, with its four Inuit members, "would involve Inuit in making decisions about which development projects should be allowed to go ahead and which should not because of their dangerous impacts."

So the public record clearly stated the intention and attitude of TFN negotiators in their dealings with the federal government. UG should have been familiar with the process and intentions of the claimants long before the Kiggavik project became a public issue. Already in 1981, a wildlife management agreement (initialed in 1986) set out clearly the aims and purposes that would later form a large part of the Agreement-in-Principle. Describing the general principle of liability, Section 6.3.1 (p. 98) of that document states: "A developer is liable absolutely, without proof of fault or negligence, for loss or damage suffered by a claimant as a result of its development activity within the Nunavut Settlement Area." Such liability applied to loss of equipment, wildlife, and present or future income from harvesting for gain or personal use. The section also discussed limits to liability and exemptions, as well as the creation of a surface rights tribunal that would not be bound by strict rules of evidence. The point is that Inuit were not going to rely on lawsuits to seek redress or allow themselves to be victimized by legal process. This attitude should have alerted the mining company of things to come, and encouraged consultation with Inuit themselves to establish the baseline standards and expectations for wildlife harvesting in the Kiggavik area.

In the fall of 1990, FEARO announced that it could be a year before the environmental review resumed, because UG might take up to a year to revise its EIS. The intervener groups were alarmed at having their government funding cut suddenly. Jim Evoy of the NWT Federation of Labour claimed that the fight was ongoing (the project was alive in the boardrooms) and would be even more challenging when hearings resumed: "I think Urangesellschaft is not going to be the fumbling, bumbling fools they have been to date. They're going to be more credible, present themselves better and understand the community better. They will be smoother in their presentation."[91]

Ironically, the Keewatin Chamber of Commerce had prepared an investment plan for the next century in August. A critical look at the economic realities in seven Keewatin communities had prompted this new strategy. Bob Leonard, Arviat businessman and president of the Keewatin Chamber of Commerce, said, "Time has all but run out for the Keewatin economy. With approximately 1500 children between the ages of 6 and 16, this group, which comprises one-third of the total population, will be needing jobs by the year 2000. Currently there are about 1400 jobs in the Keewatin, with half being provided by government." Most of these jobs were going to southerners, because Inuit did not have the level of skills needed. The Chamber was proposing to develop small-scale manufacturing of sewer tanks, boats, etc. to replace imports and offer some export potential. As the biggest problem was lack of access to capital, the strategy called for increased government funding from the Economic Development Corporation and the establishment of a territorial trust company. Leonard concluded, "In the Keewatin we are no longer facing an economic crisis. We are in an economic crisis, the economy in the Keewatin is in a mess. We are totally dependent on government spending and there's no way that can continue."[92]

SUMMATION

A more detailed look at the intervener presentations in the Kiggavik case reveals obsessive biases toward developers, and toward uranium miners in particular. The presentations also hinted at the courses of action that might have saved the project. For example, the traditional knowledge possessed by Baker Lake elders might have been used in the planning process. The *Compendium of Submissions received by the Kiggavik Uranium Mine Environmental Assessment Panel Immediately Prior to and During the Issues Scoping Workshops* (May, 1989) is revealing in this regard.[93]

The submission entitled "Northern Anti Uranium Coalition Contribution to the Draft EIC Guidelines" makes (among others) the following points:

- The public lacked trust in the FEARO process because there was no public consultation while developing the terms of reference. The Panel's conscientious adoption of the oral and written submissions now being made could help to restore this trust. (p. 90)
- There was little public education about the issues. (p. 93)
- The Inuit attitude of cynicism was exploited throughout the

contributions; the "They Lied to Us" theme, derived from nuclear weapon and weapon plant incidents, was one example. (p. 97)
- UG's continuing exploration near Deep Rose Lake, inside a caribou calving range, despite company assertions that the proposed development would affect only Kiggavik. (p. 98)
- The "woefully inadequate" social impact studies of Beak Consultants for the Kitts–Michelin inquiry, as well as the same firm's "particularly disappointing" environmental impact statement for the Warman, Saskatchewan, uranium refinery, ignored the sensitivity of the local Mennonite community. (pp. 101, 113)
- The dual involvement of government departments (both federal and territorial) in both industry promotion and regulation roles created a conflict of interest. (pp. 103, 104)
- The chummy relationship of government with the proponent and government enthusiasm for the project stood in contrast to the powerlessness of the people. (p. 117)
- UG had made no financial commitment for monitoring, compensation, and training. (p. 123)
- The company denied the existence of radon and decay-product clouds. (p. 136)
- UG claimed that strong Canadian regulations would protect the environment. (p. 139)
- Mick Stuart, UG's general manager, claimed that the review process would be as stringent as if the open pit uranium mine were located 75 km upwind from Toronto. (p. 140)
- UG lacked experience in operating such a mine in permafrost terrain.

Among other topics, in the opinion of NAUC, the Panel should have had a mandate to consider the following:

- Native land claims. UG had ignored Native land use in the inquiry that rejected the Kitts–Michelin project, but apparently learned nothing from that experience. (p. 89)
- The concept of "perceived impact." The Wildlife Compensation agreement, initialed by the TFN on 19 June 1988, recognized that proponent and people may view impact differently. This panel should have done the same. (p. 91)
- The role of bodies to be created by the final settlement with the

TFN. Would these bodies have the power to authorize, monitor, and enforce compliance? (p. 102)
- Criteria for compensation. Would the criteria of the TFN wildlife compensation agreement apply? And would UG agree that "radiological and non-radiological contamination from any aspect of the project, pre- and post-abandonment, will not be regarded as a natural phenomenon of an exceptional, inevitable, and irresistible character" and explicitly state that "the developer will be liable for any loss or damage wholly or partially the result of contamination from the normal operation of the uranium mine/mill, accidents at the mine/mill, and transportation of radioactive material from the site"? (p. 106)
- Incorporation of traditional knowledge. Would UG cooperate with the local people and incorporate their traditional knowledge to produce a mutually acceptable baseline document concerning the overall health of animals, customary breeding rates and patterns, and abundance and accessibility of game? (p. 109)
- The Keewatin reality. The social impact review should take into account the severe human resource crisis, with high unemployment, high illiteracy, low educational attainment, low labor force participation, low income levels, and heavy dependence upon social assistance. (p. 114)
- Inuit dependence on country food. UG should respond to the Inuit perception that both the quality and the quantity of country food might decline with industrial development. (p. 116)
- Stress on the community. Can the impact of fear, stress, and powerlessness be quantified and mitigated? (p. 117)

The oral and written submissions of NAUC reflected a diverse and eloquent contribution from member organizations and a vigorous research effort by Jack Hicks and others. NAUC spokesmen had made a thorough effort to investigate other uranium inquiries in Saskatchewan (Cluff Lake, Key Lake) Labrador (Kitts–Michelin) and British Columbia (Royal Commission of Inquiry) to arm themselves for the Kiggavik hearings. Their attitude also owed much to the ongoing experience at the land-claim negotiation table, where Inuit were seeking the means to control the rate, location, and nature of development in their homeland. Once approved, the mechanisms like the NIRB and participation on FEARO panels would automatically provide the Inuit with a bigger say in development decisions.

The opinions and briefs of other interveners influenced the Inuit attitudes and contributed to a pervasive sense of cynicism about Urangesellschaft. For example, the NWT Federation of Labour, through James Evoy, described the critical contribution of organized labor in securing safe working conditions for Canadian miners. The Federation also admitted the economic contribution that Eldorado Mining at Great Bear Lake had made to the North. However, no mention was made of the high incidence of lung cancer among Native workers at that mine, although this had become quite an issue in recent times. John Murray of Nuclear Free North dwelt on militarism, the abuses of nuclear energy, and the slow recognition by Canada (as compared to the United States) of the public dangers of radon in landfills and other nuclear waste disposal sites.

Gordon Edwards expanded upon his persuasive testimony to the Legislative Assembly in 1981. In particular, he described the problem of toxic wastes from uranium mining and the long-lived decay products left behind in mill tailings. He cited a *Wall Street Journal* article on February 25, 1986, which described the 220 million tons of American uranium tailings as an "ecological and financial bombshell." He contended in the Compendium that half of the 200 million tons of Canadian tailings were inadequately disposed of in unfenced ponds left open to the wind and rain, facilitating the release of radon to the atmosphere. (The cancer-causing nature of radon gas or its decay products in mines was well known, and underground ventilation had been strictly controlled in Canadian uranium mines since the 1950s.) Edwards said, "Many researchers believe that the continental toll from such radon emissions is underestimated by at least a factor of ten, even if we ignore the fallout of solid radon daughters on lichens and leafy vegetation as the radon gas passes overhead, and even if we assume that the tailings are not blown by the wind, washed by the rain, or spread through the food chain, thereby distributing the source of contamination over a much wider area" (p. 178). Edwards did not back up his statement by naming any specific researchers, yet these and many other comments about cancers made a strong impression on the Inuit.

The representatives of Nuclear Free North dwelt on the analysis and interpretation that the Panel might ignore. In addition to the so-called benefits, the real costs had to be assessed: for example, the cost of managing tailing ponds in perpetuity. Patricia Sherman contended that, considering the Keewatin annual mean wind speed of 29.5 km/h, "dust, radon, and any atmospheric contaminants will pass over Chesterfield Inlet and Rankin Inlet, as well as Baker Lake" (p. 187 of the Compendium). Another vivid statement of Nuclear Free North was that "[i]t is quite apparent that contamination of the biosphere will eventually result in contamination of the waterways. The radiation will eventually flow into the ocean" (p. 188). Ecology North, a Yellowknife environmental group, relentlessly questioned the terms of reference, the justification of the project and its global significance, the EARP process, regulatory competency, tailings treatment and disposal, decommissioning, and subsequent environmental monitoring. All these outside interveners bolstered the Inuit in their questioning of Kiggavik.

Speaking for the NWT Chamber of mines, Dave Nutter expressed confidence that the FEARO panel would make a thorough and balanced examination of the evidence. He said the project should not proceed if the proponent could not provide the necessary protection for peoples' health and safety, nor should it proceed if long-term impacts on wildlife and the environment could not be kept to acceptable levels (pp. 211–13 of the Compendium). The brief by Environment Canada dealt with standards to be expected of baseline studies and EIS reports, and pointed out some omissions in the draft guidelines, such as the protection of Chesterfield Inlet from a spill or marine accident and the all-important decommissioning plan (pp. 215–22).

David Kritterdlik, the newly elected mayor of Whale Cove, sent a late message to the EIS panel, wondering: Why it was necessary to mine for ten years and leave dangerous waste forever? Uranium was no longer needed and was mostly stored. "NWT is considered the LAST FRONTIER, and the mining company wants to take away this Canadians' pride, by taking out what they consider the valuable and leaving the Canadians to clean up the mess. This mess will never be cleaned up, and we, the Inuit, will be left exposed to its dangerous health effects" (pp. 223–24 of the Compendium).

The Inuit were influenced mightily during the Kiggavik event by the advocacy and eloquence of others who were of the anti-uranium stripe. Their own concerns about the loss of the game on which they depended and on its contamination from the radionuclide trail were genuine enough, but they were also captivated by the larger moral issue and sought to make a responsible decision for mankind as a whole.

Great thinkers have commented on humanity's dual relationship to the natural environment. We are affected by it, but also shape its future for good or ill. Over their history, the Inuit had not exerted a great impact on nature – certainly not before the fur traders provided them with rifles – but now a strong sense of responsibility had emerged, and this was well demonstrated in land-claim negotiations covering wildlife, land-use planning, and control over developers. The Inuit's mature attitude and global responsibility came to the surface in their reaction to the Kiggavik proposal.

Early in the controversy, Tagak Curley expressed why Inuit were opposed to the mine. He said he was worried about studies indicating that tailings from a uranium mine might remain dangerously radioactive for millions of years, and that radon gas emissions from the tailings could cause regional or even global radioactive contamination. He was also concerned that uranium from the Baker Lake mine might be used for military purposes, thus destroying the Inuit's reputation as a peaceful people who are careful custodians of the environment.[94]

Meanwhile, the mineral exploration industry was not boycotting the NWT or even the Eastern Arctic. Claim staking had accelerated in 1988 thanks to a buoyant economy and the Dene/Metis land freeze, with its perceived threat to land access. In the Keewatin, gold exploration proceeded on the Meadowbank Lake and Meliadine River iron formations and along the Rankin Ennadai volcanic belt. Diamond prospecting had led to staking in the central Slave province, and copper showings had attracted attention to Victoria Island. Gold exploration was vigorous in iron formations at George Lake, south of Bathurst Inlet, and between the Hood and James Rivers, on the Ulu claims. Even Urangesellschaft, in conjunction with PNC Exploration and Daewoo Corporation, performed geophysics and drilling on uranium properties west of Baker Lake.[95]

Exploration was indeed pressing on, deeply confident of the access and development rights conveyed in the Canada Mining Regulations. The mining industry appeared confident also that land-claim agreements would be realized, these being matters for others to resolve. There was a head-in-the-sand attitude, here, and a clear reluctance to learn any lessons from the struggles of Native people around them. Only a few companies showed a promising trend to deal with the people directly and make deals of mutual benefit, or build a basis of trust at an early project stage.

By 1990, the TFN land-claim negotiations had reached a plateau, symbolized by the signing of the comprehensive agreement-in-principle. A new perspective of real power and influence was now available to Inuit, so they could concentrate on the final task of selecting the lands that they wanted to own. They would bring to the negotiating table some of that exhilarating sense of power and maturity, as well as their long-standing skepticism about mining.

7. NEW OWNERS

The previous pages have described events in the lives of modern Inuit that portray struggle and achievement. After many years of preoccupation with development issues and stubborn negotiation, they finally achieved a satisfactory agreement with the federal government that accomplished most of their aims. They also engaged successfully in regional campaigns of protest against development proposals, most notably one to stop the mining of uranium in the Keewatin. An aura of confidence and maturity pervaded Arctic communities as Inuit approached the last task needed to conclude the land-claim process, that of selecting lands for themselves and becoming legal owners. They put aside other preoccupations and concentrated now on securing the most prized, economically and culturally desirable lands. This was not so much a task for strong-willed negotiators as for grass-roots community people, their leaders and elders. However, they entered a process of ratification that was framed as a negotiating and bargaining exercise, not one of arbitrary approval by the government, which I had supposed it would be.

AGREEMENT-IN-PRINCIPLE AT LAST

In 1990, the long process of consensus building and patience resulted in the signing of the Agreement-in-Principle between the Tungavik Federation of Nunavut and the federal government. Negotiators could now turn their attention to preparing the final wording of the agreement and selecting specific lands, before presenting the whole package to the people for ratification at the ballot box. Though the Inuvialuit of the Mackenzie Delta had signed the *first* territorial comprehensive land claim in 1984, their 17,000 Inuit kinfolk were on the verge of signing the *largest* claim in Canadian history. The cash compensation for their ceding aboriginal title to more land and relinquishing 82 percent of the settlement lands to the Crown – thereby rendering "certainty" for the future to the government – was a tidy $580 million (or $34,118 per capita). The settlement lands to be retained (their 18 percent) totaled 121,360 square miles of surface (or 7.14 square miles per capita) and 14,000 square miles with surface-plus-subsurface rights (0.82 square miles per capita). These lands had to be selected region by region to complete the deal, and it is to this land selection process and its outcome that we now turn.

PERSONAL NOTE

At this point the story becomes more personal. I had played no part in maintable discussions or consultations leading up to the Agreement-in-Principle, but afterward, from 1989 to 1991, I was employed by the Tungavik Federation of Nunavut (TFN) as an advisor on subsurface land selections, a post befitting my background as a professional geologist. The following pages, therefore, are extracted not from the literature, but from my own diaries and notes.[1] They convey impressions for which I alone am responsible.

My task was simply to provide geological consulting by preparing compilations of mineral prospects in the NWT using maps, reports, and assessment records filed at Yellowknife. I had done this before in connection with my work, but I had never attempted to plot the results in map notations that were easy to understand at a glance and to translate into Inuktitut. In meetings with Inuit leaders, I presented this kind of documentation in talks on regional geology. The relevant data plotted on map sheets enabled the community representatives charged with selecting lands to appraise, reject, or include subsurface rights in their selections.

I had also prepared log books on the various mineral showings and prospects, and it eventually appeared that these reference materials would be needed in the heat of negotiations. In fact, I was invited, along with my colleagues, to attend the actual sessions as part of the TFN team in case important decision-making or prioritizing had to be done on the spot. I was delighted with this challenge, which brought me into contact with many far-away communities and interesting Inuit leaders. It looked like a routine kind of job, but as the negotiations evolved, there were complications as Inuit negotiators began to treat mineral lands as bargaining tools. The more the federal side insisted on retaining particular lands (usually they were mining claims and leases, third-party rights they intended to protect or exclude), the more my Inuit employers wanted them. They were strongly motivated by a desire for quality lands with economic potential. There ensued a vigorous struggle to break the federal hold on third-party rights.

Ottawa wanted to ensure a revenue stream from mines and minerals to justify its infrastructure expenditures in the North. Inuit negotiators, for their part, were dedicated to the securing of quality lands with economic potential as well as lands for traditional usage and travel. They were obliged to relinquish aboriginal title in exchange for the lands (and other benefits of the agreement), so they intended to select well. I believe they astutely made minerals the focus of attention at times so that other selections of surface lands could be ratified without any fuss.

At any rate, the negotiations offered a fascinating display of table-pounding negotiation, oratory, and counter-argument. This was not a game, however. Inuit leaders and community representatives were engaged in a serious challenge to wrest from the Crown the very best lands for the benefit of their children and grandchildren. They knew that Crown ownership of land would still dominate their territory, so they wanted their package to be exceptional.

In this struggle, I think they succeeded. They won the largest Native land claim in history, and they secured the ownership of quality lands that form a base for eventual self-sufficiency. The passion for land and the determination displayed in the negotiations lead me to believe that Inuit will be good managers in the future, but readers will want to see for themselves how the story of this land battle (or my rendition of it) unfolded.

It all began in March 1989, when geologist Hans Bielenstein approached Comaplex Resources, my part-time employer, about bidding on a geological consulting job for the TFN. This opportunity was quite unknown to us, though we were already engaged in mineral exploration projects in the NWT. We had our nose to the grindstone, so to speak, and were not following Inuit affairs, though all of us at Comaplex had personal experience employing Native prospectors and field workers, including Inuit.

Bielenstein represented Associated Mining Consultants Ltd. (AMCL), an aggressive Calgary consulting firm that had a broad-based practice in government and nongovernment work, mostly related to the mining of coal and other sedimentary rocks, potash, and petroleum. He became aware of TFN's advertised need for a mineral consultant and knew his own ability to fulfill some of this need, including orientation, training, and advising on mining matters. He was an outgoing educator and experienced geologist, especially in structural geology, which he applied in the solution of difficult coal-mining problems, but Bielenstein was not a minerals geologist, and the AMCL firm had not pursued hard-rock exploration per se. For this reason, he contacted college acquaintance Murray Pyke, the vice-president of Comaplex Resources.

Since we (Pyke, geologist Phil Mudry, and I) had a large body of experience in the NWT and some sympathy for the welfare of northern Native people, it was agreed that AMCL and Comaplex (or rather, its subsidiary,

Wollex) would jointly bid for this consulting job. The main question, because of our private enterprise background, was whether the TFN would accept us: Comaplex was a public exploration company currently exploring in the NWT, and it held many mineral claims and permits. In fact, the TFN turned out to be pleased to have such an experienced team with in-depth knowledge of the North, but they required confidentiality in pursuit of their cause.

Bielenstein, Pyke, and Mudry were all involved or eventually became involved in the Nunavut land-selection process, but it fell to me to assemble the basic inventory of Eastern Arctic mineral resources that would guide the Inuit in land-selection negotiations. I began almost immediately to gather geological and mineralogical data and systematically prepared what we called the "mineral inventory books" for Kitikmeot, Keewatin, and Baffin Island. Delving into old reports and assessment files was exactly what I was doing anyway, since my job was largely that of research geologist. At that time, work in progress included compilations for the Nicolas Lake and Sissons Lake areas, using reference material from DIAND, the Geological Survey of Canada, the Gallagher Library at the University of Calgary, Landsat photos, and private reports.

ORGANIZING THE TEAM

The TFN land-selection team was headed by Lucassi Ivvalu from Igloolik. Our dealings, however, were exclusively with Bruce Gillies, land selection coordinator, and Jon Pierce, director of land planning. We learned that the TFN had awarded us the contract on April 15, 1989. Two days later, I talked to Bielenstein about the project and learned, by means of a newly released map, the specifics of the Dene/Metis land withdrawal in the Slave area of the NWT. This information brought us face to face with the reality of Native land claims in the North.

Dave Nutter of the Chamber of Mines was visiting Calgary to explain the land freeze to a gathering of exploration companies. He said that the mining industry had known of a possible land freeze in the Slave area since December, and people had had some time to react. This Calgary audience of claim holders was far from bitter about the land-freeze situation; they acknowledged that they could live with the changes in the mineral claim "extension" rules now adopted by DIAND, because prospectors could not be expected to work their claims unless they could also stake them. Overall, however, the industry was in a boil about this land freeze – and possibly others to follow. Rumor had it that the Inuit had already picked out their lands in the Eastern Arctic, which was probably a reference to community-based discussions on occupancy areas and quantums. In a subsequent meeting with Bielenstein to organize our thoughts, we stressed the need to keep our consultations simple and use existing maps and indexes of assessment reports in the archives of the resident geologist in Yellowknife.

On April 28, Bielenstein and I visited the TFN offices in Ottawa to discuss the project with Bruce Gillies and Jon Pierce. We talked about map scales and presentation features for a mineral-resource evaluation, noting that Inuit would need to have translations and use maps they were familiar with, preferably 250,000-scale, topographic maps. We were briefed on the agreed boundaries of Nunavut and its regions, as well as negotiation policy and strategy. I took home a huge, green map case filled with TFN maps and documents. Colleagues immediately dubbed this my "soapstone" map case, hinting that I was already getting perks in this new consulting career.

PREP WORK AND PHILOSOPHY

Right away, I began assembling data on north Baffin Island, ordering appropriate claim maps from Yellowknife and clipping property descriptions and assays from DIAND reports. I had always wanted to check into the geology of the Nanisivik area, and started plastering notes and colored dots (for zinc showings) on base maps. By May 30, we had discussed TFN title blocks and presentation matters with Bielenstein and Keith McCandlish (also of AMCL), as well as the pending confidentiality agreement. Pyke and I began to talk more about mining potential as we saw it and the projection of a future "dream" for Nunavut.

Gold and high-value base metals colored our conceptual thinking, and therefore we selected the volcanic belts of the NWT as the most desirable areas. The best course was to steer the Inuit towards volcanic belts, where there was latent and sometimes obvious potential, and to exclude granite and the like as more suitable for the larger government (or Crown) domain. We discussed the parameters of economic geology and whether we, as advisors, should extol coal measures on Ellesmere Island or iron ores on Melville Peninsula, for example. Meanwhile, I continued the interesting Nunavut compilation work, drafting map sheets with labels and annotations in a form suitable for translation. I tried to grade the mineral showings on the maps with black-dot symbols: the full dot described a mineral showing, trench in the rock, or drill hole that had yielded assays of economic importance; the half-dot or quarter-dot represented mineralization that was less explored or exhibited submarginal assay grade. Though the quantity or continuity of mineralization was often in doubt at such locations, the showings usually represented good places to conduct more exploration, and in many cases these were already staked under the Canada Mining Regulations.

The Nunavut compilation project prompted me to call or correspond with many knowledgeable geologists in industry and business – for example, Walter Gibbons of DIAND. The Yellowknife resident geologist's office was very helpful about mineral occurrences in the Eastern Arctic and Canadian Arctic Islands. Personal advice, therefore, became an important supplement to information drawn from the geological database. Besides the drafted map sheets with showings and important geological contacts of host rocks and volcanic belts, I also assembled map-referenced descriptions in a data book for each region.

THOUGHTS ON SUBSURFACE QUANTUM

The time was now approaching for the TFN to come to Calgary to discuss the work and future schedule. On July 11, we went to a meeting at AMCL attended by Bruce Gillies and Jon Pierce of the TFN. Our briefing with maps, slides, and presentations was followed by discussion on the means of attracting exploration companies and training Native prospectors. From that time on, we always carried books on mineralogy and small mineral display sets when we met with Inuit representatives.

In the afternoon, we became involved in a lengthy discussion of the subsurface land quantum, which was still unresolved at the main table. Gillies and Pierce, as team advisors themselves, needed some support to bolster the Inuit position, which had been drastically discounted by the federal negotiators. I agreed to make a rough calculation of the area of favorable geology throughout Nunavut – this meant host rocks like the Society Cliffs formation along the Nanisivik belt and volcanic belts in general. The area I came up with was around 36,000 square miles. In addition, I phoned NWT resident geologist Bill Padgham to ascertain the area of existing mineral claims – about 12,000 square miles, as it turned out. These figures, relayed to the TFN, may have assisted in the negotiations, which settled on 14,000 square miles of subsurface rights. The quantum figure was not precedent-setting in terms of per capita allotment, for the Inuvialuit and Yukon Indians had won higher, but it was easily the largest total amount ever accorded a Native group. Indeed, it would make Inuit the largest freehold owners of mineral rights in Canada, owning even more than the CPR.[2] We tended to think the quantum allotment was huge, as it exceeded the area of all the recorded claims and leases in the NWT.

The 136,000 square miles of land, of which 14,000 would include subsurface rights, represents over 18 percent of the land area of Nunavut. The Sverdrup Basin in the High Arctic, though part of Nunavut and subject to resource management provisions of the agreement, was not allowed for land selection, because Inuit could not document their occupation there. Therefore, the quantum allotment of 136,000 square miles amounted to about 25 percent of the area of demonstrated occupation.

MEETING WITH INUIT IN WORKSHOPS

Next on the agenda was a July 31, 1989 visit to Igloolik for a North Baffin workshop, where Bielenstein presented an introduction to the importance of minerals and I provided an overview of the resources of Nunavut. To facilitate this talk, I had to haul my giant "soapstone" map case to the Arctic. The trip enabled us to talk to Dave Nutter of the NWT Chamber of Mines and John Seaton, DIAND geologist at Yellowknife. Our Hawker Siddley flight across the North afforded an enjoyable glimpse of many communities, including Pelly Bay, which I remember for its wildflowers and red granite outcrops. At Igloolik, we got settled into our hotel rooms and met Gillies, the TFN staff, and the interpreters, who admitted they were nervous about the technical terms we might use to describe rocks and minerals. Our morning walk around the settlement took us by Ordovician limestone outcrops yielding a fine view toward Foxe Basin, inukshuks, and grave sites tenderly adorned with tokens and personal items. It was an active period at Igloolik: ice around the harbor was moving out, and canoes were being loaded for trips to summer hunting camps.

We set up our maps in the hamlet council chamber and met the sixteen or so representatives of the community land identification negotiating teams (CLINTs), both local and regional, as well as Paul Quassa, the TFN chief negotiator, and Lucassi Ivvalu, who was directing the whole process. Quassa started the meeting with an update of land-claim matters and discussion of hunter support, which was considered to be part of the social program and not a land-claim matter. The interpretation facilities and personnel were first-rate, enabling us to follow the discussion with ease.

The next day, Bielenstein gave a dynamic introduction to minerals and our role as advisors. It was here that someone said, approvingly, that we were needed on the team because only a shaman knows what's under the ground! Undeterred, I spoke in the afternoon on the geology of the Nunavut settlement region. The participants showed keen interest in the mineral sets we had brought and made lively comments about the local rocks, which they knew well. Malachi Arreak from Pond Inlet was quite articulate, and urged the TFN to secure copies of the mineralogy handbook we displayed.

The following day, we had an informal, shirt-sleeve session on the maps, and there was much discussion about the legend and the nature of diamond drilling, on which some of the information was based. The participants showed an evident respect for each other's opinions. Getting down to work, the group selected over nine hundred square miles for subsurface ownership in the North Baffin land-use region, covering claims, leases, and mineral trends without restriction. That left much of the 2,400 square miles of quantum unselected, to be the subject of new ideas or possibly transferred to the Keewatin district. The original pro-rated quantum was 7 percent of the North Baffin AIP entitlement of 33,230 square miles. We became aware that the required formula for all land selection was to select more than the quantum at this stage, and allow for 50 percent reduction during actual negotiation with the government. This was a process destined to cause some anxiety in the future. There was support at the meeting for an immediate Inuit land freeze, as many saw a constant whittling away of their lands. We voiced the industry's concern about that. Most of the work of the CLINTs was, amid much discussion, the plotting of surface land parcels. They were primarily coastal hunting areas distributed in more or less continuous blocks.

Our next stop was Iqaluit, with its fascinating harbor full of marine activity. We were amazed that trucks drove to meet landing craft on the sandy, rock-strewn sea bottom when the tide was out. We toured the carving-stone outlets and jewelry-making workshop, and came to appreciate the artistic production in the community.

Hans Bielenstein looked after the final collation of our mineral inventory maps and saw that they were delivered to the TFN. He said we were going to be needed next year at the actual community negotiating sessions, where the TFN teams would meet face to face with the government teams. In October 1989, we were becoming aware of the Inuit obsession about uranium and saw their anti-uranium brief, but it had very little relevance to the Baffin region.

In late October, we went to Cambridge Bay for a Kitikmeot region workshop. Lucassi Ivvalu, director of land identification, chaired the meeting. About twenty people attended, including men I would call elders; younger, articulate men like Joe Allen, mayor of Coppermine, Joe Otokiak of Cambridge Bay, and Stanley Anablak, Kitikmeot land identification coordinator; and representatives from all the other communities: Bathurst Inlet, Bay Chimo, Spence Bay, Pelly Bay, and Gjoa Haven. Bruce Gillies described the preparation work and schedule for land selection to the group. There was excellent interest in minerals and a warm exchange of views. In the afternoon, we went over detailed mineral maps and selected 2500 square miles in the first pass, thanks to the preliminary work of Anablak. For the first time, a painful tension showed up: it was necessary for the CLINTs to relinquish other selected (surface) lands to acquire the subsurface parcels that we recommended. The latter areas tended to be remote from their main activities. The land-selection process required that Inuit take the surface ownership of any subsurface lands they selected, so that they would hold them in fee simple.

Our spare-time activities included a tour of the craft shop and the up-to-date cooperative meat plant where we ordered some char and muskox, and interesting talks with participants about building komatiks and repairing ski-doos in winter. The question of a Nunavut land freeze was debated by the participants. Some thought that in the interest of time, since federal cabinet approval was needed, this measure could be omitted. We argued that if Inuit were allowed to acquire existing mineral claims and leases, this inconvenience to the mineral industry was unnecessary.

In any event, a communiqué on the subject was sent to the December Geoscience Forum in Yellowknife. Ivvalu and Gillies indicated their need to have us present as mineral advisors at actual negotiating sessions in the following year, since on-the-spot decisions about subsurface lands would have to made. So it appeared that our role as consultants for the TFN would be expanded somewhat.

Back home, I reported on the Cambridge Bay meeting to colleagues Pyke and Mudry, who were going to Rankin Inlet and Yellowknife on mining business. Their contacts there would enable us to be sensitive to the concerns of Native and exploration people on land-claim progress. We also attended the annual meeting of Borealis Exploration Ltd., environmental "bad guys" of the Arctic, with properties on Melville Peninsula and elsewhere. The residents of Repulse Bay and Rankin Inlet were under no illusions about the behavior of this company, a poor representative of the industry; legal suits by disgruntled employees and land-use authorities were to plague the company in the months ahead.

During the early months of 1990, I attended to occasional TFN duties, such as plotting new prospecting permits on working maps and collecting backup reference papers. My normal work in exploration had to do with projects in central Baffin Island and the southern Keewatin, around Cullaton Lake. I was curious about the Canadian Mining Association's (CMA) workshop on land claims, but did not attend. Murray Pyke attended the TFN's Keewatin planning workshop at the end of March. He observed that the Inuit had an unrealistic expectation that mining could be planned, not realizing that the locations of mines are largely unpredictable and their development hinges on global economic factors. The workshop participants were low-key about the anti-Kiggavik referendum, where Baker Lake voters had registered 90 percent opposition to the uranium mine. There was much criticism of Borealis Exploration and Cullaton Lake Mines in regard to oil spills at abandoned sites.

We noted that there should be some recognition of Wollex (our group) as well as AMCL, and that we might need a procedural guidebook to use during negotiations. We were all pretty vague about negotiations at this stage, and we wanted to be prepared.

Bielenstein and I went to Ottawa, loaded with our maps, at the end of April – which happened to coincide with the Igloolik signing of the Agreement-in-Principle. The meeting we attended was a working session on land matters for regional coordinators and other Inuit from key communities. We spoke on mineral industry developments during the year, and lawyer John Merritt provided a briefing on water rights.

The two main agenda items were the maps and the need for a land freeze. Those favoring a freeze said that it would impress their fellow Inuit back home and bring more pressure to bear on the federal government for speedy negotiations. Even allowing contiguous staking by existing claim holders (my amendment to the discussion, supported by Jon Pierce) was ruled out.

I still believed the interim protection afforded by a land freeze was unnecessary and ineffectual because of the high level of staking already in place. Though a land freeze would irritate industry, public relations was not a concern here, and accordingly it was decided to request a freeze of those areas selected. We concentrated on specific selections and amended the work maps around Contwoyto Lake, George Lake, Chantry Inlet, and Committee Bay, using known geology. Barry Dewar of the government's comprehensive land-claim office, who came in to address the group on negotiations, gave us a litany of do's and don'ts, including the exclusion of third-party lands. Though delivered in good humor, his comments sparked an emotional outburst afterward, led by Joe Otokiak of Cambridge Bay. All Inuit were determined to stand firm and fight hard for alienated lands. Later, I gave Keith Hay, South Baffin coordinator, some advice on subsurface selections of iron formations and volcanic rocks at Sanikiluaq. I had to cancel my projected June trip to the remote Belcher Islands, so this exchange was useful. Regrettably, the health of Bielenstein began to deteriorate around this time, and he was lost to the TFN advisory team.

NEGOTIATIONS BEGIN AT POND INLET

The first land-negotiation session took place at Pond Inlet on Baffin Island, on July 24–30, 1990. Its purpose was to settle the selections of surface and subsurface lands for the communities of the North Baffin land-use region (Pond Inlet, Arctic Bay, Clyde River), including the Melville Peninsula (Igloolik and Hall Beach) and Devon, Ellesmere, and Cornwallis Islands (Resolute Bay and Grise Fiord). By previous agreement, as stated in the AIP,[3] Inuit of this region were entitled to 33,230 square miles of land and a portion of the total 14,000 square miles in all of Nunavut that would convey both surface and subsurface rights.[4] The specified surface area (about ten times the size of Prince Edward Island) had been studied by teams headed by Lucassi Ivvalu of Igloolik during community preparation programs. The work was a combined effort over many months by community land-identification negotiating teams in the region. The teams comprised elders, hunter and trapper representatives, and younger Inuit with map and planimeter (map instrument) skills, aided by a TFN coordinator – in in this case, Malachi Arreak of Pond Inlet – and other staff specialists, especially Bruce Gillies. I was the advisor regarding minerals.

The selection process had two important features. First, it was an open process known to all community members, many of whom were touchy about the 82 percent of lands being forfeited – as if by penalty for signing the agreement. Second, the mechanics of selection required the CLINTs to identify more land on the maps than they would actually receive: 150 percent of the quantum entitlement. The federal government would endorse the stated quantum by cutting one-third of the lands identified. The government was committed to approving 75 percent of the quantum from the lands selected by the communities in case the two sides were at loggerheads, but in practice wished to approve 100 percent from those lands during negotiations. This flexible approach had a pronounced psychological effect on the Inuit communities, since the 150 percent concept was poorly understood. The local people could plainly see on large-scale maps the extent of selections made by their home representatives, who took these to the table for negotiation, only to return with smaller approved areas. To the people, this looked like a negotiation loss.

The federal team at Pond Inlet was headed by chief negotiator Tom Molloy, with staff specialists Barry Dewar and Norman Williams (one lawyer, one experienced bureaucrat) at his side. They were flanked by ten departmental specialists, representing Parks Canada, National Defence, DIAND, EMR, Surveys, Canada Wildlife, and others. The TFN team, headed by chief negotiator Paul Quassa, included Lucassi Ivvalu, coordinator of land selection; Malachi Arreak, Baffin regional leader; Tongola Sandy, his counterpart from the Keewatin; Paul Keytook of the Baffin Regional Inuit Association; twenty Inuit representatives from the communities; two interpreters; and two white advisors, Gillies and this writer.

Arriving by July 24, the Inuit team held a preliminary caucus discussion at the site of the meeting, the United Church building. Paul Quassa stressed to the community representatives that they would have to take the lead in this phase of the land-claim process, since they were the experts on the land. They should remember that the federal representatives were less familiar with the land – and they would be going home afterwards. He said, "You, or rather, we, will have to live with the choices made here forever: they are final. You should also remember that you will be giving up around 80 percent of the land to the Crown, except for what you keep." Being an old hand at main-table encounters with government, he recommended that a proud, aggressive stance would be appropriate in these negotiations. Quassa's pep talk was followed by a report from Gillies on possible sticking points at certain communities, and he urged the CLINTs to be prepared to negotiate down.

Malachi Arreak said that North Baffin was prepared to transfer eight hundred square miles of subsurface rights to the Keewatin, since these lands would be jointly owned by all Inuit. Knowing that this region had limited mineral endowment, except for the Nanisivik and Polaris mine areas, I was quite in agreement with this decision. My hope was that the settlement would leave the Inuit with a sprinkling of subsurface lands in all the known plays or prospective areas of Nunavut. I guessed that mineral rights would play a relatively small role in the negotiations here, where they amounted to 7 percent of the total package.

My first inkling of the contrary came when Allie Salluviniq of Resolute Bay said that if their region were denied its share of subsurface rights, they wanted more surface. There was also a sense of disappointment among the representatives that the government had excluded the Sverdrup basin in the High

Arctic, well known for its oil potential, from land selection by Inuit, using the convenient rationale that Inuit could not demonstrate occupation there during living memory.

I noticed that during their spare time, the Inuit participants were jolly socializers and ardent card players. The evening poker pots were up to $500, with Allie Salluviniq and Sam Palitug big winners. Perhaps this was a portent of the manner in which they would approach negotiations; searching for cards to strengthen their hand.

FIRST ROUND OF CUTS

The joint meeting of the parties began on July 25, 1990 with a prayer for success by a Pond Inlet elder. Earphones and microphones made this bilingual encounter understandable to all, thanks to the team of simultaneous interpreters ensconced in the sound room. Speaking for the government representatives, who had already studied the Inuit maps, chief negotiator Tom Molloy stated their wish to achieve ratification of the final land parcels wholly from the areas selected. He said the federal objectives were to avoid too much coastal concentration in the hands of Inuit, avoid conflict with third-party rights, and ensure travel routes or corridors to blocks of isolated Crown land. Lorne McNeith of the Surveys Branch elaborated on the need to identify all parcels of land by straight lines connected to prominent topographical features, or by lines of latitude and longitude, so that the lands successfully identified would be vested in Inuit ownership with a minimum of legal surveying. Some of the Inuit selections presented serious problems in this regard because of curvilinear boundaries. The cartographic responsibility fell to us, as the government representatives made it clear that they would not touch pen to paper during the negotiations.

Beginning first with Igloolik and vicinity, the government critiqued the Inuit 250,000-scale maps, which were displayed on the wall. The detail man, Norman Williams, went through a list of "conflicts" with third-party interests, which dealt mostly with such matters as marine navigational beacons, wharf space for the Department of National Defence (DND), and short-range radar sites. When he came to iron ore mineral leases/claims at Eqe Bay, site of a treasured summer camp and soapstone deposit on the west shore of Baffin Island, he said these selections were clearly off limits. The picture was similar around Hall Beach: the acquisition of mineral rights on the Borealis iron claims and on leases at Roche Bay (75 km to the southwest) would not be allowed.

These arbitrary government decisions to protect third-party interests were very painful to the Hall Beach people, who had historic connections to these sites. They believed that mining companies could hang onto property endlessly, and they wanted some control over that. Solomon Qanatsiag said he practically grew up at Eqe Bay, and his uncle had been the first to show iron mineralization to the whites. He was referring to Robert Blackadar's first mapping of the Precambrian iron formation and later prospecting by Murray Watts, the staker of the Mary River deposits farther north. Molloy's response to Inuit pestering on this point was simply that the government had never had any intention of granting rights to land protected by mineral claims or leases.

When the discussion turned to the Clyde River selections, the federal objections centered around inland routes, sport-fishing sites, and proposed bird-sanctuary acreage. For Arctic Bay, they focused on goose and fulmar protection areas, withdrawals for commercial travel routes (the CLINTs were to select one side only of channels south of Admiralty Inlet), public fishing in small lakes, and mineral leases totaling sixty-four square miles along the Nanisivik belt. It should be noted that Arctic Bay made no request to acquire subsurface rights in and around the Nanisivik mine area. Mike Parre of the DND spoke of the military plan to take over the Nanisivik mine site for storage and maneuvers when ores became depleted. This plan did not impress the Inuit; they preferred not to have a military base on their doorstep.

Pond Inlet selections were denied for a Co-op owned (i.e., Inuit owned) airstrip and outfitting camp, for a sea-run char fishing camp on Tay Sound, for old mining leases, and for a campsite at Mary River that was a designated site for carving stone. The government's plan to exclude Co-op lands to ensure public access was particularly irritating to the Inuit, who believed that such lands already provided a public service.

The biggest uproar centered around Bylot Island, whose beautiful, glacier-riven southern coast faces Pond Inlet. There were miscellaneous small federal needs there, but more importantly, the government, through Parks Canada, wanted to establish a huge national park preserve. Bob Gamble of Parks Canada, who had already been promoting the project among the Inuit people, presented his agency's needs and expressed a willingness to accommodate local concerns.

The communities of North Baffin continued to reduce their acreage in caucus beyond the 10 percent already chopped by the government, while trying to hold together the core of their needs. The major Inuit reaction was: did we really have any bargaining power that could be brought to bear in critical cases? The Inuit could and did state their objection to sport-fishing camps.

One, at Creswell Bay, was particularly objectionable to local residents, according to Allie Salluviniq from Resolute Bay. They also objected to low-flying jets and wanted to discourage DND expansion. The national park preserve, which was especially dear to the government, could be opposed.

In regard to subsurface acquisition, I suggested a strategy to break the impasse of applying for claims and avoiding leases. The status of Crown prospecting permits, some of which covered lands with potential, was left in doubt. The government had preconceived ideas about the way the final maps should look: the Inuit were not to control the whole coast, though they were a coastal people, and yet the Crown would control all the staked minerals.

The next session examined the land selections presented by Grise Fiord and Resolute Bay. There were no subsurface conflicts for the former, but there were federal needs to cover fuel storage, a bird sanctuary, sports fishing, and RCMP and Arctic Institute sites, as well as Eureka recreational fishing. The most concern was focused on the Inuit selection of the inland fiords and passes – obvious corridors to Crown land that had to be provided. Also, there was technical criticism over non-straight borders and skirting of ordinary, as opposed to navigable, lakes. By far the biggest row was over the mountain passes in the rugged Ellesmere Island terrain: the Inuit naturally believed they had discovered them and were their sole users.

Williams critiqued the Resolute Bay selections at great length. There were bird-protection and polar bear research needs, a DND reserve for winter maneuvers, the historical John Franklin site on Beechey Island, coastguard staging, a Suncor storage site, a Panarctic camp and staging area, private whale-watching leases, a polynya (or winter water) study site, and navigational aids to be considered. In addition, the federal negotiators excluded a number of properties held by Cominco: mineral leases and a camp on the Shields Peninsula, a mine site, and mineral leases on Cornwallis, Little Cornwallis, and Truro Islands. Altogether, the third-party subsurface conflicts totaled sixty square miles by the government's estimate, out of fifty-four square miles selected.

This concluded the first round of the negotiations. Each side had stated its position; it remained for the Inuit to accommodate the federal needs or stated exclusions while paring down their own selections to the desired quantum levels. The federal side (familiarly referred to by now as "the feds") again said that the "map fiddling" was strictly in the hands of the Inuit, some of whom were surprised at this policy or uncomfortable with the responsibility of making the cuts and having to answer for them back home. Lucassi Ivvalu urged both Resolute Bay and other representatives to "dream about the land they want to keep." As in a card game, it was useful to know your opponent's opening hand before playing your own, but the Inuit wondered if there was room for negotiation and compromise at the table. Did they have any useful cards to hold in their hand for a while, or was this strictly a payout game?

SUBSURFACE BECOMES CONTENTIOUS

The next day, Paul Quassa opened with an attack on federal policy concerning third-party rights – for example, the case of mineral claims and leases. He said the Agreement-in-Principle, already signed and sealed, allowed the Inuit to acquire ownership of "areas of known and potential mineral deposits" (Section 17.1.2.b.i), and the agreed purpose of settlement lands was to promote economic self-sufficiency through time. He asked for an explanation in the light of the high level of NWT claim-staking at the present time. The exclusion of bird sanctuaries was analogous in the case of renewable resources that the Inuit harvested. They sought to protect those areas from third parties by choosing them; federal policy implied a lack of confidence in Inuit as managers of their land. Quassa said area reductions had been made and maps had been changed. The CLINTs were giving up some lands, not because they wanted to, but because they were reasonable people and desired an agreement. He admitted being nervous that actions at this time would have significant implications for other areas, like the Keewatin.

Malachi Arreak said the North Baffin negotiators had sound reasons for their selections. Igloolik and Hall Beach representatives said surface-area reductions had been achieved; as to the subsurface, they felt that the mineral companies could sit on Inuit land forever – and they had grave concerns about that. Clyde River said there was easy access inland without the designated hundred-foot strip along waterways, and that the proposed fulmar sanctuary was already an active hunting area. Arctic Bay had concerns about mining claims and bird sanctuaries, as the birds were a big hunting resource. Grise Fiord agreed to claim some lakes and to give up the proposed bird sanctuary. They doubted that the Eureka weather station staff of eight persons needed two lakes for recreation. The big issue of public transportation routes along passes needed clarification: the exclusion of such areas would fracture their selections. Resolute Bay delivered a terse statement; they said, referring to their history as displaced persons, that they were there only for sovereignty reasons, adding, "We hope you will cooperate with us to make something of ourselves."

Tom Molloy made some general comments, to the effect that conflicting areas were relatively small, and there was lots of government activity outside those areas. He said the government considered claims and leases to be one and the same, created under the Canada Mining Regulations. The benefits of the Agreement-in-Principle, he felt, made subsurface ownership less important to the Inuit. They had a resource revenue-sharing agreement for all Crown lands, and they could take out surface rights over mineral dispositions and thereby benefit from development. As to routes of access, the general public must be able to reach Crown land, hence Inuit control of both sides of some travel routes would be a problem.

Quassa said he was not convinced, after making inquiries of the Canadian Mining Association (CMA), that mining companies preferred to operate on Crown land. He wanted to hear why third parties were being kept on Crown land, in direct conflict with the Inuit goal of eventual self-sufficiency. Molloy replied that the reason was to ensure continuity of service and land management. Privately, the Inuit leadership had heard from industry representatives that they would accept Native ownership of mineral rights. Bruce Gillies said that those opinions had been expressed at the June CMA meeting and were contingent on the continuity of rights under the Canada Mining Regulations, to which Inuit had agreed, but I myself was not aware of any accommodating public statement to this effect. Industry "preferred" to see minerals remain in Crown hands, as they worried about access protection and, when they made a discovery, might want to seek government assistance for roads and infrastructure.

(Later, after 1992, the mining industry would take direct action to become involved with Native issues by means of the multi-stakeholder Whitehorse Mining Initiative.[5] This promoted interim agreements with Native groups, such as memoranda of understanding and resource development agreements. While the latter agreements were drafted so as not to prejudice actual land-claim settlements, they contained elements that might carry over and have lasting effect on access and development terms. This Yukon action acknowledged aboriginal interest in having partnerships with developers and exercising some control. However, despite the earlier efforts of Tagak Curley, developer partnerships with government remained the preferred model in the NWT.)

During the break, I worked closely with Gillies and Arreak to define the potential mineral belts more exactly, with less overlap on granite. I thought the stated federal policy negated our entire plan of securing quality mineral holdings for Inuit ownership in every area. It required a clear change in our strategy, including more saturation and a less generalized, broad-brush coverage of mineral belts. We resolved on principle "to take all the volcanics and leave the granite to the Crown," as a measure of self-defense. We had the maps and data compilations to accomplish this, while – much to my surprise – the government was without a mineral advisor. In addition, I counseled the TFN to acquire surface rights on every last mineral claim, even if it resulted in a terrible patchwork and the consequent fragmentation created an administrative mess for the Crown.

In joint session, Arreak went to the maps to describe the changes made so far. They were significant, but they went only part way toward achieving the quantum. Travel routes needed more discussion. Pond Inlet and Arctic Bay had not cleared out of Bylot Island, so it appeared that the proposed national park would not proceed. Quassa accepted the decision to keep the Bylot Island selections, but said the communities had not formally rejected the proposed park. The exchange over Grise Fiord centered on public access along fiord travel routes, the best of which the community had decided to keep. Jaypeetee Akeorok questioned whose access was important, since only Inuit used the routes. Where there was third-party travel; why not hire an Inuit guide? Quassa concurred, saying that the Agreement-in-Principle and private agreements would take care of any problems, but Molloy contended that commercial access (to Crown land) via this popular route was not assured by the AIP, whereas Inuit had no restrictions on their travel. There was some impatience with Akeorok's foot-dragging tactics, he even added some new acreage to his bloated maps to cover the Haughton impact crater on Devon Island after I told him of its possible tourist value. Resolute Bay representatives were cooperating to a degree: they reduced their holdings on Beechey Island and Devon Island's south shore and those at Somerset Island's north-shore beluga station. The Somerset Island diamond pipes were dropped, on my advice, because they had been explored. Diamond exploration, which was later to sweep the NWT, dictated large acreage acquisition because of the non-specific nature of kimberlite intrusion. This squandering of the Inuit subsurface allotment was to be avoided. There were also minor adjustments on Little Cornwallis Island. Molloy said that a federal response would be made the next day, but he feared the government team were going to have to make decisions for the Inuit and start putting lines on the maps themselves. Quassa raised the question of mining claims as opposed to leases, stating that the Inuit did not want leases that carried deposits and mines; they wanted mineral claims under their ownership.

The next day, Molloy explained that the need for public corridors was small, but they were important, like highways down south. It was not clear how they would really interfere with Inuit use of the land. As to mineral resources, he did not "see" Inuit taking either claims or leases. In the North

Baffin region there were no exceptional circumstances, and in any event the dispositions represented small areas. He said Inuit might take surface ownership over mineral claims and leases to enjoy the benefits if there were any developments. Quassa said he could not understand why special circumstances had to be alleged – none had been brought up during their contact with the mining industry. Molloy said there never was any intention to yield staked or leased minerals during the negotiations. Asked if that included prospecting permits, he said they were not included. Norman Williams went over the subsurface selections by Igloolik and Hall Beach, and we wondered how anyone could "protect" Borealis Exploration's leases, with their insulting environmental condition, or the age-old leases of iron formation that were carried on company books because of lax NWT regulations. Since 1985, Brazilian and Australian producers had gained ascendency in the world of iron ore production. Inuit felt that their strong historical connections with the Eqe Bay property would constitute exceptional circumstances if such were needed.

Returning to our caucus, we discussed the inclusion of lakes, which were originally understood by the CLINTs to be excluded. They now believed the government was deliberately requiring them to give up land and take water – but not ocean water, of course. They were also critical of the policy that, under the guise of third-party protection, excluded the selection of land with Co-op improvements (achieved by Inuit themselves) that yielded economic benefits, such as airstrips. But the big discussion involving the whole group was over the federal holdback on mineral lands staked by third parties. The holdback entailed most of the lands slated for mineral ownership, and this mineral allotment consisted of a mere 2400 square miles, or 8 percent of the total quantum, to begin with. Quassa, admitting he had negotiated for a long time on this subject, wondered how long it would take before the Inuit could get their hands on leases. I repeated that statistics showed the Native land-claim process itself had brought on a very high level of staking, with the result that virtually all minerals with potential were staked. Mineral claims, however, were often short-lived: most would lapse after two years if no work expenditure was submitted, and the net result was an average turnover of seven to eight years. At the time of negotiation, however, a record number of mineral claims were in force. If federal policy held sway, this would prevent the Inuit from acquiring a meaningful portfolio of mineral-bearing lands. The holdout seemed an unconscionable violation of the Agreement-in-Principle. We were being driven to rely on surface coverage of claims and leases and then concentrate the subsurface quantum on strictly geological reasoning. It seemed ridiculous that we would acquire surface rights over all mineral claims and not be able to get any subsurface rights in decent areas. There was some difference of opinion among Inuit leaders at this point on tactics, whether to strive for an agreement or pull out of the negotiations, mainly over the mineral-claims issue. My earlier assumption that mineral matters would play a small part in these negotiations had been dead wrong. The federal boilerplate position of "protecting" third-party interests had touched a very sensitive nerve: it was a potential threat to the whole process because of the Inuit desire to acquire quality lands for themselves.

COMPROMISE BEGINS, IN PART

In the next session, the government heard how precious the Ellesmere Island fiords and passes were, all discovered by Inuit and dotted by their inukshuks, as described by the stubborn Grise Fiord representatives, Jaypeetee Akeorok and Laisa Ningiuk. The Co-op leases and Little Cornwallis Island subsurface leases were likewise defended. At this point, Tom Molloy relented with another offer: the government would seek written permission from the Co-ops for a switch from Crown to Inuit ownership. Quassa asked whether mining-company views could also be canvassed to see if some would opt to become Inuit land–based. Molloy said there were policy questions involved for federal negotiators – the government had to "defend" mineral claims and leases as they created them. But Quassa countered: "What if the usual thing happened, and many claims lapsed after two years? Would that not create a myriad of little gaps in our lands? And, if we took the surface of every claim, would that not create a hopeless checkerboard? In fact, didn't the Agreement-in-Principle, in section 18.1.1(j), warn against undue fragmentation of lands?" He said that fragmentation would be inevitable to meet Inuit concerns. Molloy said they would have to review the situation. He added that it was okay to take surface over mining properties, such as Mary River with its massive iron formations, subject to existing land-use permits and gravel leases.

At this stage in the piecemeal process of coming down to quantum, Malachi Arreak relinquished all the withheld land parcels on Bylot Island, thus paving the way for the North Baffin national park preserve, to which the government was prepared to commit under the terms of the Agreement-in-Principle. Bob Gamble spoke about the proposed park, complaining about a subsurface parcel (actually a bird nesting area) covering a coal occurrence, and excessive coastline taken on the west side of Navy Board Inlet. The proposed park extended far beyond Bylot Island itself and included much of the Borden

Peninsula. Gamble offered to work out a special agreement under park management to protect an Inuit archeological site, which was a helpful approach. Allowing the public free access for a hundred feet along the coast was a bit meaningless, because, as Moses Koonoo explained to Bob Gamble, you cannot land a boat there under most conditions, so access was restricted for Inuit and visitors alike. The argument here was chiefly confined to park boundaries, inasmuch as Inuit relationships – such as joint management, access and subsistence hunting rights – had previously been worked out and stated in the AIP. During the long process of negotiation by the TFN, the Inuit had come to welcome national park preserves as an important component in their struggle to gain control of their land. Such areas would officially become national parks only after ratification of the final agreement and signing of impact–benefit agreements. Inuit would not have to "own" them to exercise benefits and authority and achieve ecological conservation.[6]

Jaypeetee Akeorok was urged to straighten the boundaries of parcels around Grise Fiord. The process of cutting was now well advanced. We were down to quantum, or perhaps lower in some cases – the calculation of areas using manual planimeters had been rather tedious and slow.

Back in caucus, I listened to Solomon Qanatsiag tell about growing up around the Eqe Bay iron formations, which were well known to his uncle. I suggested preparing an open letter to the mining industry about federal policy on subsurface rights and our concern over the patchwork land arrangement that would result. Would not the Crown have to perform or require legal surveys on every mineral claim and lease upon its lapsing to ensure the integrity of land ownership, assuming we surrounded the holdout parcels? Quassa suggested that we contact major mining companies like Cominco about the question of Inuit ownership. Tongola Sandy agreed that industry might be able to help with our position. He was not concerned, as a Keewatin observer, about acreage being transferred to his region – the Baffin communities were entitled to their quantums. This statement confirmed the unselfish solidarity of Inuit people in their quest for this land-claim settlement. Ivvalu explained that the feds now wanted to break us down into small groups and go over each map in detail, but there were serious problems over subsurface rights for most communities.

Igloolik and Hall Beach still had too much subsurface. I suggested dropping the block at the west end of Fury and Hecla Strait, which covered a gabbro sill (black igneous rock) with minor copper occurrences. Regrettably, the desired leases held by Borealis at Roche Bay would be up for renewal in October. Discussion turned again to the old mess left by mining companies, for example, the HBOG camp at Mary River left in the 1970s. The Land Use Branch was said to be investigating the site (and at a later date brought charges against Borealis for violations in the Keewatin). The growing attachment to subsurface lands among the North Baffin communities meant that less would be transferred to the Keewatin than originally thought. For example, Grise Fiord insisted on taking some oil and gas acreage at the fringe of the Sverdrup basin, which was in general excluded from Inuit selection. Quassa wanted to see something accomplished at this Pond Inlet session. He said that we were close to target and sticking to the AIP: "kind of doing the feds' work for them!" He thought the community representatives should take the maps back to the communities for final choices and, if necessary, come back to the table again. At this point, Allie Salluviniq made a definite stand: his Resolute Bay people were not in agreement with the way things were going, and he would not change things by himself. Malachi Arreak then urged us not to reduce our selection any further. Grise Fiord representatives said they would continue to demand the land through the passes of southern Ellesmere Island, but they needed support in their arguments. Salluviniq said the passes, discovered by Inuit who landed there in the 1950s, had been consistently used for hunting. There were doubts, however, that fees could ever be charged for their use.

When the Igloolik/Hall Beach maps were reviewed in joint session, many people got into the argument. Paul Halli made an impassioned speech about the powerful connection with their outcamp at Eqe Bay on the Baffin Island shore, a two-hundred-kilometer journey from the two settlements. While the government's focus of attention was on subsurface ownership at the camp, the interest of the Inuit centered on the fishing, hunting, and soapstone resources of the site, as well as its historical meaning to them. Molloy restated the federal position that the government would grant mineral claims or leases to the Inuit only where special circumstances warranted, yet he would not define what those circumstances might be, seemingly constrained by federal policy. TFN's Gillies made a speech about the stated need for a mix of selections under the AIP. Harsh words were beginning to indicate that the federal representatives were straining under pressure. I volunteered that there were indeed exceptional circumstances, and the government should know this: the iron claims and leases were a holdover from the 1960s and were economically worthless now and for the foreseeable future. They conferred no present value to the government and should be given up. The world's iron ore production was now dominated by large-scale operations in Brazil and western Australia. In the end, the Igloolik andHall Beach representatives broke off for a private discussion. They decided to halt negotiations on this issue and resume discussions in the communities.

MINERAL ISSUE UNRESOLVED

This decision caused a major break in the negotiations. I was truly surprised that mineral land would be such a stumbling block, rather than just a bargaining ploy. There was good agreement on other issues, and land selections were close to the quantum numbers.

In regard to the Pond Inlet selections, the outstanding item was the Co-op airstrip on the Mary River road, which serviced a sport fishing camp. One of those present was Joe Inuk, director of the Co-op, and he addressed the issue effectively. Inuit had managed this business in the past and wanted to continue. The government contended that where the public wants access, as it does here, the government wants control. So this bit of old road, with empty fuel drums along its sides to denote a landing strip, would be denied Inuit ownership. The Arctic Bay maps were reviewed by Moses Koonoo, who was not, in the opinion of the government, able to identify any special circumstances, so subsurface parcels were denied. Molloy admitted that our side had not requested the Nanisivik property, but suggested they put the matter in abeyance until contact could be made with third-party owners and the government could decide the matter. Asked if government negotiators had had any prior contact with industry, Molloy replied that they had not, except to transmit the content of the AIP. The Inuit response was that the government representatives needed to read the AIP, as well as transmitting it.

Quassa expressed the view that these negotiations had become a one-sided affair, with benefits accruing only to the federal side. Lazarus Aola of Arctic Bay said they needed to go back to their communities – and people were going to be disappointed. They already had a concern (in giving up 80 percent of the land) that the government was going to restrict their movement. Molloy denied this and suggested moving on to solvable issues – a mark, I thought, of his long experience as a negotiator. Aola nevertheless contended that this one-square-mile parcel (the airstrip) would hold up the agreement, and the government would not be back for years. He said he was worried that exploration would disturb caribou in their calving grounds. This certainly was a reflection of Native ambivalence or mixed emotions impinging on mineral ownership. I thought Inuit mineral ownership should be enhanced at Arctic Bay because Inuit residents who were working at Nanisivik might be interested in becoming prospectors. What better place to launch an Inuit exploration company? However, this "special circumstance" was never voiced by the Inuit, and I never heard of Nanisivik training Native prospectors either, despite the growing need for more ore reserves.

Salluviniq asked: are you not in a position to do anything, in these so-called negotiations? We at Resolute Bay already are the victims of special circumstances. *We* cannot claim the ocean bottom, while the *mining companies* can put claims there! He meant the Inuit could not apply for surface rights along the strike extension of the Polaris zinc mine, because it runs under the sea. The discussion was getting acrimonious. Quassa said, "First, you take the Queen Elizabeth Islands [Sverdrup basin] from us, and then 80 percent of the rest, and now you are citing special circumstances – a term not even used in the AIP. The negotiating process is nothing more than a federal land-grab."

Quassa concluded by expressing his view that we had essentially reached a deadlock in North Baffin, adding that he hoped the government would do some homework before we met again. As a last offer, Grise Fiord representatives said they would open up an alternative route to the interior, namely the one used by the RCMP patrols.

PERSONAL IMPRESSIONS

Thus ended the first round of land-selection negotiations under the terms of the TFN Agreement-in-Principle. It was an eye-opener for me, and no doubt for some of the Inuit participants as well. I could see now that this was a task of "negotiating" through a sea of federal hard-line positions, not the least of which was their stout defense of third-party interests. That the Inuit had already agreed to honor existing regulations, such as the Canada Mining Regulations, did not seem to matter. Perhaps now the federal negotiators had a new perspective and were beginning to form the same conclusion about the process that Tom Hoefer, general manager of the NWT Chamber of Mines, would reach in 1992: "In the eastern Arctic, the Inuit have bargained hard to acquire areas of high mineral potential, including existing mining claims and leases. Their land-claim agreement is quite specific in its requirement for developers to pass benefits of resource development on to Inuit residents."[7]

After the meeting, Gillies and others thanked me for my supportive efforts. On my way home, I visited Bob Ginn and Tony Andrews of the PDAC, who were very interested in the Eastern Arctic land claim. They were still preparing a brief for government on the June Canadian Mining Association workshop (attended by Gillies and others of the TFN); they said they might include some comment jointly with the CMA for the minister's attention. I was still determined to aim for a creditable acquisition of mineral rights for Inuit in all known plays and districts – that is, to get for them a good mix of subsurface lands across Nunavut. I had come to appreciate that Inuit in the communities somehow regarded the one-third of selected lands they had to give up as negotiable. They were likely to continue using the "mineral card" in the negotiations, and that was all right with me.

NEGOTIATIONS CONTINUE AT PANGNIRTUNG

The next meeting I attended was the joint negotiating session for the South Baffin region, which began on November 27, 1990, at the historic settlement of Pangnirtung. Before the meeting, the TFN had secured interim withdrawals of Crown lands and had them displayed on DIAND's latest mineral claim maps. No progress had been made at the main table on the question of mineral-rights acquisition, and the future of negotiations seemed in doubt.

My preview of the community selections before leaving Calgary had made me wish for better aeromagnetic survey coverage of Baffin Island to guide us. This was high-grade metamorphic terrain, suggesting that, in the absence of better information, we should confine the subsurface parcels to mafic volcanic and intrusive igneous rocks with a possible potential for nickel and copper mineralization. I got some advice from Don Hogarth at the University of Ottawa, a mineralogist acquaintance with field experience in Baffin Island. Hogarth was something of an expert on local soapstone and lapis lazuli occurrences, and he was currently a member of the Smithsonian expedition to explore Martin Frobisher's 1577 gold pits at Kodlunarn Island. These excavations, located near Iqaluit, were the site of Canada's first prospecting scam.

To our deep regret, Hans Bielenstein, our original colleague on this project, was now seriously ill. Murray Pyke and I visited him, however, to talk about hydrocarbons in the Arctic and the feasibility of coal-fired power generation at Rankin Inlet. He was not optimistic about High Arctic coal exploitation in the near future (including the Pond Inlet seam) or, for that matter, about the mining of oil shales on Southampton Island, for which he had some good references. Our purpose was to become as knowledgeable as possible, not knowing when the federal team would show up with their own geological experts.

At Pangnirtung, the TFN team described previously was bolstered by Mary Crnkovich, a lawyer, and Terry Fenge, regional planner and research director

Photo 1 Malachi Arreak with planimeter,
Pangnirtung, December 1990
Photo Credit: M. W. Pyke of Calgary

Photo 2 Mary Crnkovitch, Bruce Gillies, Lucassi Ivvalu, Malachi Arreak, Murray Pyke, Bob McPherson, Coral Harbour, March 1991
Photo Credit: M. W. Pyke of Calgary

Photo 3 Interpreters Mary Nashook, Veronica Curley, Looee Okalik (Editor Inuktitut), Martha Flaherty, Coral Harbour, March 1991
Photo Credit: M. W. Pyke of Calgary

Photo 4 Presentation of Federal requirements by Tom Molloy Coral Harbour, Mach 1991
Photo credit: M. W. Pyke of Calgary

Photo 5 TFN caucus discussion on Baker Lake, Coral Harbour, March 1991
Photo credit: M. W. Pyke of Calgary

Photo 6 Agreement reached with Repulse Bay, Coral Harbour, March 1991
Photo credit: M. W. Pyke of Calgary

Photo 7 Government and TFN negotiating teams, Coral Harbour, March 1991
Photo credit: M. W. Pyke of Calgary

Photo 8 Paul Quassa, Louis Pilakapsi, David Alagalak, Tagak Curley, Rankin Inlet, April
Photo credit: M. W. Pyke of Calgary 1991

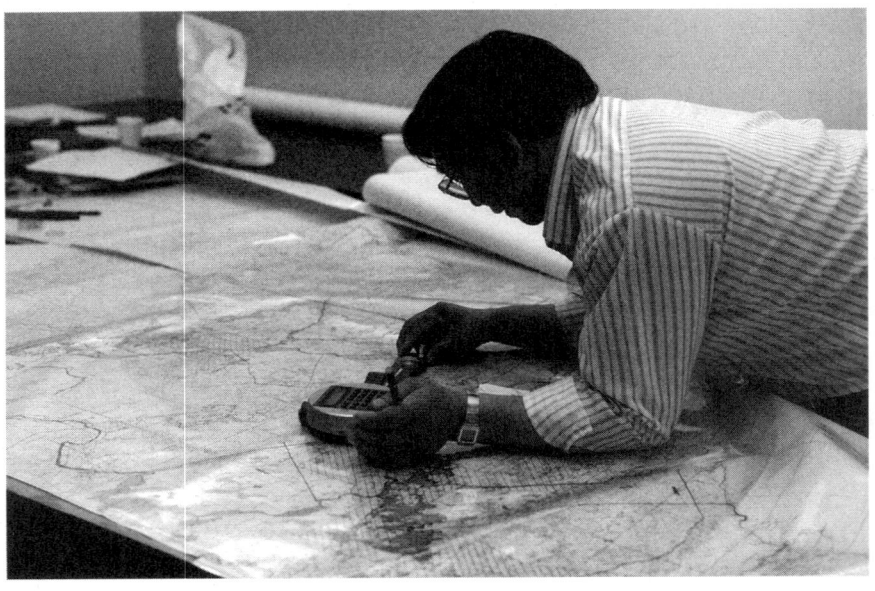

Photo 9 Malachi Arreak, David Aglukark, Tagak Curley, Robert Tatti and others, Rankin Inlet, April 1991
Photo credit: M. W. Pyke of Calgary

Photo 10 Anthyme Kadjuk at maps with planimeter, Rankin Inlet, April 1991
Photo credit: M. W. Pyke of Calgary

Photo 11 TFN presentation at the map wall,
Coppermine, July 1991
Photo credit: M. W. Pyke of Calgary

Photo 12 Norm Williams and Tom Molloy
presenting Federal position,
Coppermine, July 1991
Photo credit: M. W. Pyke of Calgary

Photo 13 Paul Quassa, Terry Fenge, Bob Kadlun and Bruce Gillies enjoying break, Coppermine, July 1991
Photo credit: M. W. Pyke of Calgary

Photo 14 Prospector Noel Avadluk receiving plaque, Coppermine, July 1991
Photo credit: M. W. Pyke of Calgary

Photo 15 Tom Molloy tries the drum, Coppermine, July 1991
Photo credit: M. W. Pyke of Calgary

Photo 16 Tom Molloy, Paul Quassa, Ross McKinnon, Iqaluit, 1993. Copyright by Hans Blohm

for the TFN, as well as Keith Hay, the South Baffin coordinator. The site of this meeting was in a spectacular fiord, and our accommodations at the Auyuittuq Lodge were comfortable. The Inuit reps assembled from Pangnirtung, Iqaluit, Cape Dorset, Lake Harbour, and Broughton Island formed an animated and good-natured group that brought our numbers to thirty. The federal/territorial team of sixteen had able support, including Dave Perrin, a land and survey specialist from the comprehensive-land-claim office in Yellowknife.

The Pangnirtung meeting was interesting because the surface land issues were hotly debated, but there was little subsurface controversy owing to the small endowment of minerals in the region. During the five days of negotiations, serious questions arose over the surface selections of the communities. The most contentious pertained to overselection along the coast, which Tom Molloy wanted to restrict to 30 percent of the coastline; access routes to Crown lands; and selections in and around communities, territorial parks, wildlife preserves, and the existing (since 1974) Auyuittuq national park preserve. Parks Canada acceded to the wishes of Broughton Island and gave up acreage to them, amounting to 12 percent of the park preserve, for sport hunting of polar bear and other commercial purposes. Inuit rights in parks had been agreed to previously and set out in the AIP, which included the provision that national parks would not be formalized until impact and benefit agreements with local residents were in place after final ratification of the land-claim agreement.[8]

The government relinquished its attempt to exempt outpost camps, since the vast majority were Inuit. However, they insisted on Crown parcels near communities that could be purchased by non-Inuit residents for summer cottages and recreational purposes. This was felt to be a niggardly way of depriving the Inuit of some modest real-estate profits in the future.

Some chinks appeared in the otherwise solid armor of the TFN as a result of a previous agreement on disproportionate quantums. Pangnirtung's was larger than the others, probably because of that settlement's natural aggressiveness and Johnnie Mikes' strong personality. The squabble was solved by some shifting of acreage during the meeting. Both Lucassi Ivvalu and Paul Quassa tried to explain to participants that the land would be held in the name of all Inuit irrespective of location, but that view was regarded as a "hard sell" back home in Lake Harbour. In fact, the misconception about the negotiable nature of the land selections that exceeded the quantum was a sore point there and in all communities. Coming back home with a smaller bundle of land parcels was worrisome to the CLINTs, and they expressed concern about whether they could sell the whole land-claim agreement to their constituencies. The final agreement and the land package were to be voted on by all Inuit before ratification. This was going to require a lot of persuasion on the part of Inuit leaders.

The minimal amount of potential for minerals in the whole South Baffin region – and the consequent absence of mineral claims and leases – had their effect on subsurface selections. In the first place, our team ruled out uranium showings in granites and migmatites, which they considered unimportant. Cape Dorset selected about 330 square miles of iron formations, volcanic and ultra-mafic igneous rock on the basis of the GSC mapping that we had compiled. Others had similar but more modest selections. Broughton Island selected some coastal Tertiary basalts and a portion of iron-bearing greywacke sandstone with possible lead–zinc potential in the Piling sedimentary basin, where exploration permits existed. The biggest fuss arose over the Lake Harbour lapis lazuli deposit: it happened to be within the proposed Soper River territorial park, which would render it unrecoverable. My strong feelings about the economic potential of this semiprecious stone infected the carving-minded group, and there was a vigorous fight to hang on. The lapis parcel was reduced in area in the hope of its being excised totally from the boundaries of the park, but the GNWT rejected that plan in favor of a special agreement allowing the Inuit to have park access to exploit the resource. We took pains to ensure that the agreement envisaged a fully commercial venture, with mining machinery and power tools, not just the hand tools that were typical of carving-stone exploitation. I felt the lapis beds would really need underground mining. The efforts of Terry Fenge and Mary Crnkovich were instrumental in hammering out the agreement.

From conversation with Dave Perrin and others, I learned that the federal negotiators had come to realize the enormous surveying obligation inherent in their "no claims or leases" position. They were moving towards a policy of allowing representative selections of third-party lands in the Keewatin, a crucial region for minerals. They would also strive to see that subsurface-block boundaries did not cut through actual claims. In these negotiations, the South Baffin representatives agreed to donate around four hundred square miles of subsurface quantum to the Keewatin region.

BREAKTHROUGH AT RESOLUTE BAY

I wrote a working paper on subsurface land strategy during January 1991, in preparation for the upcoming negotiations in areas with much more industry interest and activity. The 14,000 square miles of subsurface rights allotted to the Inuit amounted to 2 percent of the Nunavut area, so the Crown with its 98 percent of mineral lands could afford the risk of low-potential land covered by granite. We were therefore resolved to guide the Inuit selections so as to conform with the best possible geology, such as volcanic belts and gold-bearing iron formations. We hoped this kind of selection strategy would provide for long-term gain. Nevertheless, the federal "no claims" policy in this era of high claim-staking would leave the Inuit with lands that were only peripheral to the high-potential areas with showings and deposits, despite the idealism of the AIP. So it appeared that the Inuit might not achieve any short- or medium-term mineral opportunities. Consequently, our parallel tactic was to surround the existing mineral claims denied to us and secure surface rights over them — thereby controlling exploration and development access to them except where Crown land-use permits existed. I was convinced that the federal government would not get any applause from industry for this state of affairs, because mineral explorers would face an administrative mess as well as a disenchanted Native population. The latter would be uninvolved, having no stake in minerals themselves, but they would nevertheless "hold the hammer," so to speak. The government would probably have to conduct legal surveys of all mineral claims, an unheard-of procedure, to ensure the integrity of the Crown domain. Given the rapid turnover of most mineral claims, the resulting checkerboard of interests would boggle the mind and possibly deter prospecting activity. These tactics were received and appreciated by the TFN machine. They were less enthusiastic about my compromise suggestion that we avoid mineral leases if we could choose mineral claims.

Engaging sympathetic members of industry to put in a good word for us was an idea suggested by TFN leaders. Cominco was a logical choice because we would soon be going to Resolute Bay, a community that was close to the Polaris mine. My own public-relations effort with Cominco endeavoring to secure a letter of consent proved useless, however. I finally talked by telephone with Vice-president Dave Johnston to describe the Inuit land-claim status. He showed little interest in getting involved, and thought the Cominco people would pay no attention to the Resolute Bay meeting. He wanted Inuit to recognize mining risks and not expect rewards without investment. Perhaps it was his knowledge of unsuccessful discussions with the Inuit Development Corporation a decade earlier, on the subject of participation in Polaris, that colored his comments.

Before the Resolute Bay meeting, a side issue arose when federal bureaucrats belatedly discovered a conflict over the Bylot Island national park preserve. On January 23, 1991, I flew to Pond Inlet, along with Charlie Jefferson and Wayne Wagner of the Geological Survey of Canada and the other TFN advisors. The government geologists were along to discuss the Bylot Island park preserve, not as consultants to the federal negotiators. The Department of Energy, Mines and Resources and the Government of the Northwest Territories had called the meeting to discuss a modification to the park boundary with the people of Pond Inlet. The modification had been recommended by the intergovernmental Mineral and Energy Resource Assessment (MERA) committee; it was a requirement of Parks Canada policy that such an assessment be undertaken before any lands were withdrawn and designated under the Act.[9] Jefferson, Wagner, and Don Brown proposed to exclude the South Mala River drainage from the park, as it contained a structural basin of the ore-hosting Society Cliffs formation. This deeply buried geology was an extension of the rock formation hosting the Nanisivik mine to the southwest, but no substantive information about new discoveries or mineral claims was brought forward. Their position was that the basin comprised favorable geology that might need to be explored in the future. Indeed, the MERA committee had a good point: national parks should not include mineralogically prospective areas. But they were tardy in coming forward with the finding, owing to their learning of the park proposal only recently, or to a failure in internal communications.

The Inuit negotiating team, which included the TFN reps, the mayor, and the fish-and-game committee, rejected the federal proposal to trim the park and terminated the meeting summarily. Actually, local Inuit had previously discussed the issue with Bob Gamble of Parks Canada, who had arrived in Pond Inlet several days earlier. The motivation for their decision derived from the AIP, which allowed for a later change to negotiated park boundaries with land options for the Inuit. They were not prepared to re-open the negotiation of land ownership at this time. The incident highlighted the inferior liaison that mineral people, including the GSC, have with Native communities in comparison to that enjoyed by wildlife and park representatives. The Pond Inlet people had tried in land negotiations to acquire an old Petro-Canada lead–zinc prospect in this general area, but the land was simply swallowed up in the Bylot park preserve.

We flew on to Resolute Bay, enjoying, though we were engulfed in winter darkness, the marvelous views of open water leads along Lancaster Sound. But no one spotted any polar bears on the ice as we stared from the Twin Otter windows.

The major work of the Resolute Bay meeting was to resume negotiations on outstanding items left on the table in July. The federal–GNWT and TFN negotiators were present, along with CLINTs from Resolute Bay, Arctic Bay,

Pond Inlet, Igloolik, and Hall Beach. The items on the agenda mostly revolved around the issue of subsurface quantum selections. Since this was a stubborn and vital issue, the federal team included Dave Nutter, director of DIAND's Minerals and Economic Analysis Branch. The Inuit team was headed by Bob Kadlun, vice-president of TFN from Coppermine, and a long-standing maintable negotiator. We had the usual complement of advisors, including Terry Fenge, Bruce Gillies, Mary Crnkovich, and me.

Tom Molloy opened the discussions with a statement on the federal negotiating policy, which included a number of new criteria to guide the Inuit acquisition of third-party mineral interests. He admitted that criteria were still being developed, as they might apply to "improved properties" in general. Here, their criteria related specifically to the producing lead-zinc mine at Polaris. They declared they would exclude from selection any abutting and adjacent parcels that included ore reserves, to sustain the mining operation. In addition, they would exclude "supporting properties" nearby that might include truckable ores. He urged Inuit representatives to define their blocks so that they contained all the local claims in order to eliminate survey complications. If complications did arise, the Inuit were expected to abide by the dispute mechanisms of the Canadian Mining Regulations. This was a real breakthrough for mineral-rights acquisition, although the government was obviously determined to retain the known ore bodies in Crown hands – a potential point of conflict. This precedent would be of enormous value in the Keewatin and elsewhere.

Norman Williams dealt with the substance of the federal position by referring to specific maps on the wall. Selection problems in other contested areas were eliminated, except for an iron-ore claim at Roche Bay near Hall Beach, contentious because it covered sea floor, and one out of three iron-ore leases at Mary River – the one containing the ore body and old camp. The critical Resolute Bay area was different. The government excluded Cominco leases at Truro Island and Eclipse, as well as nearly all the properties along the north and west coasts of Cornwallis Island (and also some on the northern peninsulas) because they lapped onto sea floor. Inland properties on Cornwallis Island, including some recorded in September 1990, were deemed acceptable.

The instant reaction of the Inuit, eloquently expressed by Bob Kadlun, was that Inuit were only granted the leftovers. I had to agree that the new criteria were onerous and left us only the crumbs in terms of known or potential minerals, scarcely any of the favorable Thumb Mountain formation. According to my calculations, of all the Cominco claims and leases in the area, 52 percent had been added in September 1990 in areas of Inuit selection interest.

No wonder the mining people were not coming to the meeting! Some youths, however, came in to observe the discussions, no doubt at the urging of their schoolteachers. At this stage, one of them said to me, when I inquired about her interest, that "land claims suck." I thought they would have their eyes opened yet to the importance of land claims.

In the next round, during which we felt quite depressed about our prospects, the TFN held onto the Eclipse property and everything else except Truro Island and even transferred some useless acreage from Little Cornwallis Island to cover the Rook group and a new block on Cornwallis Island. As to the Mary River leases on Baffin Island, we proposed to switch blocks to get the main ore body for ourselves.

There was much rhetoric exchanged in the next joint session, and we could appreciate the energy and dedication of Bob Kadlun and others. The discussion revolved around our mere 2 percent and the greed exhibited by the feds, with their 98 percent, in hanging onto the best properties. We said that, in view of their known size, the satellite mining properties would not generate any royalties: they would be mined out during the existing three-year royalty holiday for new mines; they therefore had no present value in that sense to the federal government. Whether we, as owners, could escape the royalty holiday while honoring the Canada Mining Regulations was not brought up. We stressed the fact that due to our small entitlement, we were impelled to go after known mineral occurrences. Those occurrences would be the focus of activity in the near term and of the discovery of new deposits in the medium term. Of course we were motivated with our own kind of acquisitiveness. In this and other exchanges, I had to respect the professionalism and lack of rancor of the federal team: they were tenacious, but they walked the moral high ground.

Finally, in a much-awaited breakthrough, the government granted our major selections on Cornwallis Island except for the Rook claims, where a previously agreed corridor would have been violated – provided we would stand off from the Eclipse deposit. They also acceded to our demand for a switch of blocks at Mary River, as their only concern was the cleanup of the old camp, not the ore body.

At this point, the Resolute Bay people, headed by Allie Salluviniq, reconsidered their position in caucus. The Eclipse deposit was a key item, and they had strong feelings about it. My advice was that even without it, Inuit now had a strong land position within their grasp. It included known base-metal showings on Cornwallis Island along an important geological trend, linking Thumb Mountain properties, that Cominco was actively exploring. Returning to the table, they agreed to the scheme, providing they were allowed to transfer the Eclipse acreage and consolidate around other claims, and also retain the Rook by allowing a narrow corridor south of these claims. They also insisted on retaining surface rights over the Eclipse deposit and Truro Island. At this point Tom Molloy, seemingly a good poker player himself, negotiated for a slight opening up of two nearby blocks of subsurface to allow public access to Crown land in the interior of Cornwallis Island. The two combatants stood up and shook hands. We had a general celebration, knowing that Inuit had made a major breakthrough here that would enable them to achieve big things in Keewatin and Kitikmeot. The Inuit were jubilant and called for an impromptu dance. The student I had spoken to earlier said she had never seen anything like it! We all knew that nothing could have been gained without a fight and without the firm determination of Allie Salluviniq.

Salluviniq drove me from the village back to the motel, pointing out the vacant place along the shore where the original group of displaced Inuit had landed in 1952. I asked if the government put up some housing there before the families were brought. He said, "Are you kidding? We were dumped off in October, when there wasn't even good snow for igloos. If we hadn't searched the garbage at the military base, we wouldn't have made it through the winter." There was something very special about Salluviniq and his fellow survivors at Resolute Bay. He did not come across so much as a victim but as a very proud man. I felt happy indeed for their sakes that victory had come at the land-claim table: they had established the principle that Native claimants could acquire ownership of third-party staked ground, and they themselves had acquired for the Inuit about 523 square miles of mineral rights in the central Arctic's prominent lead–zinc mineral district. The other community participants in this process were Ekaksak Amagoalik, Simeonie Amaroalik, and Paul Amaroalik.

The whole experience at Resolute was gratifying – it strengthened my interest in the North and its people. It was a springboard for launching a productive campaign of mineral-rights selections in the rest of Nunavut, one that would challenge my skills. As for my impressions of Resolute Bay, I had never experienced such biting cold or tramped over such cement-like snow drifts. It was a fitting place for winter military maneuvers, to bring our young soldiers up to the standards of the Inuit. I found out that the hunters and trappers' society there had a good stock of arctic char, so I bought some to remind me of this place. But I delayed my departure to join Dave Nutter on a trip over to Polaris, where we were given a plant tour by mine manager Tony Keane. The next day, we went under ground with geologist Bob Sharp and marveled at this rich, carbonate-hosted lead–zinc deposit, a model for Arctic mining conditions. Besides geology, we had a chance to talk about Native hiring, apprentice programs with more lenient Territorial requirements, and land-claim matters in general.

NEGOTIATIONS CONTINUE AT CORAL HARBOUR

The next encounter took place at Coral Harbour on Southampton Island, on March 4–11, 1991. Both Murray Pyke and I attended this series of negotiations between the federal and TFN sides to ratify the lands selected in the North Keewatin district. This included the communities of Coral Harbour, Repulse Bay, Baker Lake, and Chesterfield Inlet, each of which had made its preliminary selections to 150 percent of quantum values. The meeting was a crucial one because it saw the appearance of full federal guidelines intended for the selection of mineral properties. There was apprehension on the government side, lest the Inuit secure all the good prospects. Also, the GNWT was concerned about the dwindling of the Crown resource-base, which it might inherit once the planned devolution of powers took place. So, in addition to the regular team of seven, there were two geologists present: Steve Goff of DIAND and Michael Cunningham of the GNWT's Department of Energy, Mines and Petroleum Resources. Ross McKinnon was at the table as GNWT advisor on parks and municipal matters.

The twenty-three-member TFN team, headed by Paul Quassa and Lucassi Ivvalu, included a number of outstanding and experienced Inuit leaders, as well as talented community representatives. There were Donat Milortuk from Repulse Bay, Michael Amarook from Baker Lake, Tagak Curley from Rankin Inlet, David Aglukark from Arviat, and Raymond Ningeocheak from Coral Harbour. Tongola Sandy was there as Keewatin land-identification coordinator, as was his counterpart from North Baffin, Malachi Arreak. Bruce Gillies, Mary Crnkovich, Murray Pyke and I were among the support staff as advisors, along with interpreters and a writer.

Initially, there was much discussion of federal criteria for appraising third-party mineral interests. The government was essentially retaining "the mines, adjacent properties and advanced properties," each characterized by exploration work, reserves, feasibility studies, government infrastructure, and location. Right away, Paul Quassa and Tagak Curley charged that these criteria were a violation of the AIP, which mentions nothing of the kind. Tom Molloy assured the gathering that these were simply rules for their own guidance. I wondered whether the criteria would have much application here. Were the feds constructing a plan for the bigger battles in South Keewatin and Kitikmeot?

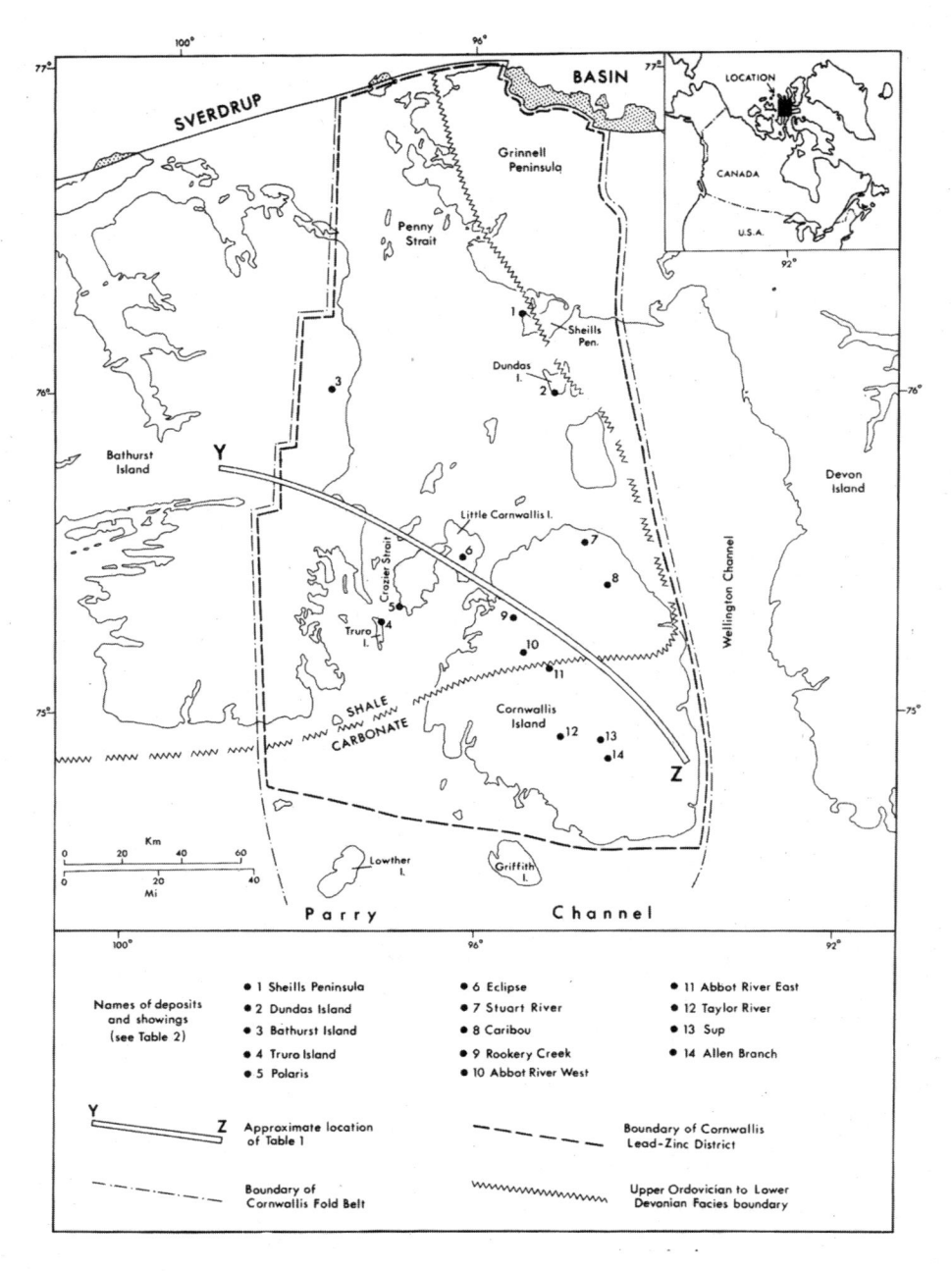

Map 1 Mineral deposits and showings in the Cornwallis lead-zinc district. After J. W. Kerr, Canadian Journal of Earth Science, Vol. 14, 1977.

244 · *New Owners in Their Own Land*

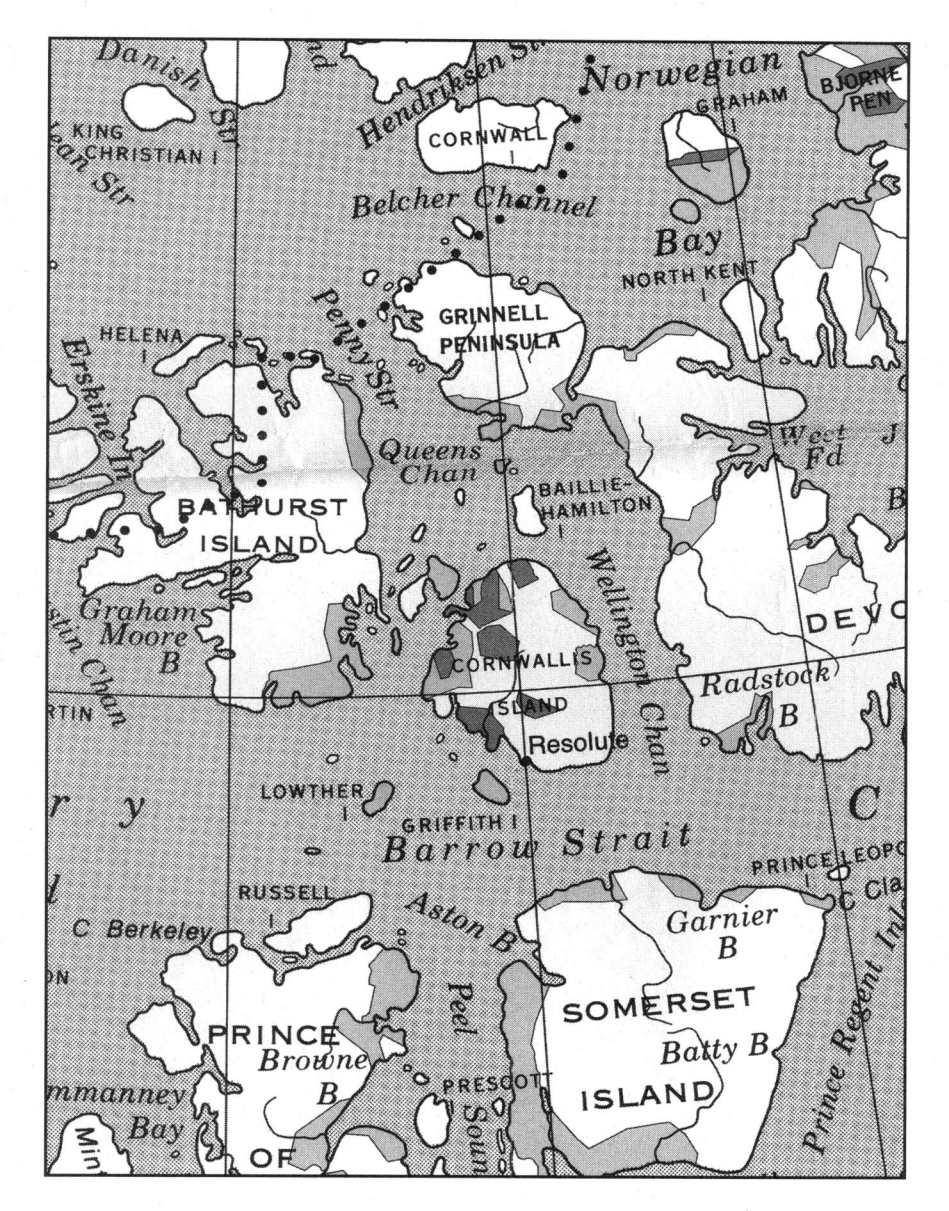

Map 2 Inuit-owned lands near Resolute, showing surface only (lighter shade), surface and subsurface (darker shade), from map of Nunavut Territory courtesy Nunavut Tunngavik Incorporated.

The first response by the government was almost wholly a criticism of the solid-block nature of the Inuit selections covering the coast and the important waterway of Chesterfield Inlet. Tom Molloy called for a major opening by the TFN – it was too big a job for the feds to say just where – and he threatened to invoke the 75 percent rule if they did not; this was the negotiated rule that binds the government to accept that much of the quantum from within the area of interest selections, though in the past they had always affirmed they would accept 100 percent. After much talk about this in the TFN caucus, we decided, as with the subsurface criteria problem, that we had better wait and see.

The CLINTs set to work and made partial reductions from their 150 percent area-of-interest maps, some of which I thought were done rather hastily. In response, the government produced its own map to link up Crown parcels and provide generous avenues for access, basing the configuration on the first reductions of the Inuit. This led to some consternation over the way the Inuit had set a pattern for government action in this chess game of land selection. Nevertheless, the representatives proceeded with the chipping-away of their areas of interest in ways that ignored the federal map. With each phase, the feds took a long time in their own caucus discussing the maps.

In the final stages, the critical items for the respective communities were as follows. The Coral Harbour people were mostly upset about the Crown blocking the westward growth of the town and denying Inuit any real estate opportunities. Pyke and I advised them to acquire seventy-five square miles of subsurface rights on the Boas oil shale. This extensive sedimentary deposit of Paleozoic-age bitumen outcrops west of the community. We did not concern ourselves here with the question of whether oil shales were legally part of the surface or not. Pyke and I had previously secured the latest information and maps on these rich bituminous deposits from Fari Goodarsi of the GSC in Calgary. The oil shales were uneconomic by today's standards, but a potential energy resource of local importance for the future. I made the suggestion, for what it was worth, that a future open pit to recover the oil shale might serve as a repository for the enormous pile of World War II fuel drums that still scarred the landscape near Coral Harbour. In these subsurface decisions of the Inuit, there was an uncomfortable trade-off: they were constrained in all such cases to acquire surface rights to the same area. This meant switching their focus of attention away from areas that were valuable to them for hunting, traditional, or other reasons.

Repulse Bay people were intent on two parcels of land fronting on sea-like Wager Bay, the source of popular interest and considerable eco-tourist potential. The government wanted these parcels removed to make way for a national park preserve. The preserve had been proposed in 1978, but the government had refused formal guarantee of the park. The Inuit persisted, however, and managed to retain these rich areas, with a caveat to consider a land exchange later if a park should materialize. Even if it did, some of the Inuit had doubts that they could call it their own. The same group selected and won about three hundred square miles of subsurface rights along the Penryn Precambrian rocks on Melville Peninsula, which had known lead–zinc potential. To secure this, they had to assume cleanup responsibility at the Borealis Exploration camp, with which they were familiar. Norman Williams said Land Use had been pursuing Borealis for two years and would make another effort to pressure them to clean up. He suggested that a cleanup deal might possibly be arranged with the Inuit over this matter, as certain company assets, like all-weather tents, had been left at the site.

Chesterfield people were mainly concerned with some key locations along the navigable water of Chesterfield Inlet. One parcel that was dropped was quickly re-taken by neighboring Baker Lake, which infuriated Barry Dewar and other federal negotiators. Finally, after much dickering by Andre Tautu over land at the entrance of the Inlet, the selections and broad access ways were adjusted to the acceptance of both parties. Along the 150-km Inlet, Cross Bay was most appreciated as a possible deepwater port to service potential zinc ores to the south (Sand Hill property was acquired by Inuit) to be shared between Inuit and government. The strategic importance of Cross Bay for trans-shipping originally came to light during the Kiggavik uranium mine feasibility study.

Baker Lake, the last community to settle, held four or five critical areas on the lake and along Chesterfield Inlet: the Christopher Island–Cross Bay section. Much time elapsed, and phone calls were made to their elder constituency back home about giving up these blocks along the shore. A bottom-line speech was made by Michael Amarook, formerly president of ITC, whose eloquence and demeanor were riveting. He explained that the Inuit claimed the Kiggavik uranium mine because of grave sites and occupation history, referring to the hunting camps downwind, along the Pointer Lake–Judge Sissons drainage, as described in the 1990 environmental hearings. He stressed their attachment to and dependency on these sites and others along the Inlet once the white fox prices were on the decline. Stories were then related of encounters with polar bears and consequent graves of forebears. Pyke and I were touched and deeply convinced that some of these lands were so treasured by Inuit they should not be traded off for any block of subsurface rights. The case in point was the

ML31 block west of Baker Lake, covering the PNC uranium property, a sister deposit of the infamous Kiggavik deposit of Urangesellschaft. Throughout the negotiations, Baker Lake really held onto the Kiggavik block with a view to relinquishing it in order to keep ML31. Now the latter became a bargaining chip in the negotiations, although Tom Molloy disavowed any horse-trading in these proceedings.

Molloy maintained, however, that no consideration of the Kiggavik area would be made until five concessions were wrested from the Inuit to provide access corridors into Crown land north, south, and west of the lake, and also to Kazan Falls. These precise areas were specified as a bottom-line condition for the government. The Baker Lake representatives, after making a telephone call to elders back home, played their cards adeptly by reducing some critical blocks of surface, for example near Cross Bay, along the Inlet, and at the west end of the lake. These were cautiously made, on condition that ML31 went to them. They succeeded in this, so we felt elated with the result, and astonished, too, that they had flipped the ardently contested land for a mineral property. Baker Lake also secured most of the Archean rocks with gold potential and surface rights around the uranium mine, as well as the Thelon dolomite occurrence that might be mined for lime if Kiggavik ever went into production. I felt pleased that they acquired the important subsurface block around Meadowbank and Tehek Lakes, which Pyke knew about personally from gold exploration on the area's iron formations. Having a stake in mineral rights there would encourage the Inuit to be more receptive to an access road from Baker Lake, which they had unanimously rejected for the Polar Gas pipeline proposal fifteen years earlier.

These results, especially the procuring of an advanced uranium property and a substantial holding of surface rights, will astonish members of the mining industry who are familiar with the decades of opposition by Baker Lake people (and Keewatin residents in general) to mineral exploration and uranium mining. Their anti-nuclear stand was primarily responsible for frustrating the development of Kiggavik at a time when that project appeared economically feasible. Why would they now acquire ownership of another uranium deposit? Amarook laconically said his people wanted to be in on the action if the mine ever went ahead. Another consideration was the desire to have a monitoring window on industrial activity, even if for the time being that was just government environmental studies around the site. Perhaps the negativism of the past was mellowing with the realization that Inuit were gaining real power and more confidence in environmental processes achieved through their land-claim agreement. While I heard anti-nuclear sentiment being expressed from time to time, preventing development through ownership was never a goal.

This series of negotiations witnessed a see-sawing of positions that ended with both sides essentially getting what they were after. Pyke and I were again impressed with how mineral lands (totaling some 1,150 square miles) became bargaining chips in negotiations. Both sides understood this, and the federal side did not seem to begrudge what they were giving up because our demands had been reasonable. The Inuit quickly learned the ins and outs of the negotiating process. They took pains to understand what it was the other side really wanted and to use this in their bargaining strategy. However, there was one new lesson learned here about opening moves that might prejudice the ultimate result: the government fixated on early Inuit concessions in order to build the substantial system of corridors between Crown lands. Inuit negotiators needed to be wary in future of setting up a negative trend. The federal position on mineral lands previously staked by third parties certainly evolved, from one of exclusion to one of sharing resources. This was opening a clear opportunity for Inuit in the Keewatin and Kitikmeot regions, in our opinion. It was going to challenge our professional skills to settle on the right priorities.

It was a fine, clear morning when we flew out of Coral Harbour, on a route that took us through Repulse and other communities on the way to Rankin Inlet. We talked with Tagak Curley, who pointed out some features on Wager Bay, a place that was dear to him and always worthy of a Ski-Doo trip. Time allowed us to visit Scott Mitchell at the Land Use office in Rankin Inlet. He briefed us on proposals for the cleanup of nickel tailings in town, necessary to make way for community expansion. He also showed us a video of the World War II barrel dumps near Coral Harbour, perhaps the biggest environmental mess in the Arctic.

NEGOTIATIONS BREAK DOWN IN RANKIN INLET

The first South Keewatin meeting was staged at Rankin Inlet during the period April 16–23, 1991, just after a flight-delaying blizzard and successful search for some missing persons. Phil Mudry and I attended as mineral advisors to the TFN. Though Mudry was already knowledgeable about the Keewatin region from years of exploration, we had gone over the economic geology thoroughly and studied the recent activity and claim stakings. A very strong team assembled on behalf of the Inuit communities of Rankin Inlet, Baker Lake, Whale Cove, and Arviat. Some of the notable leaders were Paul

Quassa, Thomas Suluk, Tagak Curley, Bob Kadlun, Michael Amarook, David Aglukark, Louie Pilakapsi of the KIA, and Paul Kaludjak, the mayor of Rankin Inlet. Tongola Sandy coordinated the twelve community (CLINT) representatives. The women present included Bernadette Makpah; Robert Tatty's wife, Annie; writer Looee Okalik; interpreters Martha Flaherty and Mary Nashook; and lawyer Mary Crnkovich. Terry Fenge and Bruce Gillies were also present. Altogether, the Inuit team numbered thirty-five persons. The government team, led by Tom Molloy and Ross McKinnon of the GNWT, numbered eleven persons. Norm Williams and Dave Perrin provided technical support, and Michael Cunningham gave geological advice.

The purpose of the meetings was for the communities of this subregion to negotiate for their respective land quantums, totaling 15,800 square miles of land including 4060 square miles with subsurface rights. It should be noted that the ratio of subsurface to surface for Nunavut as a whole was 10 percent, but here in the Keewatin the figure was 25 percent, owing to transfers from elsewhere. Therefore, we faced a daunting task of selecting high-potential subsurface areas without antagonizing the other side.

The first TFN caucus was a strategy session and review of the numbers. Mudry and I presented our own map of the whole region so everybody could appreciate the broad picture. We pointed out all of the deposits and important mineral showings present in this well-endowed region. The Kaminak volcanic belt was compared to the Abitibi belt of Ontario and Quebec in terms of its gold potential. We suggested a strategy of land elimination in the first round to prevent the government from achieving any strong land corridors that would block important parcels of subsurface from Inuit acquisition. The Meliadine River gold belt was a good example, lying as it did along the Rankin–Baker Lake transportation route that they would no doubt have their eyes on. There was good support for this idea from Tagak Curley, Tongola Sandy, and others, and we were urged to keep the CLINTs informed of high-potential areas as the negotiations progressed.

In the first joint session, after Paul Kaludjak and Louie Pilakapsi had welcomed all, Paul Quassa noted the close proximity of the communities and their overlapping interests, the prominence of the much-used coastal area, and the large subsurface quantum to fill. Tom Molloy, whose side had already viewed the preliminary 150 percent area-of-interest maps, said there were many obstacles to meeting government needs during the reductions necessary to reach the quantums. As to subsurface lands, he observed that Inuit had selected two existing mines and virtually all mining claims and leases. He felt the mines and the majority of claims should be held by the Crown, but the Inuit selection should be "representative" of this high-potential area. The Crown would need to have a presence along the coast and corridors more than a few miles wide around the communities, to link Crown lands. If need be, the 75-percent rule would be invoked. Molloy also stated the government intention of protecting Saskatchewan Indians' hunting rights around Ennadai Lake; he obviously touched a sensitive nerve, as many Keewatin Inuit are especially attached to that ancestral land and regularly travel there for wolf hunting.

The government's own corridor map was presented and defended by Norm Williams. It was bitterly attacked by the TFN, as it blocked most of the good subsurface parcels and garnered 50 percent of the coastline. Tagak Curley said, "This scenario is segregational – we don't want to be put on reserves!" Bob Kadlun referred to it as a pure land-grab by which the Inuit, a coastal people, would be pushed inland and separated from any economic potential. Paul Quassa said the government was expressing no faith in the principles of the AIP and urged CLINTs to ignore the federal map totally (which they did). Nevertheless, the first round of surface cuts seemed a little over-zealous to me. Rankin reduced its selection by 32 percent, Whale Cove by 38 percent, and Arviat by 21 percent. Louie Bruce and others insisted that Arviat retain the Heninga base-metal deposit along a proposed corridor, but the Cullaton Lake gold-mine site and leases were given up. (The mine had previously been worked out and closed.)

The federal reaction on April 19 dealt with subsurface selections in a significant way. Molloy said that their side needed some economic potential. He suggested that in the Baffin region, a ratio of one-third/two-thirds (one-third for us, and two-thirds for them) would be satisfactory, and our side could have the first choice. This drew a lot of Inuit heckling, since the fact that this split was not allowed us in North Baffin at the start had seriously affected the result. Molloy admitted that some of the land corridors outlined here were designed to retain Crown mineral lands. The government was intent on preserving possible royalty income from mines, although it was pointed out that revenue comes primarily from income taxes. We had already made some quality concessions, but the federal negotiators questioned the merits of some subsurface parcels. Tweaked by our knowledge of greenstone belts in the Keewatin, they asked to borrow our map. What followed was a federal subsurface overlay map suggesting blocks to decide upon. They were apparently leaving the first choice to us, so long as we fell within their ratio parameters. Mudry and I thought we could work with this and select the best ground, given a free choice of the first one-third.

Dealing first with Arviat, there was a lot of dickering about the boundaries of corridors in an attempt to retain traditional travel routes as Inuit land. The federal side wanted more surface coverage north of Heninga Lake but did not quibble about our retaining mineral leases on the south side of the lake. As these were "water claims," I felt that an important concession had been made, here. (Remember that water claims were specifically excluded from Inuit ownership in North Baffin.) We said that, in the light of the government's wanting more surface to the north, we would be interested in a subsurface position on both sides of the lake, thus ensuring us the 6 million ton zinc–copper ore body. Actually, these corridor lands became a bargaining chip in negotiations over surface blocks west of the community and at Padlei, the location of outpost hunting camps. Third-party lots at Ennadai Lake were a concern: though this area was distant, the Arviat people had sentimental and historical attachment to it. The feds were bent on excluding any land acquisition there to forestall lawsuits that might be launched by the Chipewyan Indians. By previous agreement, both the Inuit and the Indians had hunting rights in that area of overlapping occupation.

Speeches were made by Luke Suluk and others about the economic importance of subsurface entitlements and the goal of economic self-sufficiency proclaimed in the Agreement-in-Principle. Molloy said he wanted us to have a preferred choice in the matter, provided we remained within an acceptable range. Mudry and I had observed this range to be around 18 percent of existing claims, judging from their own map of the day before, and we were using this number in plotting selections for Arviat. The government agreed to these selections of 1070 square miles, but this was short of quantum, necessitating our return to uncontested areas, such as those south of Cullaton Lake. In the end, Arviat acquired subsurface rights to 1257.4 square miles.

Next in line for detailed discussion was Rankin Inlet, where the federal citations were rather devastating, especially along the coast, on Marble Island, and along the Ferguson River. Since the CLINTs were already close to quantum, they decided to shift a block of surface for a better location on Kaminak Lake. Quassa described the change as regrettable but necessary. Molloy adopted an angry tone over the change, and said we were denying the government any coastline and blocking their needs at Kaminak Lake. At this juncture, Paul Quassa, our accommodating and polite leader, had to leave the meeting to catch a plane. The negotiations continued, and even intensified, without him.

In caucus, there was a long discussion of the traditional value and quasi-religious significance of Marble Island and portions of the coast. To this day, as Robert Tatty and Jose Kusugak attested, there was a tradition among visiting Inuit who ascended the beach at Marble Island, to go on their knees and elbows to mimic the starving whalers who perished there long ago. Mudry and I showed the CLINTs an alternative routing for a corridor that would preserve some subsurface, including the desired parcels on Kaminak Lake.

When we met in joint session, Curley and Kadlun both stated the importance of coastal lands to community elders, and John Arnalukjuark made a stirring speech about life on the land. Molloy said we had started to address government needs and consequently they revised their position. He meant they would back off east of Rankin Inlet, along the coast, but needed a land concession to the west to maintain linkage with the Cross Bay route to Chesterfield Inlet. There was a slight modification of federal demands south of Rankin, too, but a flood of speakers came forward to object. Elders and women who fish the Diana River and hunt near town raised their voices in opposition. The Rankin people drew a narrow corridor to access Cross Bay that did not relinquish subsurface, and they decided to exchange a parcel for a string of islands and a piece of coast at the mouth of the Diana River. The latter created an unacceptable break in the otherwise continuous strip of coast retained by the Crown, they said they wanted the Bissett Lake block, which Rankin shared with Baker Lake, to consolidate Crown holdings and access Kazan Falls. A tenuous deal seemed close at hand as we broke for caucus.

In the next session, Paul Kaludjak suggested trading the Bissett block for a piece of coast south of town, recognizing the "grey power" now being exerted by the community elders. Their intervention was a new element in the negotiations, brought about by a realization that their interest (in nearby fishing) was being overlooked. The Diana River mouth parcel of land, which the Inuit had previously given up, had become a make-or-break item in the negotiations. Molloy said the blocks suggested by Kaludjak were not tradable, but he was prepared to give up Marble Island – his last chip. The Rankin people said their backs were to the wall, but the feds were adamant about retaining an uninterrupted strip of land along the coast. Molloy said he feared negotiations were slipping off the rails. He had looked at possible ways of breaking the impasse and suggested splitting the parcel, to give the Inuit the east side of the river connected to the islands, but he insisted on restoring the south side of the block to its original, expanded configuration. He urged Rankin Inlet people to take a hard look at his proposal. Both Kadlun and Curley objected to the proposal, so they suggested adjournment and discussion of the Whale Cove selections the following day. Molloy indicated that no agreement seemed possible, and the government side would be leaving in two days. He seemed to be feeling a sense of impatience over the transfer of parcels between communities, and perhaps frustration because he had played his cards too soon.

The next day, Molloy noted the critical wedge of Whale Cove land between the other two communities. He said the Crown wanted one-third of the coast and the mineral lands with potential, and called for an immediate response on this basis. Mudry and I saw no problem meeting the subsurface requirement and wanted only to apply some of the credits to cover the important Meliadine gold trend back in the Rankin Inlet area. To achieve the overall four-thousand-square-mile quantum, we would also have to select some of the subsurface allotment from areas with geological potential at Kaminuriak and Cullaton Lakes, where we already had surface rights.

Inuit spokesman Andy Kowtuk appeared confident but firm. He was getting a fax message from some elders at Whale Cove on the coastal question. A busy caucus meeting took place, and then we reconvened to present the results. Coastal concessions had been made and they merged with a hundred-square-mile Crown corridor running inland. Admittedly, it was configured in such a way as to protect the Kaminak Lake parcel that had been retaken earlier and caused so much debate. There was now one-third open coastline, including the peninsula comprising the community land transfer. I had never myself heard the feds say that municipal lands were not to be included in their coastal demands, but certainly without these the Inuit offer fell short of the required one-third. Molloy looked over the map and requested confirmation of the latest deletions, which was provided to him. He and his team abruptly left the table with the words: "You have my telephone number!"

That the meeting broke up at this point was quite a setback for the negotiations, in my opinion. The Inuit seemed reconciled to it and firm in their resolve not to be rushed into decisions with such lasting consequences. Unfortunately, Paul Quassa was not there to suggest a possible compromise, to take a second look at things.

As for the future, I could see no difficulty in acquiring a first-class subsurface entitlement for the Inuit of the Keewatin. If the new federal stance on minerals had existed from the beginning, the North Baffin region – and Arctic Bay in particular – would have fared better. In fact, some of their subsurface credits might not have been transferred out, to put more pressure on the highly encumbered Keewatin mineral resource lands. One could say that the original stance was counterproductive from the federal perspective; logically, they should have promoted a wide distribution of subsurface selections throughout Nunavut.

Inuit gains had indeed reflected the tenets of the Agreement-in-Principle, that Inuit lands should promote economic self-sufficiency through time and are expected to include areas of value related to nonrenewable resource development, including areas of known or potential mineral deposits. Even a casual look at the whole Keewatin subsurface portfolio has to impress the observer. The Inuit acquired substantial blocks of the following: Woodburn komatiite volcanics around Tehek Lake, including the Three Portage (Meadowbank Lake) gold deposit; the Amer Group around the fringe of the Thelon sandstone, including one of the major (PNC) uranium deposits; the Meliadine River fault trend, including several high-grade gold occurrences in iron formations; the Cross Bay greenstone belt, including the Sandhill zinc–copper discovery; Dubawnt sandstone near Yathkyed Lake, including the YU pitchblende (uranium) vein; the Kaminak volcanics, including the gold veins on Pork Peninsula, copper showings at Carr Lake, gold showings at Turquetil Lake, and base-metal deposit at Heninga Lake; Hurwitz Group quartzites, including known gold showings around Cullaton Lake. Altogether, these Keewatin holdings of over five thousand square miles implied that there would be a huge involvement of the Inuit with mine developers in the future.

NEGOTIATORS TRY AGAIN AT RANKIN INLET

The same group reassembled at Rankin Inlet two months later, on June 23, 1991. I met Phil Mudry at the Siniktarvik Hotel and found him ruddy-faced from his field work on the nearby Meliadine gold property. Paul Quassa was not in attendance at first, leaving Bob Kadlun, Louie Pilakapsi, and Tagak Curley to lead the charge. The first day of the negotiations showed little progress, with both sides laboring to justifying their positions. The federal side under Tom Molloy was not withdrawing specific lands from the Inuit land selection process, but they said they wanted one-third of the whole area between Whale Cove and Pistol Bay, and they spelled it out that municipal land was *not* to be included in Crown-owned coast. The feds were clearly concerned about the geographical distribution of lands, and wanted to achieve a representative share that would show favorably on the map. They were less concerned about the specifics. The Whale Cove people were, unfortunately, squeezed between the blocks already selected by the other two coastal communities; they had limited options because of their position as the last party to negotiate. We talked to Andy Kowtuk and others about widening the Crown corridor in such a way as to minimize the subsurface loss. Otherwise, this change could leave us short of our quantum. One idea was to trade a wider Crown corridor for a better position on the south side of the Copperneedle River.

The proceedings allowed us some free time and enabled me to take a good look at the nearby nickel-bearing serpentinite. No mention had been made of the old North Rankin nickel property, because this remained entirely within the settlement boundaries. Since all municipal lands had mineral rights vested in the Crown, Crown ownership of this parcel would not change its subsurface status. We made a short pleasure trip to Kujarik Park on the Meliadine River in Tongola Sandy's jeep. This was now a popular camping spot, where people enjoyed char fishing in the shallow rapids, and apparently it had been a popular spot for a long time, because we saw numerous tent rings and ancient pits that made up this archeological site. Rankin Inlet was not a very old community, but ancient Inuit seal hunters had sometimes camped at these inland rapids.

The Rankin Inlet people made some progress at the table: they accepted the Diana River compromise, and they also agreed to let Whale Cove draw some land from their area. The feds made some minor moves, but held to their basic position. On June 26, Kowtuk presented a new corridor map, asserting that this was their bottom line. Molloy rejected the proposal rather abruptly, and also took issue with the reassigned parcel south of Copperneedle River and the acquisition of a portage block across the Kazan River at Yathkyed Lake.

Mudry and I busied ourselves trying to assign the four hundred square miles of subsurface still undecided. We were in danger of losing more, since Inuit claims now stood at 30 percent, by exceeding the third-party rule-of-thumb (originally one-third for us, two-thirds for them). The subsurface selection numbers stood at 1,450 square miles for Arviat, 1,080 for Rankin, and 1,275 for Whale Cove. We were trying to minimize the taking of claims in the Whale Cove area by selecting from open, unencumbered ground, but we found ourselves within the contested surface area. The feds were taking their time to consider our Whale Cove plotting, and Mike Cunningham was trying to pick our brains on the geology that lay behind our choices. During the night session, the federal side offered to raise the mineral claim ratio to 33 percent, but insisted on removing the Copperneedle block from the deal. Our side was hesitant and decided in caucus to play for 38 percent, which the federal side correctly interpreted as indicating that we were in a "subsurface-for-surface" mode.

They came back with a comprehensive plan that would meet all their needs in exchange for a figure of 35 percent. There was then some agonizing and speech-making about resource entitlement to benefit children and grandchildren. Thinking that the Inuit were going to hold out for 36 percent, I tried to say that the offer would yield a fine subsurface endowment and, to be helpful, suggested a transfer of some surface acreage to the McConnell River in the Arviat area. Thanks to the cool heads of Quassa, Curley, and Pilakapsi, we accepted the 35 percent, as long as we had a free hand in the subsurface selections. There was much relief on both sides, attended by applause, smiles, and handshakes. The squeeze-play involving Whale Cove had been resolved with even more generous acquisition of mineral prospects in the Keewatin. Mudry and I thought that acquiring 35 percent of existing mineral claims, with first choice, was a very satisfactory way to end the negotiations. Afterwards, we had a conversation with Andy Kowtuk at the airport. He said that he did not want the Copperneedle block anyway – it was only a creek! Playing tough at the bargaining table was not just personal stubbornness on his part: he wanted to do the best possible job for his people back home.

NEGOTIATIONS CONCLUDE AT COPPERMINE

Some testy questions were still unresolved by the TFN, such as the Contwoyto Lake area boundary dispute with the Dene being mediated by John Parker, former NWT Commissioner, and also final agreement on land selections in South Baffin around Pangnirtung. These matters were concluded, and the land-selection process resumed for the remaining Kitikmeot region, the one with which we mineral consultants were least familiar.

I was completing some field work in central Baffin Island and left Iqaluit en route to Coppermine on July 27, 1991. It was a long but interesting flight through Igloolik, Spence Bay, and Cambridge Bay. Jack Kupeuna picked me up and drove me into Coppermine, where I joined the TFN caucus meeting. My colleague Murray Pyke was already there. The Inuit team, led by Paul Quassa and Bob Kadlun, was composed of Stanley Anablak, Malachi Arreak, James Eetoolook of Spence Bay, Joe Otokiak of Cambridge Bay, Joe Allen Evaigotailak of Coppermine, and many more community representatives, as well as Terry Fenge, Bruce Gillies and Mary Crnkovich. Edna Elias (a local educator), Martha Flaherty, and Mona Tiktalik were our capable interpreters. The federal team included Tom Molloy, Ross McKinnon, Barry Dewar, Norm Williams, Dave Perrin, geologist Michael Cunningham, Bernie Karpen of Mineral Lands, and others.

James Eetoolook, president of TFN, made the opening remarks of the Kitikmeot negotiations, the last of such meetings on the schedule. Paul Quassa reviewed the TFN selection record to date, pointing out that the government access routes and coastal requirements had been huge impediments elsewhere but were not likely to be big concerns here. The main items of controversy were going to be Contwoyto Lake and the boundary issue. The newly defined boundary separating Dene and Inuit territory, made by retired Commissioner John Parker, did not sit well with Inuit familiar with this long-debated issue. Bob Kadlun elaborated: The Parker line, which overrode the old 1986 line, was unsatisfactory. Inuit rights had been trampled upon, and the TFN was going to have trouble explaining this to the people. A lot would depend on what interests the Inuit acquired in the Contwoyto Lake area. Tom Molloy stated that the federal team recognized the importance of this area and the crucial role the boundary would have, but they believed an agreement could be achieved on all land selections within the week. The government, which had been pleased with the final resolution of the long-standing boundary issue, announced that an additional 220 square miles of selections would be allowed in the area as a bonus.

The main federal concerns were blocking out a window to Contwoyto Lake for future road access to the Arctic Ocean coast, and maintaining a hundred-foot strip on either side of the boundary line. Some of the coastline concerns were at Chantry Inlet, NE Victoria Island, Cambridge Bay, and Bathurst Inlet. If community lands were to be expanded, for example, at Bloody Falls, the government should be informed now. Specific exclusions for park, conservation, military, and other purposes would be detailed later. Molloy reiterated the government's earnest desire to conclude an agreement subject to certain policy demands that had to be protected. As to subsurface selections, it was expected that the TFN would secure an equitable share of known minerals (existing claims and leases), but the government expected to hold the larger portion. Also, the Crown wished to hold onto existing mines and adjacent property. Quassa said the TFN wanted to acquire 4,100 square miles of subsurface rights in the Kitikmeot region.

In response, Kadlun raised the question of DND destruction of Inuit camps at Anderson Bay, where no compensation was paid. Transportation routes to the Arctic coast were fine, but did the government really have a road-building strategy, or was this just a pipe dream that was controlling policy? He questioned why the majority of land around Contwoyto Lake had to go to the Crown. Molloy said they would get back on the Anderson Bay matter, and pointed out that there were already communications concerning Frey Inlet and Healy Lake, where additional lands would be made available. Joe Trinka, from the policy division of DND, tried to be helpful by reviewing the history of Dew Line operations and the department's present plans for cleanup and downsizing of manned and unmanned sites. He said backup airfields with paved, extended runways would be in use thirty-five days a year for military exercises that would not include any low-level flying. Quassa questioned why the Inuit could not own the small, unmanned radar sites, since they would not hinder maintenance of the facilities. Such ownership would be consistent with Section 17.1 of the Agreement-in-Principle, which stated that the purpose of Inuit settlement lands was to promote economic self-sufficiency through time. This apparently was an old argument, which had reached an impasse because of the government's insistence on ownership. What economic benefits would accrue to Inuit from the military was uncertain, and the details of cleanup were vague and unavailable because they depended on future negotiations between Canada and the United States. The Inuit insisted that the radar sites be included in the Dew Line cleanup, to be sure it was done properly. These matters were on their minds because of their close proximity to radar stations at Cambridge Bay and elsewhere.

In discussion of details, Norm Williams explained the opening of coastal area required by the government around Pelly Bay. He said the exact size of the proposed radar site was uncertain, but the new, automated sites would be much smaller than the original sites. The easements previously agreed to for a transportation route from Repulse Bay to Pelly Bay would affect six or seven parcels, but the government was not proposing a precise route now. For the Spence Bay area, exclusions were described that would allow for more coast, navigational aids, third-party sites, and a wildlife sanctuary in the Rasmussen Low Lands – an area for co-management. Gjoa Haven people would have restrictions in coming down to quantum. Cambridge Bay would have to contend with some coastal restrictions, tourism leases, and DND installations. The Hope Bay subsurface blocks would be acceptable, as there were access lands to the northeast. Bathurst Inlet/Bay Chimo negotiators were told that tourist camps, Back River surface lease, and river access in general were concerns from the appearance of their preliminary maps. In the Coppermine area, surface leases around Dismal Lake, Bloody Falls, and the Lupin mine site at Contwoyto Lake were concerns, as was the routing for winter and potential all-weather roads to the lake and outpost camps. The question of a road to deep water on the Arctic coast and a corridor of exclusion prompted a great deal of discussion. There was no route study or environmental impact study to put on the table. One Inuit participant responded, "The mining industry with lots of bucks has been trying to get our land for years; now we are prohibited from getting it by the government." As the roads would follow different routes, Terry Fenge pointed out that easements might be the solution: actual rights of way could be negotiated in the future, if and when the roads were undertaken.

In summing up, Molloy said that these were initial responses to the Inuit selections as portrayed on the maps. They should be kept in mind as each negotiating group pared down its area of interest to meet the quantum figures. No outright prohibition of land parcels was suggested here. Nevertheless, Kadlun thought the federal policies implied that Inuit lacked integrity or responsibility in managing rivers, unmanned radar sites, and third-party lands, thereby defeating the intent of the AIP to encourage self-sufficiency. Molloy insisted that the government was working in the long-term interests of the Inuit but just wanted to define reasonable rules for the future. Strong opinions were expressed about the government's lack of control and responsibility in matters of the environment, without which all the hideous Dew Line cleanup would not now be necessary. Some said that the strongest argument for Inuit control of mineral lands, as opposed to government control, was that it would ensure adequate cleanup. Molloy countered that this control could be achieved through surface, not necessarily subsurface, ownership.

In caucus, the Inuit representatives were cautioned not to give up too much in the first phase of reductions, and generally to ignore the federal comments about travel routes. The Bluenose Lake national park proposal was a recent initiative[10] and not a firm commitment from Parks Canada. The area should be treated like Wager Bay and placed under contingent agreement. The Queen Maude bird sanctuary, likely to be the largest in Canada, should be established as a co-management deal with the Canadian Wildlife Service, allowing Inuit to manage the land and CWS, the birds. There was doubt about excluding water gauge sites at this time because of the impending Arctic Environmental Strategy, and about "certificates of title" like the old Coppermine River Ltd. lease that was assigned to the Coppermine Development Association. Much of this advice came from advisors Terry Fenge and Mary Crnkovich. In addition, Malachi Arreak had some comments about the surface area required for a radar site, usually in the order of fifteen square miles.

During a break, Murray Pyke had a good opportunity to talk to Terry Fenge about future mineral administration policy of the Inuit, thinking that he might be their advisor during implementation of the final agreement. Pyke was particularly interested in procedures for holding and working Inuit-owned subsurface lands. To his surprise, he found that Fenge advocated offering exploration rights to the highest bidder who would agree to Inuit terms. These would include back-in provisions, allowing the owner to regain equity participation rights after the fact. Of course, Fenge was strongly in favor of measures to ensure secondary benefits through expediting and provision of services and education of mine workers. This was our first inkling that the future Inuit minerals regime was not going to resemble the Canada-wide, free-entry system for hard minerals that we had come to know, and which had produced a prosperous mineral industry. The Inuit were going to operate as a major freehold owner (like the CPR), which they were destined to be. Fenge was a tough-minded person who would act strictly in the best interest of the Inuit as he saw it, and with whom we might have a future opportunity to debate the issue. (Actually, he did not become their land advisor at all, but left the TFN to become executive director of the Canadian Arctic Resources Committee, where he could work on behalf of northern Native people environmentally.)

There was much concern in caucus about the government's subsurface policy for the Kitikmeot, and this took the form of insistent questioning in the next joint meeting. We wanted to know how the 35/65 percent formula for sharing would work. Was this split to be averaged over the better prospects (i.e., mineral claims and leases) as well as mineral belts in general? We wanted to keep whole properties and own them 100 percent rather than chop them up. They were accommodating, and wanted only a 65 percent position that would justify and conceivably fuel their future investment in mine roads and infrastructure. They declared that the Lupin mine would remain Crown-owned because of these considerations, but they did not mention the Izok property, which had the best undeveloped base-metal ore body in the NWT. We knew of this and many other smaller deposits scattered around the Kitikmeot mineral belts. These deposits would need major improvements in the way of roads to the coast and infrastructure investment before they could be economic, and realistically, the key to such a wave of investment would be the giant Izok copper deposit. We continued to be cautious on this point. The government carried a list of the mineral belts, thanks to geologist Michael Cunningham, and we thought he might suggest a way of dividing them up. At bottom, we were confident, because we knew the locations of most of the core claims that carried the best mineral showings in the Kitikmeot. If the 35-percent rule became restrictive, we were prepared to take only the core claims of properties.

Stanley Anablak and Malachi Arreak outlined the initial reductions made by the communities. Quassa and others made speeches about Inuit concerns, especially the inequity of the Parker Line and the exclusions around Spence Bay. There was a plea to achieve a balance. Williams specified that around Pelly Bay, the government required more shoreline for access and a number of sites for microwave installations, radar facilities, and radar cleanup, plus a caveat to cover the overland route to Repulse Bay. At Gjoa Haven, exemptions were needed for lodge leases and government reserves. The CLINTs agreed to reduce a large block of their subsurface selection along the east shore of Chantry Inlet, retaining only the portions covering volcanic rocks that contained rare-element pegmatites. Spence Bay people made progress toward a co-management agreement with the Canadian Wildlife Service and a lessening of DND easements.

Pyke and I discussed Noranda's new prospecting permits and recent stakings in the Coppermine area that had caused considerable interest locally. We thought there was a possibility of over-reacting and pledging too much acreage to this current copper play.

The evening of July 28 afforded me an opportunity to savor local traditions at Coppermine. A drum dance took place at a private home, where a female elder of the community provided the main entertainment. She was revered for her knowledge of songs and dances and repeatedly urged on by those gathered around. One particular original song caught the group's fancy, and I found out afterward that it was about our own meetings, with the refrain, "Oh, the land-claim process – how long it takes!" She was obviously referring to the twelve-odd years that negotiations had dragged on. Others eventually took their turn with the drum, always egged on by the spectators. I liked the spirit shown by Joe Otokiak of Cambridge. He danced a sort of jig, drummed a lot, and gave a whoop. Everybody was in a party mood, in a wonderful humor, drinking strong tea and chewing caribou.

Pyke and I made a presentation to our caucus on the mineral potential of the west Kitikmeot region. Like the Keewatin, it held exciting potential, but the occurrences were more scattered and had special transportation needs. Ocean access was going to be essential, and this called for government participation to achieve full development of the base-metal deposits. Some of the gold deposits, perhaps a third of the total, had a chance of development without major investment in infrastructure. We had to assess the government's attitude toward the Izok copper property, which might be assumed to be the lynchpin of their portfolio and justification for an Arctic coast road-building program. We should consider our options in the light of this, such as going for all of the remaining advanced properties. To this end, Quassa requested further elucidation of the government's subsurface intentions or rationale when we met with them on July 28. Molloy reserved the Lupin gold mine of Echo Bay as a special case: he did not want the TFN to select any of the Echo Bay property on Contwoyto Lake. As for the existing claims and leases of other properties, a portion (around 35 percent) could be obtained in all of them to guarantee that economic benefits accrued to Inuit. The government's major interest of 65 percent was deemed appropriate to ensure the provision of services. Molloy furnished a list of mineral belts or trends where the Crown wanted to hold a proportionate share, though not necessarily 65 percent. The varying interests of high potential should add up to a 35/65 split. In caucus, we surmised that the government would look seriously at our proposal, and we might even push the 35-percent limit. Mary Crnkovich had doubts whether Izok was included in the "proportionate share of properties," as it was perilously close to the Parker line.

We were treated to speech making by some interested Inuit spectators, including women, who implored the negotiators: "Think of future generations and don't give anything away. The companies don't let us use their land." One man urged the team to look at the best areas and forget about the percentages. Quassa said that we wanted to get the best from the 2,684 square miles of entitlement for the west Kitikmeot, considering that we were not selecting the Lupin mine. The Inuit would need to get some economic benefits in the future from this region with so much potential. Pyke and I were trying to determine whether the sum total of priority properties in square miles was less than 35 percent of the total. If so, we might feel comfortable giving up Izok for first choice of the remainder. Fenge thought we should hold out in this region for a 50/50 split. Our original package and bargaining position was about 23 percent of the total as prime properties and 45 percent overall.

The federal team responded well to the TFN's initial submission of sub-surface selections. The mood was good, and they seemed conciliatory. Molloy noted that the Lupin mine was omitted and reiterated the government's need for royalties and other income to justify infrastructure. The Inuit would benefit from the AIP's royalty sharing agreement (50 percent of the first $2 million and 5 percent of the remainder). He argued that this agreement was predicated on the government's holding a large interest in prime mineral properties. Again, he said that the surface holdings of the Inuit should be maximized in mineral areas to provide job opportunities, although he did not explain the exact mechanism that would allow this to happen.

The government now wished to examine each community selection separately from a fair-share point of view. At the end of the process, the Inuit were allowed to retain the Anialik gold belt, Gondor, Hood, Hackett River, Flood and Hope Bay deposits, Noranda's Victoria Island copper claims, Contwoyto Lake east, as well as the Turner or the Pistol Lake gold property. The government wanted, essentially, to retain George Lake, Izok, Kennecott, and Turner or Pistol Lake.

Kadlun summed up the situation: the government wanted 100 percent of the best gold mine in Canada, the Lupin; the richest base-metal deposit in the NWT, the Izok; the second-best copper deposit in the NWT, the Kennecott at High Lake; and the top gold prospect in the NWT, George Lake. He declared that the TFN would have a difficult time selling the Parker Line to the communities if Inuit held no interest in the Izok deposit. Moreover, the Hope Bay silver camp left the Inuit too many environmental liabilities. The Inuit did not wish to grandfather these offenders when the responsibility lay with the government.

After a break, the federal team came in with a reply intended to meet the TFN concerns. Molloy offered to split George Lake, giving us the eastern gold occurrences; in the Anialik belt, the Arcade property would go to the TFN and four southern claims to the Crown; the TFN could keep the Turner *and* Pistol, as well as the Muskox platinum property and the rest. It looked as though there was a basis here for agreement, but Pyke and I had to work out some specific block-boundary problems, and surface selections that interfingered with subsurface had to be confirmed. In short order, the government came up with some restrictions around mining properties, including Izok, where a 1/4-mile surface lease already existed. In my diary of July 29, I recorded: "We were thrilled that we had won all our demands except Izok, where we will take surface rights for political reasons, as this is very near the Parker Line."

Quassa pressed the government to undertake a cleanup of industrial waste in the Kitikmeot in approximately seven years. Molloy agreed to convene meetings with the appropriate departments on this matter, but could not speak for the government at this time. He thought the cost was going to be extremely high. Mostly on environmental grounds, the TFN decided not to claim the Ida Point property and old silver operations at Hope Bay. The unfulfilled quantum was then taken up by some careful selections of open ground. In this exercise, we regretted our lack of data on the Ulu gold property, which we had heard contained an exciting gold find. The Inuit, however, would acquire this property later on, during a final subsurface mop-up, at which time we had more definite information.

On July 30, 1991, we were able to make our full quantum presentation. Bay Chimo had transferred a block to Cambridge Bay, enabling us to cover more of the "red bed" copper play in north Victoria Island. We were even concerned that we had taken too much of the High Lake volcanic belt in the interest of diversification, but this proved erroneous. Our figures showed that 981 square miles of subsurface (30 percent of quantum) in the west Kitikmeot (Coppermine, Cambridge Bay, Bathurst, and Bay Chimo) had been selected, covering 36.5 percent of existing mineral claims and leases. The open-ground selections were 1467.6 in the first round and 691.7 in the second, accounting for 52.7 percent of quantum. To this was added 974.6 square miles in east Kitikmeot (Pelly Bay and Gjoa Haven), for 26.2 percent of quantum. The total came in at 4,115 square miles, very close to the quantum number. Molloy said they reviewed the submission and found it acceptable: we had an agreement on subsurface lands. There was a lot of applause at this point, and some appreciation that professionalism had paid off in the negotiations.

The weather was fine on that day, and we were all the more thrilled to see Coppermine kids swimming at the community beach in the Arctic Ocean. The view over island-studded Coronation Gulf at the mouth of the Coppermine River was beautiful indeed. I was able to enjoy the festivities of the evening, which included a feast of country food, with the whole community in attendance. For me, the fish chowder was the best. Later, I had the honor of presenting a plaque to Noel Avadluk, the famous Inuk prospector who twenty years earlier had made significant discoveries near Bathurst Inlet, including the Pistol Lake gold occurrence. I did this with much help from interpreter Edna Elias, commenting on Noel's career and the value to mining companies of an individual who would stay on the job until early winter, long after white crews had left for home. I also credited the perseverance of Inuit all over Nunavut in securing a good subsurface entitlement for themselves. Noel Avadluk was there with his now-disabled wife, who had also been his prospecting partner. He said he had enjoyed his career of silver- and gold-prospecting, and urged young Inuit to learn mineralogy. When I first mentioned Avadluk's reputation to Joe Allan Evaigotailak at these negotiations, he said, "Well, he hasn't anything to show for it!" since Avadluk was now a humble senior citizen but Avadluk had worked for wages in his day, and none of his discoveries had ever reached production. He carried no grudge over his treatment.

Another Coppermine resident, a woman speaking through her son, told me that her husband was denied any benefit after telling prospectors about a mineral showing in the late 1960s. She was referring to a copper sulphide showing (now the Hood claims of Kidd Creek Mines Ltd.) at the south end of Takijuq Lake, reputed to contain nearly a million tons of ore. The showing was at the site of an Inuit outpost camp frequented by her husband. As the Hood claims were acquired by the Inuit at these negotiations, I decided to check into the story later, using the mining recorder's archives at Yellowknife. Sure enough, the ground was staked by two prospectors during the Coppermine rush of the 1960s. However, no assessment work was done, and the claims lapsed after two years. Later, Kidd Creek Mines rediscovered the mineralization when they flew an airborne electro-magnetic survey in the region. It was explored and drilled between 1974 and 1982. I wrote the Coppermine woman about the history of the property and sent a copy to Kidd Creek Mines.

The negotiations on surface quantums continued, with skirmishes over military lands, navigational aids, surface leases, specific wording related to the management of the Queen Maude bird sanctuary, and the Bluenose park contingency plan, but the main debate of the last two days of the meeting centered on surface rights around Contwoyto Lake in the vicinity of the controversial Parker Line.

The federal team insisted that a certain block of land be left open to afford a travel route for Dene wishing to enter the lake at Fry Inlet. This block was chosen because it was little used by Inuit. However, the Coppermine people were adamant that this was their historic hunting territory, and the elders were furious. Quassa asked the government to produce documentation to show Inuit lack of use of the area. The government acknowledged that this was disputed territory, but a resolution of the issue was going to depend on accepting it as a Crown-owned travel corridor for the Dene. John Parker's boundary proposal, and the minister's approval of it, served the purpose of forestalling a Dene lawsuit over the issue and bringing the Inuit land-claim process to a halt. The decision was said to be fair to all Canadians – a consideration that would be important in getting ultimate parliamentary approval for the Nunavut land claim. Moreover, the Inuit were being allowed 230 square miles over and above their quantum as a bonus.

The exact land parcel to provide the Indians with access to Contwoyto Lake was said to lie along a traditional trail of theirs. Peter Kamingoak said it was a trail from nowhere; if there was a trail, we should be able to find it on the ground. Joe Niptaniatiak said that if the government could not provide facts, the Inuit could not accept this decision. Molloy replied that the Inuit still had acquired the areas of highest use by their people. If there were a counterclaim to be made, positive information would have to be brought forward to convince other people, not those at this table. Many stories by Walter Topilak, John Magsagak, and others poured forth, describing the late arrival or nonarrival of the Indians there. The speakers demanded to know the names of the Indians. Noel Avadluk, who had taken in many of the sessions, said in his experience there were never any Indians in the area, whereas the Inuit had lived there continually. Indians had not been there since he frequented the area in 1942; otherwise, their trails and camp sites would be in evidence. Other speakers also related their experiences over the years and swore they had never seen any Indians. One story recounted how, in 1988, Indians were heard on the side-band radio who did not know how to get to Contwoyto Lake, so an Inuk picked them up and showed them the way. This land was evidently treasured by the Inuit elders in attendance, and their chorus of protest was bringing the discussion to an impasse. Joe Otokiak from Cambridge Bay said, "I wouldn't want to be in your shoes. We've tried to convince you with fact. You will have to produce more fact than you have on the map there to convince our side. We from Cambridge are not signing until Coppermine is satisfied." At this point, the federal side tried some trading of lands, but to no avail.

The government did not defend the Dene position on the boundary issue, but the negotiators wanted to avoid future grievances and promote acceptance of the claim, knowing full well that the parties would never agree. Clearly the Inuit, in order to compensate for the northward and 20,000-square-mile wrongful displacement of the Parker Line, had flooded the area with Inuit surface selections. I could see on the map which TFN parcels had been turfed out to make the Contwoyto window for the Dene, and how the Healy Lake block was shifted to replace them. I regretted that we had no trump cards, like the mineral lease at Baker Lake, that might afford a trade-off. We should have held onto the Izok mineral property right to the end, or perhaps we should demand it now! My last impression of the discussions was of Molloy insisting that he had no mandate to change the boundary, and of the Inuit clinging to their ancestral lands, as they should.

Pyke and I left the negotiations to prepare for our flight to Yellowknife on Ptarmigan Airways, but around noon, an agreement was reached, and tears of joy were flowing. Apparently the TFN had offered a two-hundred-foot easement around the access lake, but the feds had refused this. Finally, the offending corridor block was traded for another, already dropped, block near Contwoyto Lake. Molloy, Quassa and others rushed out and caught the same plane, full of smiles.

By popular demand, since no one had been able make the trip during the week, the pilot banked steeply, so we could all see the historic Bloody Falls on the Coppermine River. Mary Crnkovich and I were enthusiastic about this place, and thought we should canoe down the river next year to attend the final signing of the TFN agreement, which we giddily assumed would take place at Coppermine.

During the flight, I spoke with Quassa and Gillies, who would now be turning their attention to the difficult task of selling the agreement to the Inuit communities. They engaged me to write a final report on Inuit minerals to assist them.[11] It would not be hard to do that, and in glowing terms. The Inuit settlement lands and mineral entitlement would make the Inuit the largest freehold owners in Canada, larger even than the CPR. And those lands, thanks to patience and determination in the negotiations, contained some of the best mineral occurrences in the Northwest Territories.

8. LAND BASE ACHIEVED

SUMMATION

Mineral ownership was hailed as an important aspect of the Nunavut land-claim agreement. Bob Kadlun, TFN vice-president, noted that "Inuit are to gain ownership of many proven mineral deposits, currently under lease to third parties, some of which may be developed in the 1990s. This agreement insures that Inuit will receive worthwhile and lasting benefits if and when mineral development occurs in the Kitikmeot region."[1] Terry Fenge observed that the boundary issue was particularly contentious. It was resolved by an additional concession of fee simple ownership in the vicinity of the boundary. Paul Quassa, TFN president, also pointed to the acquisition of 240 square miles adjacent to Bluenose Lake, an area where the government was considering a national park. He said the TFN would soon be consulting every community in Nunavut in preparation for a vote to ratify the Nunavut Final Agreement.[2] The Final Agreement, with all its terms and benefits, would be given in exchange for the psychologically potent surrender to the Crown of aboriginal title to lands, waters, and the offshore.

Bruce Gillies, as assistant project director, provided some final comments upon the conclusion of land ownership negotiations. He mentioned that one hundred ten Inuit representing their communities had worked since 1988 to decide which 18 percent of Nunavut's landmass they should retain: "Given that Inuit still *intensively* use 50 to 60 percent of Nunavut in land-related activities, having to choose and successfully negotiate for 18 percent of it was, to say the least, demanding." In spite of many negotiating pressures, Inuit retained approximately one thousand parcels from the land they selected, outside of municipal land. They had to be patient and hold out at the table many times to get them. Gillies said, "In spite of the government tabling the initial position that Inuit needn't own any lands under mineral claim or lease, Inuit held out and retained 14,000 square miles of some of the best mineral land in Nunavut, including 50 percent of the uranium potential at Kiggavik." Gillies felt that all of the lands acquired, if managed properly, would go a long way in meeting economic and conservation needs. Everything possible under the negotiating conditions "was fought for until it was obtained."[3]

In a later article, Tom Molloy described the view from the government's side, as he had experienced it since 1982: "The land selection process itself was done on a community-by-community basis in 1991 and 1992 [sic], and added a whole new dynamic to the negotiations. A number of people not previously involved in the negotiations were added to both the Inuit and Government teams.... Government had to carefully consider the consequences of agreeing to the proposed land selections, including what might occur on the lands in the future, as well as the impact that Inuit ownership would have on third parties. Government also had to provide for access corridors, and areas for recreation, and to insure that areas of high mineral potential were shared equitably." Speaking of the long and difficult process ending with the Agreement, he said, "After over a decade of long days and nights – first working within government and thereafter with TFN – our team could point to a final agreement that both the Government of Canada and TFN were proud to be part of. Unlike a great many aboriginal peoples world-wide, the Inuit did not have to resort to litigation to have their rights acknowledged. That alone made the time and effort worthwhile. The Nunavut Agreement is a testament to good faith. It's an agreement about people living and working together – not only for the benefit of Inuit, but for all Canada."[4]

After many years of debate and negotiation Inuit (and Inuvialuit) recognized the great importance of national parks, as described in respective agreements, in the furthering of their goals of conservation, joint management of resources, and wildlife harvesting. They cooperated energetically during land-ownership negotiations to further the park system, and even badgered the government to complete its northern Green Plan (a 1990 cabinet plan for five new parks in Nunavut) under this forum. However, the government failed to make the necessary commitments because of financial constraint or limitation of negotiating mandate and, in the opinion of Terry Fenge, lost a fine opportunity.[5] The national park story shows how Inuit were amenable to side deals outside the land-claim process, if properly structured with benefits and legally not prejudicial to comprehensive land-claim negotiations. Auyuittuq was established as a national park preserve in 1974, for example.[6] It took a full twenty years for the mining industry, in cooperation with government and aboriginal peoples, to devise a similar formula for interim advancement of their interests, as expressed in the Whitehorse Mining Initiative of 1994.

Many far-reaching benefits were achieved in the landmark TFN agreement, including the eventual establishment of Nunavut Territory covering the

Eastern Arctic. The patience and pragmatic skill of Inuit leaders finally produced handsome results, clearly a fruition of the community-development process set in motion years before to encourage Native people to better themselves. In terms of land, the Inuit themselves became legal owners of 137,450 square miles of surface, or 87,968,000 acres. They also gained ownership of 14,475 square miles with subsurface mineral rights included, or 9,264,000 acres. That makes them very large owners indeed.

In the nineteenth century, the CPR was awarded land grants to build the transcontinental railroad totaling 36 million acres along the right-of-way. Sale of these lands was instrumental in financing the project. By 1959, the CPR had only 1,200,000 acres of land left. However, from 1905 onward, the company had retained the mineral rights when they sold the land. The CPR merchandised these mineral rights separately, leaving about 800,000 acres in their own hands in modern times. The mineral rights portfolio of the CPR enabled the establishment and success of its resource subsidiaries, PanCanadian Petroleum Limited and Canadian Pacific Oil and Gas. In 1969, Canadian Pacific Oil and Gas transferred its coal and mineral rights to Canpak Minerals Limited, which became Fording Coal Ltd. in 1977. These companies, as well as subsidiary Cominco Ltd., a major developer of Arctic minerals, have been active participants in the petroleum and mineral business. The Inuit are succeeding the CPR as the largest freehold landowner in the country, and very possibly in time they will spawn their own land-management and development companies.

CONCLUDING NOTE

In subsequent years, the inclination of the Inuit led them to reach out to resource lawyers and experienced administrators to establish their own brand of land administration, just as they had hired us as mineral specialists to aid them in selecting appropriate parcels to fulfil their negotiated quantum of 14,000 square miles. We were private operators, that is, explorers and prospectors, and could not guide them without bias toward a tenure system that satisfied their need for control, planning, and immediate return. So our work ceased with implementation of the Nunavut Final Agreement.

Murray Pyke, who has remained continuously involved in mineral exploration in the Eastern Arctic, said this in 1996 about the evolving situation: "Nunavut is now a stable jurisdiction that has a positive pro-mining philosophy. It's exciting to witness the tremendous effort of the Inuit people, as their doors of business open wider and wider for global mining opportunities." In an interview with Robert Smith for the *Canadian Miner*, Pyke described many recent developments, such as workable access and acquisition rules, partnership ventures, and mining forums to promote mutual understanding between exploration participants and Nunavut communities.[7]

A sense of unease resides in the mining community over having to deal with Inuit landowners, who are powerful now and have a strong bent for planning and control, with the result that they are more demanding than other governments in their administration of lands. Unless the companies restrict themselves to Crown land (where they operate under the Canada Mining Regulations), they must deal with Nunavut access provisions, anticipate the movement of their prospecting crews in advance, and negotiate for exclusive exploration rights. These procedures conflict with the long-accustomed privacy and freedom of movement allowed by provincial and federal regulations, but the mining industry will simply have to get used to the new, limited-entry regime if it wants to explore or expand onto Inuit-owned mineral lands, which are now massive in extent and quality.

Industry developers have brought this situation upon themselves, in more ways than they realize. Over the years, the developers supplied the irritations that helped to motivate Inuit and propelled them toward such a far-reaching agreement. They showed the Inuit that they were tenacious intruders (though perfectly legal ones) in their land, and disclined to spread the benefits around. They provided a negative message that became ingrained in the aboriginal mind, fostering an anti-development bias that saw its expression in protests, intervention, and dogged pursuit of a claims agreement.

Industry communicated another, positive message, however: the resources of the land were worthy of extravagant investment, and they could yield wealth and benefits. This message also impressed the Inuit, who came to realize that they and their children could reap some of the rewards. Of course they wanted to gain a measure of control, a means of effecting guarantees and securing a piece of the action. They fought tenaciously for these rights, in the faith that the resulting benefits, employment, and royalties would ensure the survival and prosperity of Nunavut. They are now poised to be partners in development, to cooperate with industry in achieving production, to bring their land and their capital into new ventures, and to achieve real social benefits for the people of Nunavut.

I feel that the Nunavut land-claim success has a special significance for minerals, given the passion for land of the Inuit and their faith in cooperation, planning, control, and modern technology. For example, the profitable use of nonrenewable resources, such as carving stone and semi-precious materials (possibly lapis lazuli), will benefit from their sophistication and from science education.

On the practical side, Inuit have proven themselves adaptable and trainable. They have been miners and tradesmen; they can become able prospectors and explorers who are observant, persistent, and well-trained. They could form their own community-based exploration groups and enter joint-venture agreements with well-funded private companies, initially offering only Inuit land to the deal while the companies furnish the risk capital. Thus, the Inuit could install themselves in a powerful position in future eras of development, gaining a bigger stake, a piece of the action in whatever happens.

The management role will increasingly fall to the younger, educated generations, and their expertise and practice will have important implications. Success in managing the largest block of mineral lands in the country will encourage Inuit to a larger dream – the responsibility for all Crown lands in Nunavut – if Ottawa can be persuaded to devolve it. Inuit will then become the true custodians of the North for the benefit of all Canadians.

NOTES

INTRODUCTION

1. Barry J. Barton, *Canadian Law of Mining*, Canadian Institute of Resource Law, 1993, p. 149.
2. Ian Thompson and Susan A. Joyce, S.A., *Mineral Exploration and the Challenge of Community Relations*, Communiqué of the Prospectors and Developers Association of Canada, 1997.
3. Ibid., p. 7.
4. Murray Dobbin, *The One-and-a-half Men: The Story of Jim Brady and Malcolm Norris, Metis Patriots of the Twentieth Century*, Vancouver: New Star, 1981.

1. KEEWATIN EVENTS

1. *The Cantley Economic Report on Eskimo Affairs*, Department of Resources and Development, 1950.
2. Frank J. Tester and Peter Kulchyski, *Tammarniit (Mistakes): Inuit Relocation in the Eastern Arctic, 1939–63*, UBC Press, 1994, p. 48.
3. Farley Mowat, *The Desperate People*, Little Brown and Company, 1959.
4. Tester and Kulchyski, p. 55.
5. Ibid., p. 94.
6. Ibid., p. 98.
7. Frank G. Vallee, *Kabloona and Eskimo in the Central Keewatin*, Department of Northern Affairs and National Resources, 1962, p. 41.
8. Tester and Kulchyski, p. 99.
9. Farley Mowat, *People of the Deer*, Little Brown and Company, 1952.
10. Tester and Kulchyski, p. 100.
11. John Dryborough, "A Nickel-Copper Deposit on Hudson Bay," *Transactions, Canadian Institute of Mining and Metallurgy* 34 (1931), p. 157.
12. J. C. Browning, "Rankin Inlet Nickel Mines," *Canadian Mining Journal* 74, no. 12 (December 1953), pp. 56–57.
13. Mine Staff, "North Rankin Nickel Mines," *Canadian Mining Journal* 78, no. 8 (August 1957), pp. 93–97.
14. "Rankin meeting," *The Northern Miner*, March 12, 1953, p. 9.
15. "Nickel abundance brings difficulties for North Rankin," *The Northern Miner*, March 6, 1958, p. 5.
16. "Annual Review Number – 1962," *Canadian Mining Journal* 84, no. 2 (February 1963), p. 76.

17　Robert G. Williamson, "Eskimo Underground: Socio-Cultural Change in the Canadian Central Arctic," Occasional paper 2 (Institute for General and Comparative Ethnography, Uppsala University, Sweden), 1974, p. 92.
18　Ibid., p. 93.
19　Robert C. Dailey and Lois A. Dailey, *The Eskimo of Rankin Inlet: A Preliminary Report*, Ottawa: Northern Co-ordination and Research Centre, Department of Northern Affairs and National Resources, 1961.
20　Mine Staff, "North Rankin Nickel Mines," *Canadian Mining Journal* 78, no. 8 (August 1957), pp. 93–97.
21　Dailey and Dailey, p. 83.
22　"Hours and Earnings of Wage Earners," *Canada Year Book* 1963–64, Dominion Bureau of Statistics, Queen's Printer, p. 723.
23　Ibid., p. 721.
24　Dailey and Dailey, p. 83.
25　Mine Staff, "North Rankin Nickel Mines," *Canadian Mining Journal* 78, no. 8 (August 1957), pp. 93–97.
26　Williamson, 1974, p. 116.
27　Ibid., p. 137.
28　Dailey and Dailey, p. 79.
29　Farley Mowat, *The Desperate People*, p. 284.
30　Dailey and Dailey, p. 93.
31　Ibid., p. 94.
32　Williamson, 1974, p. 149.
33　Ibid., p. 152.
34　Ibid., p. 127.
35　Ibid., p. 132.
36　R. G. Williamson and T. Foster, *Eskimo Relocation in Canada*, Institute for Northern Studies, University of Saskatchewan, 1973, p. 111.
37　Ibid., p. 109.
38　Ibid., p. 110.
39　Williamson, 1974, p. 135.
40　Vallee, 1962, p. 51.
41　Ibid., p. 56.
42　Ibid., p. 59.
43　Ibid., p. 130.
44　Ibid., p. 132.
45　Ibid., p. 145.
46　Ibid., p. 195.
47　Ibid., p. 202.
48　Williamson, 1974, p. 153.
49　Vallee, 1962, p. 212.
50　Williamson, 1974, p. 178.
51　Ibid., p. 189.
52　Tester and Kulchyski, p. 104.
53　Ibid., p. 194.
54　Ibid., p. 204.
55　Ibid., p. 211.
56　Ibid., p. 215.
57　Farley Mowat, *Desperate People*, p. 215.

58 Tester and Kulchyski, p. 221.
59 Ibid., p. 222.
60 C. S. Lord, "Operation Keewatin, 1952: A Geological Reconnaissance by Helicopter," *The Canadian Institute of Mining and Metallurgical Bulletin*, April 1953, pp. 224–33.
61 Ken. E. Eade, personal communication, 1995.
62 Tester and Kulchyski, p. 234.
63 Ibid., p. 236.
64 Ibid., p. 296.
65 Ibid., p. 297.
66 Ibid., p. 339.
67 Ibid., p. 346.
68 Ibid., p. 59.
69 Ibid., p. 351.
70 Ibid., p. 352.
71 Ibid., p. 353.
72 Abraham Okpik, "What Does It Mean to Be an Eskimo?" *North*, March–April, 1962, pp. 2–4.

2. EXPLORATION PRESSURES

1 Ian Bickle, *Turmoil and Triumph: Controversial Railway to Hudson Bay*, Detselig Enterprises, 1995, p. 102.
2 Farley Mowat, *Tundra: Selections from the Great Accounts of Arctic Land Voyages*, McClelland and Stewart, 1973, p. 270.
3 L. J. Weeks, "Mistake Bay Area, West Coast of Hudson Bay, North West Territories," *Report of Activities*, Geological Survey of Canada, 1930, p. 172E.
4 Richard Pearce, "First to fly across Canada's barren lands," *The Northern Miner*, September 20, 1928, pp. 1–3.
5 L. J. Weeks, "Rankin Inlet Area, West Coast of Hudson Bay, North West Territories," *Report of Activities*, Geological Survey of Canada, 1932, p. 40C.
6 Pearce, p. 1.
7 "First ski planes out of bay," *The Northern Miner*, November 29, 1928, p. 8.
8 "Picked up in the far north," *The Northern Miner*, September 20, 1928, p. 20.
9 This information is extracted from issues of *The Northern Miner*, usually from the Industry Reviews published in March.
10 This information is extracted from issues of *Canadian Mining Journal*, usually from the Industry Reviews published in February–March.
11 "Big, active, constructive convention sparkling success," *The Northern Miner*, March 10, 1966, p. 2.
12 Les MacDonald, "Royal Commission on Taxation," *The Canadian Encyclopedia*, Hurtig, 1985, p. 1788.
13 *Annual Report 1965–66*, Department of Northern Affairs and National Resources, p. 14; *Annual Report 1966–67*, Department of Indian Affairs and Northern Development, p. 41.

14   "Prospectors' role remains important in 'U' exploration," *The Northern Miner*, March 9, 1967, p. 4.
15   Ibid.
16   "New mineral sources need urgent, major AIME meet assured," *The Northern Miner*, February 29, 1968, p. 5.
17   "Bigness in thinking as well as capital needed for success in changing industry," *The Northern Miner*, March 7, 1968, p. 37.
18   "Junior firms face extinction without a mining exchange," *The Northern Miner*, March 14, 1968, p. 12.
19   Ray Price, *Yellowknife*, Peter Martin and Associates, 1967.
20   Frank Rasky, "Murray Watts: A Man With a Vision," *North/Nord* 20, no. 3 (May–June 1972), pp. 22–25.
21   Murray Watts and Doris K. Megill, "Iron Ore on Baffin Island: Development Work by Baffinland Iron Mines Ltd," *Canadian Geographic Journal* 73, no. 5 (November 1966), pp. 157–65.
22   Graham Rowley, "A Long-Neglected Arctic Island Enters the Modern Age," *North* 15, no. 3 (May–June 1968), pp. 52–55.
23   James Woodford, *The Violated Vision: The Rape of Canada's North*, McClelland and Stewart, 1972, p. 86.
24   Ibid., p. 88.
25   Peter J. Usher, *The Bankslanders: Economy and Ecology of a Frontier Trapping Community*, Vol. 3, *The Community*, Department of Indian Affairs and Northern Development, 1971.
26   Peter A. Cumming, "Why Eskimo fear rights not protected," *Globe and Mail*, August 1, 1970, p. 7.
27   Peter A. Cumming, "Canada: Native Land Rights and Northern Development," Document 26, International Working Group for Indigenous Affairs, *Northern Perspectives, No. 7*, Canadian Arctic Resources Committee, 1973.
28   Jim Lotz, "The Myth of the Rich North," *Canadian Forum*, January 1968, pp. 217–19.
29   Woodford, p. 124.
30   Ibid., p. 120.
31   Pat Carney, "Panel V – Social Responsibility," *Proceedings*, National Northern Development Conference, October 31–November 2, 1973, Edmonton, Alberta, p. 141.
32   François Paulette, "Comments to Panel V – Social Responsibility," *Proceedings*, National Northern Development Conference, October 31–November 2, 1973, Edmonton, Alberta, p. 144.
33   Ibid., p. 143.
34   D.M. Dickinson, "Northern Resources: A Study of Constraints, Conflicts, and Alternatives," *Northern Transitions* 1, Canadian Arctic Resources Committee, 1978, pp. 253–315.
35   Ibid., p. 283.
36   Ibid., pp. 281–83.

3. INUIT POLITICAL ORGANIZATION

1. *Indian Eskimo Association Newsletter* 1, no. 1 (March 1960), pp. 4–5.
2. "Second Conference on Eskimos," *Indian Eskimo Association Newsletter* 1, no. 2 (May 1960), p. 2.
3. Clare Clark, "Why an Indian-Eskimo Association?" *Indian Eskimo Association Newsletter* 1, no. 5 (December 1960), pp. 1–2.
4. Frank Vallee, "Regional Planning for Community Development," *Indian Eskimo Association Newsletter* 6, no. 2 (March–May 1965), p. 4.
5. Helen Buckley, "Dear Editor –," *Indian Eskimo Association Newsletter* 6, no. 3 (June 1965), p. 4.
6. "Legal Research Project to be Finished by Fall," *Indian Eskimo Association of Canada Bulletin* 10, no. 2 (April 1969), pp. 1–2.
7. Peter Cumming, "Arctic Power!" *Indian Eskimo Association Bulletin* 11, no. 4 (October 1970), pp. 1–3.
8. "United Voice Now Eskimos' First Aim," *Indian Eskimo Association Bulletin* 12, no. 1 (1971), p. 1.
9. "CASNP Seeks Publications Director," *Canadian Association in Support of Native Peoples Newsletter* 16, January 1978, p. 1.
10. "Eskimo Worker Hire," *Indian Eskimo Association Bulletin* 12, no. 1 (1971), p. 6.
11. "Inuit leaders meet to form organization," *Indian Eskimo Association Bulletin* 12, no. 2 (1971), p. 1.
12. Letter from Jeela Moss-Davies, Secretary of the Eskimo University in Frobisher Bay. *Indian Eskimo Association Bulletin* 12, no. 3 (April 1971), p. 5.
13. Peter A. Cumming, N.H. Mickenberg, and K.R. Aalto, eds., *Native Rights in Canada*, Indian-Eskimo Association of Canada, 1972, and P.A. Cumming et al., *Inuit and the Law*, translated into syllabics by Jeela Moss-Davies, Inuit Tapirisat of Canada, 1974.
14. "The Inuit Plan for Sane Progress," *Canadian Association in Support of Native Peoples Bulletin*, July 1973, p. 10.
15. Ibid.
16. Peter Cumming, "Native Land Rights and Northern Development," *Northern Perspectives*, No. 7, Canadian Arctic Resources Committee, 1973.
17. Ibid.; also reprinted in CASNP *Bulletin* 14, no. 4 (December 1973), p. 5.
18. Usher, P.J., "The Committee for Original Peoples' Entitlement," [unpublished report distributed by COPE] Ottawa, April 25, 1973, p. 22.
19. Ibid., p. 44.
20. "Land Claims and Land Selection," [enclosure in] *Inuvialuit Nunangat: The Proposal for an Agreement-in-Principle to achieve the Settlement of Inuvialuit Land Rights in the Western Arctic Region of the Northwest and Yukon Territories between The Government of Canada and The Committee for Original Peoples Entitlement*, May 13, 1977, p. 4.
21. D. Pimlott, D. Brown, and K. Sam, *Oil Under the Ice*, Canadian Arctic Resources Committee, 1976, p. 20.
22. T.R. Berger, *Northern Frontier, Northern Homeland*, The Report of the Mackenzie Valley Pipeline Inquiry: Volume 1, 1977, p. 167.
23. Ibid., p. 169.

24 Ibid., p. 170.
25 Ibid., p. 171.
26 Ibid., p. 172.
27 Ibid., p. 176.
28 Ibid., p. 177.
29 Ibid., pp. 177–78.
30 Ibid., p. 179.
31 Keith Crowe, "A Summary of Northern Native Claims in Canada: The Process and Progress of Negotiations," *Etudes/Inuit/Studies* 3, no. 1, 1979, pp. 31–37.
32 Frank J. Tester, ed., *Reflections of the People: The Potential Social, Economic and Environmental Impacts of the Proposed Polar Gas Pipeline on the District of Keewatin, Northwest Territories*, Vol. 1, Department of Indian Affairs and Northern Development, 1978.
33 Ibid., p. 25.
34 Ibid., p. 249.
35 Ibid., p. 35.
36 *Inuvialuit Nunangat: The proposal for an Agreement-in-Principle to achieve the Settlement of Inuvialuit Land Rights in the Western Arctic Region of the Northwest Territories and Yukon Territories between The Government of Canada and The Committee For The Original Peoples' Entitlement*, May 13, 1977.
37 B.A. Hochstein, *New Rights or No Rights? COPE and the Federal Government of Canada*, M.A. thesis, University of Calgary, 1987, p. 108.
38 Ibid., p. 113.
39 Ibid., p. 114.
40 The COPE/Government Working Group, *Joint Position Paper on the Inuvialuit Land Rights Claim*, 1978, See Preamble, pp. ii–iv.
41 Hochstein, p. 115.
42 Ibid., p. 118.
43 Canadian Arctic Resources Committee, "Inuvialuit Land Rights: Summary of Joint Position Paper," *Northern Perspectives* 6, no. 4 (1978), p. 5.
44 Ibid., pp. 1–4.
45 Keith J. Crowe, "Claims on the Land," *Arctic Circle* (November–December 1990), pp. 14–23.
46 H. Herchmer, "Caribou Eskimos v. The Canadian Legal System," *Northern Perspectives* 7, no. 3 (1980), p. 2.
47 *Proceedings*, Coppermine Conference of Arctic Native People, July 14–18, 1970, p. II-5. (These *Proceedings* comprise the taped minutes of discussions held at Coppermine, with translations provided by Peter Ernerk and Mary Cousins. The Indian-Eskimo Association arranged financing for the conference, which was attended by thirty-three delegates and four resource people.)
48 Ibid., p. IV-2.
49 James Arvaluk, "President's Report," *1976 Annual Report*, Inuit Tapirisat of Canada, pp. 1–2.
50 Herchmer, p. 4.
51 Interdisciplinary Systems Ltd. *Effects of exploration and development in the Baker Lake area*, 2 vols. Report prepared for the Department of Indian Affairs and Northern Development, Winnipeg, Manitoba: Interdisciplinary Systems Ltd., 1978, 309 pp. and maps.
52 Herchmer, p. 4.

53   "Keewatin Land Issue Brews," *News of the North*, Yellowknife, March 12, 1979, cited in *Musk-Ox* 24 (1976), p. 76.
54   Herchmer, p. 5.
55   Frank Calder, a Nisga'a elder and member of the B.C. legislature, brought a case on extinguishment of aboriginal title before the Supreme Court in 1973. It was held that aboriginal title depended on traditional use and occupation. The decision was inconclusive, as three judges said there must be direct extinguishment, and three said extinguishment could be implied from historical circumstances. Henderson, William B., "Aboriginal Rights," "Sources of Aboriginal Rights," and "Extinguishment of Aboriginal Rights," *Canadian Encyclopedia*, Hurtig, 1985, pp. 3–4.
56   "Native Rights Secondary to Mining," *Oilweek*, November 26, 1979, pp. 58–59.
57   Herchmer, p. 7.
58   Ibid., p. 8.
59   Frank J. Tester, "And the first shall be last: Some social implications of the Baker Lake decision," *Northern Perspectives* 8, no. 3 (1980), pp. 9–12.
60   Ibid., p. 10.
61   Ibid.
62   Ibid., p. 12.

4. THE NANISIVIK MINE

1   Robert G. Blackadar, "Precambrian Geology Northwestern Baffin Island, District of Franklin," *GSC Bulletin* 191, 1970, p. 85; and "Geological Reconnaissance of Admiralty Inlet, Baffin Island," GSC Paper 55–6, 1956, pp. 22–23. *See also*: Richard Fish, "The place where people find things: Nanisivik Mines in Canada's High Arctic," *Canadian Mining Journal* 100, no. 9 (September 1979), p. 35.
2   Diamond Jenness, "Household Furniture/Fire Making," In: *Canadian Arctic Expedition 1913–18*, vol. 16, Material Culture of the Copper Eskimo, 1946, p. 55.
3   Margery Hinds, *High Arctic Venture*, Ryerson Press, 1968, p. 148.
4   Blackadar, "Geological Reconnaissance," pp. 1–2.
5   Ibid., p. 24.
6   Robert Blackadar, personal communication, 1995.
7   Robert Blackadar, "Geological Reconnaissance."
8   "The Kidd Creek sensation," In John Carrington, G. Lonn, L.C. Brown, and P. Tremblay, eds., *Risk Taking in Canadian Mining*, Pitt Publishing Company, 1979, pp. 127–28.
9   "The world's most northerly mine," In John Carrington, G. Lonn, L.C. Brown, and P. Tremblay, eds., *Risk Taking in Canadian Mining*, Pitt Publishing Company, 1979, p. 133.
10 *Annual Report 1984*, Mineral Resources International Limited, Calgary.
11 Robert B. Gibson, *The Strathcona Sound Mining Project: A Case Study in Decision Making*, Background Study No. 42, Science Council of Canada, 1978, p. 31.

12   Ibid., p. 32.
13   *Canada Year Book* 1975, Information Division, Statistics Canada.
14   "The world's most northerly mine," p. 134.
15   Hugh Brody, *The People's Land: Eskimos and Whites in the Eastern Arctic*, Penguin, 1975, p. 123.
16   Ibid., p. 165.
17   Ibid., p. 229.
18   "The world's most northerly mine," p. 134.
19   Frank Agar, personal communication 1994.
20   Information Canada, Ottawa, Policy Statement on Northern Development, March 1972.
21   Gibson, p. 33.
22   Ibid., Appendix B.
23   Ibid., p. 38.
24   D. Bissett, *A Background Paper on the North Baffin Communities in Relation to the Strathcona Sound Project*, DIAND Northern Policy and Program Planning Branch, 1974, p. 15.
25   Brody, p. 226, quotes Father Guy Mary-Rousselier's editorial in the Summer 1974 issue of *Eskimo*.
26   Brody, p. 229.
27   Gibson, p. 60.
28   Ibid.
29   Ibid., p. 79.
30   Ibid., p. 255.
31   Letter from Levi Kalluk, Chairman of the Settlement Council, to Commissioner Hodgson, "Arctic Bay Protests Strathcona Townsite," *Inukshuk*, March 26, 1975, p. 3.
32   Gibson, p. 86.
33   Ibid., p. 91.
34   Peter J. Usher, *The Bankslanders: Economy and Ecology of a Frontier Trapping Community*, Department of Indian Affairs and Northern Development, 1971.
35   Milton M.R. Freeman and Linda M. Hackman, "Bathurst Island, NWT: A Test Case of Canada's Northern Policy," *Canadian Public Policy* 1, no. 3 (1975), pp. 402–14.
36   Gibson, p. 105.
37   Bissett, pp. 8–9, 11.
38   Gibson, p. 255.
39   Diamond Jenness, "The Hour of Crisis," chapter 2 of *Eskimo Administration*: V. *Analysis and Reflection*, Technical Paper No. 21, Arctic Institute of North America, 1968, pp. 42–43.
40   Brody, p. 229.
41   Gibson, p. 47.
42   "Nanisivik: Canada's First Arctic Mine," Department of Indian Affairs and Northern Development communique, June 18, 1974.
43   *Annual Report 1977*, Mineral Resources International Limited, Calgary, p. 2.
44   *Annual Report 1984*, Mineral Resources International Limited, Calgary, p. 5.

45 Baffin Regional Inuit Association (BRIA), Charles Hobart, principal investigator, *Socio-Economic Impacts of the Nanisivik Mine on North Baffin Region Communities*, BRIA: November 1979, pp. 3–4.
46 Robert G. Williamson, "Eskimo Underground: Socio-Cultural Change in the Canadian Central Arctic," Occasional paper 2 (Institute for General and Comparative Ethnography, Uppsala University, Sweden), 1974.
47 BRIA, *Socio-Economic Impacts of the Nanisivik Mine*, p. 217.
48 Ibid., p. 278.
49 J. R. Ames, R. B. Gibson, R.F. Keith, and G.C. Michalenko, *Analysis of the Polaris Project Proposal: Arvik Mines Ltd.*, Draft report for Canadian Arctic Resources Committee, 1980, p. 80.
50 "Arctic mine sets world record," *Oilweek*, December 10, 1979, p. 8.
51 Graham Farquharson and James Marshall "The Mining Town: An Endangered Species?" *Canadian Institute of Mining and Metallurgy Bulletin* 89, (1996), pp. 74–76.
52 Ibid., p. 76.
53 Baffin Regional Inuit Association, "Programs," *The Inuit North*, July 1979, p. 19.
54 BRIA, *Socio-Economic Impacts of the Nanisivik Mine*, p. 123.
55 Hon. Allen Blakeney, "How to Effect Change," *Proceedings*, Ninth National Northern Development Conference, October 27–29, 1982, Edmonton, Alberta, pp. 154–59.
56 Tagak Curley, "Industry and Native People: Time for a New Approach to Economic Co-operation," *Proceedings*, Ninth National Northern Development Conference, October 27–29, 1982, Edmonton, Alberta, pp. 145–48.

5. THE ROAD TO NUNAVUT

1 Mel Watkins, ed., *Dene Nation: The Colony Within*, University of Toronto Press (for the University League for Social Reform), 1977, p. 16.
2 C.M. Drury, *Constitutional Development in the Northwest Territories*, Report of the Special Representative to the Prime Minister for Constitutional Development in the Northwest Territories, Ottawa: Supply and Services Canada, 1980.
3 "Arctic residents support proposal for Inuit province," Saskatoon *Star-Phoenix*, October 2, 1979, cited in *Musk-Ox* 25 (1979), p. 99.
4 "Royalties could pay for Nunavut government," *The Inuit North*, January 1980, pp. 63–64. (Report of an interview with John Amagoalik.)
5 *Edmonton Journal*, November 13, 1979, cited in "Mining is still Backbone of Industry in the North," *Musk-Ox* 25, 1979, p. 104.
6 John Todd, "Eastern Arctic," Panel IV, Economic Development Priority Policies – Regional Perspectives, *Proceedings*, Eighth National Northern Development Conference, November 14–16, 1979, Edmonton, Alberta, pp. 117–19.
7 Message from Michael Amarook, president, *Annual Report 1979–80*, Inuit Tapirisat of Canada, pp. 3–4.

8   The text of Ittinuar's speech was published on pages 69–71 of *The Inuit North* for January 1980, under the rubric "For the Record."
9   J.R. Ames, R.B. Gibson, R. F. Keith, and G.C. Michalenko, *Analysis of the Polaris Project Proposal: Arvik Mines Ltd.*, Draft Report for Canadian Arctic Resources Committee, May 1980.
10 Ibid., p. 91.
11 Ibid., p. 63.
12 *Annual Report 1979–80*, Inuit Tapirisat of Canada, p. 10.
13 "No Mississaugas on Little Cornwallis Island" [editorial], *The Northern Miner*, November 22, 1979, p. 6.
14 J. E. Barrett, *Employment of the Inuit at Polaris: Little Cornwallis Island*, A Feasibility Study Requested by Arvik Mines Ltd., Personnel Development Services, Cominco Ltd., Trail, B.C., 1973.
15 Outcrop Ltd., and DPA Consulting Limited, *Potential Socio-Economic Impacts of the Polaris Mine Project*, June 1980, pp. 167–70.
16 "Changing lifestyles problem for Eskimos," Saskatoon *Star-Phoenix*, August 12, 1980, cited in *Musk-Ox* 28 (1981), p. 104.
17 Toronto *Globe and Mail*, December 1980, cited in *Musk-Ox* 28 (1981), p. 106.
18 "ITC charges IDC with hampering land claim negotiations," Yellowknife *News/North*, February 27, 1981, cited in *Musk-Ox* 28 (1981), p.106.
19 "Yukon land claim battle heats up," Saskatoon *Star-Phoenix*, May 27, 1981, cited in *Musk-Ox* 28 (1981), p. 108.
20 Terry Fenge, "Political Development and Environmental Management in Northern Canada: The Case of the Nunavut Agreement," *Etudes/Inuit/Studies* 16, no. 1–2 (1992), p. 127.
21 Thomas Suluk, "Nunavut Land Claims," *Annual Report 1979–80*, Inuit Tapirisat of Canada, pp. 14–16.
22 Ibid., p. 16.
23 "Negotiations to Start on Land and Resources in Nunavut," *Nunavut Newsletter* No. 2, July 15, 1982, pp. 1–2.
24 Ibid., p. 5.
25 "Inuit Not in Strong Negotiating Position," *Nunavut Newsletter* No. 5, October-November 1982, p. 11.
26 "TFN Board Meets with Federal Negotiator," *Nunavut Newsletter* No. 5, October-November 1982, p. 14.
27 Tagak Curley, "Industry and Native People: Time for a New Approach to Economic Cooperation," *Proceedings*, Ninth National Northern Development Conference, October 27–29, 1982, Edmonton, Alberta, pp. 145–48.
28 Ibid., p. 148.
29 *Nunavut Newsletter* No.3, August 15, 1982, p. 5.
30 Hon. Allen Blakeney, "How to Effect Change," *Proceedings*, Ninth National Northern Development Conference, October 27–29, 1982, Edmonton, Alberta, pp. 154–59.
31 Murray Dobbin, *The One-and-a-half Men: The Story of Jim Brady and Malcolm Norris, Metis Patriots of the Twentieth Century*, Vancouver: New Star, 1981.
32 John Parker, "Northwest Territories Today," *Proceedings*, Ninth National Northern Development Conference, October 27–29, 1982, Edmonton, Alberta, pp. 27–31.

33 Hon. David H. Searle, Q.C., "How a Canadian Views the Process," *Proceedings*, Ninth National Northern Development Conference, October 27–29, 1982, Edmonton, Alberta, pp. 128–31.
34 Walter Kupsch, "Governments and Northern Development," *Proceedings*, Ninth National Northern Development Conference, October 27–29, 1982, Edmonton, Alberta, pp. 140–44.
35 John Todd, "Eastern Arctic," Panel IV, Economic Development Priority Policies – Regional Perspectives, *Proceedings*, Eighth National Northern Development Conference, November 14–16, 1979, Edmonton, Alberta, pp. 117–19.
36 Simon McInnes, "The Policy Consequences of Northern Development," in Michael Atkinson and Marsha Chandler, eds., *The Politics of Canadian Public Policy*, University of Toronto Press, 1983, pp. 249–65.
37 B. A. Hochstein, *New Rights or No Rights? COPE and the Federal Government of Canada*, M.A. thesis, University of Calgary, 1987, p. 150.
38 Keith J. Crowe, "Claims on the Land," *Arctic Circle* (November–December 1990), pp. 20–21.
39 Vic Humphries, "Which Way Pacific?" *Oilweek*, November 27, 1978, p. 3.
40 "Syncrude Opening," *Oilweek*, December 18, 1978, p. 15.
41 J. Hugh Faulkner, "Increased Northern Development in 1979," *Oilweek*, January 8, 1979, p. 19.
42 "Panarctic Sketches Northern Vision," *Oilweek*, November 19, 1979, pp. 24–25.
43 Keith J. Crowe, "A Summary of Northern Native Claims in Canada: The Process and Progress of Negotiations," *Etudes/Inuit/Studies* 3, no. 1 (1979), pp. 32–34.
44 Allen Maghagak, "Chief Negotiator Reports," *Nunavut Newsletter* (May/June 1983), pp. 2–5.
45 Terry Fenge, "Inuit Surprised by Minister's Lamentable Decision," *Nunavut* 5, no. 8 (August 1986), p. 4.
46 J.U. Bayly, *COPE/ITC Submissions*, Compilation of Final Argument for the Berger Inquiry, 1976, p. 10.
47 Federal White Paper, *In all Fairness*; see also *Outstanding Business: A Native Claims Policy*, Department of Indian Affairs and Northern Development, 1982.
48 "A Good Track Record: says Chief Negotiator" [interview with Bob Kadlun]. *Nunavut*, October 1986, pp. 4–5.
49 Donat Milortuk and Bob Kadlun, "What's the Purpose of Settling Land Claims?" *Nunavut*, October 1986, p. 3.
50 "Special Report: An Open Letter to the People of the Northwest Territories," *Nunavut* 5, no. 3 (March 1986), pp. 5–7.
51 Ibid., p. 6.
52 Murray Coolican, Chairman, *Living Treaties, Lasting Agreements*, Report of the Task Force to Review Comprehensive Claims Policy, Department of Indian Affairs and Northern Development, 1985.
53 "The Task Force Report: TFN Supports Major Policy Recommendations," *Nunavut* 5, no. 5 (May 1986), pp. 2–3.
54 Rebecca Dufton, "Offshore Bill May Complicate Land Claims," *Nunavut* 5, no. 5 (May 1986), p. 4.

55   Donat Milortuk, "Petroleum Resources Act Prejudices Claims," *Nunavut* 5, no. 11 (November/December 1986), pp. 6–7.

56   Guarston Dacks, "The Case Against Dividing the Northwest Territories," *Canadian Public Policy* 12, no. 1 (1986), p. 10.

57   John Amagoalik, "Nunavut is Focus of Shared Identity," *Nunavut* 5, no. 11 (November/December 1986), pp. 10–11.

58   "Backgrounder: Five Years of Frustration are Over," *Nunavut* 5, no. 7 (July 1986), p. 2.

59   "TFN and Dene/Metis Agree on Boundary," *Nunavut* 5, no. 6 (June 1986), p. 3.

60   "Inuit will Share Benefits of Development," *Nunavut* 5, no. 2 (February 1986), pp. 3–4.

61   "Chief Negotiator Explains Claim to Geo-Science Forum," *Nunavut* 6, no. 1 (January 1987), pp. 5–6.

62   Ibid., p. 6.

63   Peter Ernerk, "Offshore Resources Become Food on our Tables," *Nunavut* 6, no. 1 (January 1987), pp. 6–7.

64   "No Room to Move," *Nunavut* 6, no. 5 (May 1987), pp. 2–4.

65   Ibid., p. 4.

66   Rebecca Dufton, "Policies are Inconsistent, says TFN," *Nunavut* 6, no. 6 (June 1987), p. 6.

67   Terry Fenge, "Political Development and Environmental Management in Northern Canada: The Case of the Nunavut Agreement," *Etudes/Inuit/Studies* 16, no. 1–2, 1992, p. 136.

68   "Land ID Workers Visit Two Communities," *Nunavut* 5, no. 1 (January 1986), p. 6.

69   "Land Ownership is Topic at Negotiations," *Nunavut* 6, no. 2 (February 1987), p. 4.

70   "Who owns what? TFN Needs Answers Now to Crucial Questions," *Nunavut* 6, no. 7 (September 1987), p. 3.

71   "TFN, Gov't Initial Agreement on Entry and Access – Inuit will Control Who Goes on Their Land," *Nunavut* 6, no. 8 (October 1987), pp. 2–5.

72   "Inuit Finally Feel Policy's Impact – Government Approves Federal Mandate for TFN Claim," *Nunavut* 7, no. 1 (January 1988), pp. 2–3.

73   J. Merritt, T. Fenge, R. Ames, P. Jull, *Nunavut: Political Choices and Manifest Destiny*, Canadian Arctic Resources Committee, 1989, p. 17.

74   "Inuit get Landmark Agreement on Offshore," *Nunavut* 7, no. 3 (March 1988), pp. 2–3.

75   "Agreement Reached on Land Identification Process – TFN and Government Cross Major Hurdle," *Nunavut* 7, no. 4 (April/May 1988), pp. 2–4.

76   Edi Casimiri, "Cheezie urges cooperation on land ID," Yellowknife *News/North*, January 18, 1988, p. 11.

77   "Maps represent memories for Dene elders," In John Bayly, "On the bias," Yellowknife *News/North*, April 18, 1988, p. 5.

78   Rosie Oolooyuk, John Maksagak, David Aglukark, "… describe the year that was 1988!" *Nunavut* 7, no. 9 (December 1988), p. 3.

79   Ibid.

80   "Sixteen-year Fight by Claimants against 'Extinguishhment' Was Not in Vain," *Nunavut* 8, no. 2 (Spring 1989), p. 3.

81 "Negotiations Round-up," *Nunavut* 8, no. 3 (September 1989), pp. 6–8.
82 "NIRB Will Screen and Review Development Projects," *Nunavut* 8, no. 1 (Winter 1989), pp. 2–7.
83 Terry Fenge, "Political Development and Environmental Management in Northern Canada: The Case of the Nunavut Agreement," *Etudes/Inuit/Studies* 16, no. 1–2 (1992), p. 128.
84 Ibid.
85 Melanie Paniaq, "TFN Seeks Arbitration in Shared Boundary Debate," *Nunavut* 8, no. 1 (Winter 1989), p. 12.
86 *Agreement-in-Principle Between the Inuit of the Nunavut Settlement Area and Her Majesty in Right of Canada*, Tungavik Federation of Nunavut and Department of Indian Affairs and Northern Development, 1990, 372 pp.
87 Ibid., Schedule 19–1.
88 Ibid., Schedules 20–1 to 20–6.
89 Ibid., Schedule 29–7.
90 Ibid., Schedule 19–3.
91 "Paul Quassa, "Signing of the Nunavut Land Claim Agreement-in-Principle Speech, Igloolik, April 30, 1990," *Nunavut* 10, no. 1 (Winter 1991), pp. 2–4.
92 Saali Peter, "Opinion" Is TFN's Agreement-in-Principle Outdated? *Nunavut* 10, no. 1 (Winter 1991), p. 15.
93 Terry Fenge, "National Parks in the Canadian Arctic: The Case of the Nunavut Land Claim Agreement," *Environments* (University of Waterloo. Faculty of Environmental Studies) 22, no. 1 (1993), 21–36.
94 J. U. Bayly, *COPE/ITC Submissions*, Final Arguments Submitted to the Berger Inquiry, 1976, p. 6.

6. EXPLORATION PRESSES ON

1 *The Globe and Mail* (Toronto), November 16, 1979, cited in "Inuit given title at Baker Lake, but mining to continue," *Musk-Ox* 25 (1979), p. 103.
2 *Edmonton Journal*, November 13, 1979, cited in "Mining is still backbone of industry in the North," *Musk-Ox* 25 (1979), p. 104.
3 "Pan Ocean expanding horizons beyond uranium: Active in exploration in four far-flung countries," *The Northern Miner*, March 2, 1978, p. B14.
4 "Panarctic's interest in Bankeno reduced," *The Northern Miner*, March 17, 1978, p. 22.
5 "B.C.'s uranium clamp stuns industry" (editorial), *The Northern Miner*, March 6, 1980, p. A8.
6 William Padgham, "U leads peak exploration pace in NWT," *The Northern Miner*, March 6, 1980, p. C6.
7 Ibid., p. C7.
8 B.E. Owen and W.J. Kops, *The Impact of Policy Change on Decisions in the Mineral Industry*, Centre for Resource Studies, Queen's University, 1980.
9 Saskatoon *Star-Phoenix*, April 8, 1981, cited in "Inuit want new province, share of oil, gas wealth," *Musk-Ox* 28 (1981), p. 104.

10 Saskatoon *Star-Phoenix*, August 14, 1980, cited in "Gulf between Inuit, whites in North said widening," *Musk-Ox* 27 (1980), p. 87.
11 *The Globe and Mail* (Toronto), December, 1980, cited in "Inuit group to provide $25 million loan to finance Cullaton gold mine opening," *Musk-Ox* 28 (1981), p. 106.
12 John Parker, "Northwest Territories Today," *Proceedings*, Ninth National Northern Development Conference, October 27–29, 1982, Edmonton, Alberta, p. 30.
13 H.M. Giegerich, "Resource Development, Planned and Possible," *Proceedings*, Ninth National Northern Development Conference, October 27–29, 1982, Edmonton, Alberta, p. 53.
14 Ibid., p. 59.
15 John Abbot, "Resolutions of the Ninth National Northern Development Conference," *Proceedings*, Ninth National Northern Development Conference, October 27–29, 1982, Edmonton, Alberta, p. 162–67.
16 Saskatoon *Star-Phoenix*, August 28, 1981, cited in "Key Lake mine lease signed; uranium operation proceeds," *Musk-Ox* 28 (1981), p. 108.
17 R.T. Willan, "Uranium uranium uranium," *Canadian Mining Journal* 102, no. 2 (February 1981), p. 150.
18 Ibid., p. 151.
19 Bill Padgham, "Report on NWT mineral activity for 1982–83," *Exploration Overview, 1982–1983*, Yellowknife, Northwest Territories: Department of Indian Affairs and Northern Development. (The *Exploration Overview* is published yearly in November in advance of the December Mining Exploration Forum.) pp. 1–4.
20 Ibid., p. 4.
21 Ibid., p. 4.
22 R.T. Willan, "Uranium," *Canadian Mining Journal* 107, no. 2 (February 1986), pp. 83–85.
23 Ibid.
24 R.T. Willan, "Uranium," *Canadian Mining Journal* 109, no. 3 (March 1988), pp. 73–74.
25 James Borland, "Exploration on the increase in Keewatin District," *The Northern Miner*, March 6, 1989, p. B6.
26 Ibid., p. B16.
27 Bill Padgham, "Introduction to Overview," *Exploration Overview 1991*, Yellowknife: Indian and Northern Affairs Canada, November, 1991, pp. 1–4.
28 Kathryn Welbourn, "Mine review may wait," Yellowknife *News/North*, October 17, 1988, p. 9.
29 Kirsty Jackson, "Mine proposal raise [sic] concern," Yellowknife *News/North*, November 7, 1988, p. 15.
30 Brad Heath, "Dene-Metis and mine avoid confrontation," Yellowknife *News/North*, November 14, 1988, p. 1.
31 Hon. David H. Searle, Q.C., "How a Canadian Views the Process," *Proceedings*, Ninth National Northern Development Conference, October 27–29, 1982, Edmonton, Alberta, p. 131.
32 Yellowknife *News/North*, November 14, 1988, p. A13.
33 Kirsty Jackson, "DIAND's land use permit system criticized," Yellowknife *News/North*, December 5, 1988, p. 4.

34 Peter Kusugak, "Review of mine critical, says candidate" [letter], Yellowknife *News/North*, November 14, 1988, p. 7.
35 Jim Shirley, "Keewatin residents debate proposed uranium mine," *Nunatsiaq News*, January 20, 1989, pp. 9, 10, 12, 13.
36 Ibid., p. 10.
37 Ibid., p. 12.
38 Ibid., pp. 11–12.
39 Tagak Curley, "Committee concerned about FEARO process" [letter], *Nunatsiaq News*, February 3, 1989, p. 6.
40 Peter Ernerk, "Ernerk questions intent of committee" [letter], *Nunatsiaq News*, February 17, 1989, p. 5.
41 Donat Milortok [*sic*], "GNWT Silence is 'unwarranted,' 'disturbing'" [letter]. *Nunatsiaq News*, February 17, 1989, p. 4.
42 Jim Shirley, "Kiggavik: A test case for future development" [commentary], *Nunatsiaq News*, February 24, 1989, p. 4.
43 Jim Shirley, "Mining sessions poorly organized – KUMICC," *Nunatsiaq News*, March 17, 1989, p. 3.
44 Chris O'Brien, "Only half the story?" [letter], *Nunatsiaq News*, March 17, 1989, p. 6.
45 Jim Shirley, "Consultants paint bleak picture of uranium mining," *Nunatsiaq News*, April 28, 1989, pp. 10–12.
46 Ibid., p. 11.
47 Matthew Spence, "FEARO scoping workshops to go ahead as planned," *Nunatsiaq News*, April 28, 1989, p. 2.
48 Jim Shirley, "Baker Lake residents organize to fight Kiggavik," *Nunatsiaq News*, May 5, 1989, p. 13.
49 Jim Shirley, "FEARO workshops charged with emotion," *Nunatsiaq News*, May 12, 1989, pp. 1–2.
50 "Mine concerns students" [Open letter by Grade 9 Social Studies Class, Qitiqliq School, Arviat], and other letters, *Nunatsiaq News*, June 2, 1989, p. 5.
51 Jim Shirley, "Uranium mine opponents asking for more funding," *Nunatsiaq News*, July 7, 1989, p. 18.
52 Matthew Spence, "Hicks quits over performance review," *Nunatsiaq News*, July 7, 1989, p. 12.
53 Paul Williamson, "Politicians should not try to cloud the issue" [open letter], *Nunatsiaq News*, July 21, 1989, p. 5.
54 "UG will offer employment to Northerners," *Nunatsiaq News*, July 21, 1989, p. 15.
55 Jim Shirley, "UG starts public relations campaign in Keewatin," *Nunatsiaq News*, July 28, 1989, p. 8.
56 Jim Shirley, "Vision becomes reality for Nunatsiaq MP Jack Anawak," *Nunatsiaq News*, August 11, 1989, p. 9.
57 Jim Shirley, "Kiggavik opposition regroups after losing Hicks," *Nunatsiaq News*, August 25, 1989, p. 16.
58 Jim Shirley, "Residents voice concerns about uranium," *Nunatsiaq News*, November 17, 1989, p. B2.
59 Jim Shirley, "Anawak tables Kiggavik petitions," *Nunatsiaq News*, November 3, 1989, p. B4.

60 Jim Shirley, "FEARO group hears Saskatchewan residents complain about tailings spill," *Nunatsiaq News*, September 15, 1989, p. 3.
61 Jim Shirley, "Urangesellschaft finds new ore body near Kiggavik site," *Nunatsiaq News*, September 29, 1989, p. 6.
62 "The year in review," *Nunatsiaq News*, January 5, 1990, p. 6.
63 FEARO, *Environmental Impact Statement Guidelines and Government Information Requirements*, Kiggavik Uranium Mine Environmental Assessment Panel, 1989.
64 Jim Shirley, "Residents voice concerns about uranium," *Nunatsiaq News*, November 17, 1989, p. B2.
65 "The year in review," *Nunatsiaq News*, January 5, 1990, p. 8.
66 John Murray, "Government siding with proposed uranium mine" [letter], *Nunatsiaq News*, December 8, 1989, p. 5.
67 Jim Shirley, "Government officials to meet over Rankin tailings issue," *Nunatsiaq News*, December 8, 1989, p. 1.
68 Janet Smellie, "Wray 'nervous' about Kiggavik," *Nunatsiaq News*, December 29, 1989, p. B1.
69 John Bayly, "Jury still 'out' on Kiggavik," *Nunatsiaq News*, January 12, 1990, p. B2.
70 Matthew Spence, "Kiggavik environmental assessment study late," *Nunatsiaq News*, January 12, 1990, p. B1.
71 Jim Shirley, "Keewatin groups prepare response to EIS," *Nunatsiaq News*, February 16, 1990, p. 10.
72 Janet Smellie, "Patterson denies GNWT support of Kiggavik," *Nunatsiaq News*, February 16, 1990, pp. 1–2.
73 Ibid., p. 2.
74 Louis Pilakapsi, "ICC opposes uranium mining" [letter], *Nunatsiaq News*, February 23, 1990, p. 5.
75 Jim Shirley, "KIA demands GNWT take stand on Kiggavik," *Nunatsiaq News*, March 2, 1990, p. 13.
76 Jim Shirley, "Major gold discovery announced near Baker Lake area," *Nunatsiaq News*, February 23, 1990, p. 13.
77 Jim Shirley, "Mining official says acid tailings under control," *Nunatsiaq News*, March 2, 1990, p. 14.
78 Jim Shirley, "KIA members find UG presentation condescending," *Nunatsiaq News*, February 23, 1990, p. 7.
79 Jim Shirley, "KIA demands GNWT take stand on Kiggavik," *Nunatsiaq News*, March 2, 1990, pp. 13, 16.
80 *Kiggavik Uranium Project, Environmental Assessment*, Summary Report by Urangesellschaft Canada Limited, 1990, p. 23.
81 Jim Shirley, "Baker Lake prepares for plebiscite on Kiggavik," *Nunatsiaq News*, March 9, 1990, p. 6.
82 Janet Smellie, "Motion opposing Kiggavik stalled by Legislative Assembly," *Nunatsiaq News*, March 9, 1990, p. 1.
83 Peter Ernerk, "Ernerk's motion repects wishes of constituents" [letter], *Nunatsiaq News*, March 23, 1990, p. 6.
84 Jim Shirley, "Anawak addresses issues," *Nunatsiaq News*, March 30, 1990, p. 18.

85 Jim Shirley, "BL says 'no' to Kiggavik," *Nunatsiaq News*, March 30, 1990, p. 1.
86 Jim Shirley, "Mayor wants UG to abandon Kiggavik," *Nunatsiaq News*, April 20, 1990, p. 15.
87 Jim Shirley, "Company asks BL for permission to explore for uranium," *Nunatsiaq News*, June 8, 1990, p. 1.
88 Jim Shirley, "Environmental assessment – a democratic process," *Nunatsiaq News*, March 23, 1990, pp. 19–20.
89 Janet Smellie, "Urangesellschaft puts uranium mine on hold," *Nunatsiaq News*, July 13, 1990, p. 8.
90 FEARO, *Compendium of Submissions Received*, May 1989, p. 140.
91 Janet Smellie, "FEARO says Kiggavik review will be stalled for a year," *Nunatsiaq News*, October 12, 1990, p. 7.
92 Janet Smellie, "Keewatin businesses plot own future with economic strategy," *Nunatsiaq News*, November 23, 1990, p. 17.
93 FEARO, *Compendium of Submissions*, May 1989, pp. 90–117.
94 "Mine opposed by Inuit," *Canadian Energy News* 5, no. 29 (March 1, 1989), p. 230.
95 Bill Padgham, "Introduction to Overview," *Exploration Overview 1991*, Yellowknife: Indian and Northern Affairs Canada, November, 1991, pp. 1–4.

7. NEW OWNERS

1 Bob McPherson, personal diaries and notes, 1989–91.
2 Keith Crowe, "Claims on the Land," *Arctic Circle* (November–December 1990), pp. 14–23.
3 *Agreement-in-Principle Between the Inuit of the Nunavut Settlement Area and Her Majesty in Right of Canada*, 1990, Schedule 20–1.
4 Ibid., Schedule 20–7.
5 Mining Association of Canada, *Land Access Issue Group - Whitehorse Mining Initiative, Final Report*, Mining Association of Canada, 1994, pp. 9–11.
6 Terry Fenge, "National Parks in the Canadian Arctic: The Case of the Nunavut Land Claim Agreement," *Environments* (University of Waterloo, Faculty of Environmental Studies) 22, no. 1 (1993), pp. 21–36.
7 Prospectors and Developers Association of Canada, *Exploration and Development Highlights*, March 1992, p. 17.
8 Fenge, 1993, pp. 30–31.
9 Ibid., p. 32
10 Ibid., p. 33.
11 Bob McPherson, *TFN Land Claim: Handbook of Nunavut Mineral Potential*, Wollex Exploration, Geological Consultants, Calgary, Alberta, 1992.

8. LAND BASE ACHIEVED

1. Terry Fenge, "Land selection negotiations completed Nunavut wide," *Nunavut* 10, no. 3 (September 1991), p. 6.
2. Ibid.
3. Bruce Gillies, "Conclusion of Land Ownership Negotiations – Some Final Comments," *Nunavut* 10, no. 4 (November 1991), p. 8.
4. Tom Molloy, Q.C., "Negotiating the Nunavut Agreement: A View from the Government's Side," *Northern Perspectives* 21, no. 3 (Fall 1993), pp. 9–11.
5. Fenge, 1993, p. 35.
6. Ibid., p. 1.
7. Robert D. Smith, "Nunavut Set To Boom" [feature interview with Murray W. Pyke], *Canadian Miner* (Flagship Edition, January 1997), pp. 18–21.

INDEX

A

Aboriginal rights, 52, 59, 78, 83, 85, 140, 144, 149, 153, 159
Aboriginal title, i, 153, 203, 269
Activism, 95, 172
Adaptation, 113
Advocacy, 30, 62
Aeromag survey, 41
Agar, Frank, 94
Aglukark, David, 153, 243
Agreement in Principle (AIP), 155, 157, 203, 213, 219, 225, 238, 243, 255
Agreement on subsurface lands, 265
Agreement was reached, 267
Aircraft, 35, 37, 46, 72, 85
Akeorok, Jaypeetee, 149, 220–23
Alaska, 65, 158
Alaska Native Claims Settlement, 68
Alcohol, 14, 31, 72, 102
Alienation, 21, 60, 78, 80, 136, 145
Allan, Victor, 52, 65
Allen, Joe, 211
Allmand, Hon. Warren, 81
Allooloo, Titus, 155
Alurut, Peter, 131
Amagoalik, Ekaksak, 242
Amagoalik, John, 69, 82, 86, 121, 125, 131, 145
Amaroali, 90
Amaroalik, Paul, 242
Amaroalik, Simeonie, 242
Amarook, Michael, 72, 82, 87, 124, 243–48, 250
Amer group, 161
American Institute of Mining Engineers (AIME), 46
Amok, 49
Anablak, Stanley, 211, 257, 261
Anawak, Donat, 104, 186
Anawak, Jack, 179, 181, 191, 194
Andre, Robert, 68
Andrews, Tony, 226
Anguhadluk, Mary-Rose, 177
Anvil, 45
Aodla, Walter, 131

Aola, Lazarus, 225
Arctic Bay, 89, 96, 102, 108
Arctic Canadian sovereignty, 22
Arctic frontier, 48
Arctic Pilot Project, 128, 163
Area of interest or occupancy, 152, 156, 246, 250
Arnaan'aaq, Ruby, 152, 188
Arnalukjuark, John, 253
Arreak, Malachi, 210, 213, 218, 222, 243, 257, 260
Arvaluk, James, 69, 82
Arviat, 64. *See also* Eskimo Point.
Asarco, 38
Asbestos Hill, 14
Assimilation, 19, 31, 59, 68
Associated Mining Consultants Ltd. (AMCL), 205, 212
Athabasca basin, 164
Atkinson Point, 65
Atomic Energy Canada (AEC), 184
Atomic Energy Commission, 45
Atomic Energy Control Board (AECB), 190
Attagutsiak, Isaac, 91, 97, 106, 109, 111
Avaala, John, 72
Avadluk, Noel, 42, 265
Axel Heiberg, 40
Ayaruark, John, 29, 31

B

BA Oil, 47
Back River, 16
Baffin Island, 131, 206, 209, 213, 227
Baffin Regional Inuit Association (BRIA), 70, 112, 114, 117
Baffinland Iron, 41, 50
Baker Lake, 17, 26, 30, 33, 54, 85, 124, 159, 169, 171, 174, 248
Baker Lake court case, 83
Ballantyne, Mike, 190
Bancroft, 48
Bankeno, 43
Banks Island, 51, 60
Barber, Lloyd, 51, 60
Bargaining, 74
Barrett, J. E., 127
Barringer, 44

Base metals, 39, 165
Bateman, J. D., 42
Bates Royal Commission, 160, 165
Bathurst Inlet, 40, 42
Bathurst Island, 42
Bayly, John U., 152, 158, 185
Beak Consultants, 169, 178, 188
Beaufort Sea, 77, 128
Beaverlodge, 47
Belcher Islands, 3, 22
Belcher Mining Corp., 38
Bell, MacIntosh, 40
Benefits, 148, 185
Berger, Justice Thomas R., 66, 69
Berger Inquiry, 77, 121, 123, 158
Bernier, J. E., 89
Bertell, Rosalie, 175
Bielenstein, Hans, 205, 209, 227
Billiton, 99
Bissett Lake, 164
Blackadar, Robert, 90, 216
Blair, Robert, 121, 123
Blakeney, Hon. Allan, 117, 135
Blanchett, G. H., 37
Blizzard Mountain property, 165
Blondin, Ted, 170
Boothia Peninsula, 40
Borealis, 132, 166, 211, 221, 247
Borland, James, 167
Bouchard, Lucien, 173
Boundary issue, 146, 155, 257, 267, 269
Brady, Jim, xxi, 136
Bray Lake, 24
Brinex, 47, 124, 165
British Ungava, 40
Brody, Hugh, 95, 100, 107, 114
Brown, Lavinia, 177, 239
Browning, J. C., 6
Bruce, Louie, 251
Bruk, John, 123, 159
Buchanan, Judd, 81
Buckley, Helen, 59

C

Calder, Frank, 59
Calder decision, 85
Cambridge Bay, 210

Cameron, Donald, 25
Camp life, 19
Campbell, Neil, 41
Canada Mining Regulations, xix, xxiv, 39, 169, 201, 208, 226, 241, 272
Canada Petroleum Resources Bill, 145
Canadian Arctic Resources Committee (CARC), 52, 63, 116, 126, 186
Canadian Association for Adult Education (CAAE), 57
Canadian Association for the Support of Native Peoples (CASNP), 61, 63
Canadian Bechtel Ltd., 42
Canadian Broadcasting Corporation (CBC), 21
Canadian Coalition for Nuclear Responsibility, 175
Canadian Institute of Mining and Metallurgy (CIMM), 161
Canadian Mining Association, 42, 212, 226
Canadian National Railway (CNR), 33
Canadian Pacific Railway (CPR), xxi, xxiv, 271
Canadian Wildlife Service (CWS), 82, 260
Cantley, James, 1, 4, 22
Cantung tungsten mine, 15, 43
Caribou, 23, 30, 55, 75, 82, 85, 115, 159, 177, 182, 189, 225
Caribou Protection, 167
Carney, Pat, 53
Carter Commission, 44, 46
Casual labourer, 108
Central Electric Generating Board (CEGB), 172, 175
Certainty, 144, 150, 203
Cesium, 182, 185
Chamber of Mines, 141, 171, 177
Chambers, Luther, 159
Chesterfield Inlet, 2, 7, 13, 33, 172
Chrétien, Hon. Jean, 51, 60, 64, 89, 97, 109
Churchill, 33
Citizens' organization, 58
Claim holdings, 182, 219
Clark, Allan, 58
Cluff Mining, 166
Coast, 215, 236, 246, 252, 259
Collins Bay, 166

Co-management, 70, 132, 142, 147, 154, 259
Comaplex Resources, 188, 205
Cominco, 39–41, 43, 54, 83, 115, 126, 163, 223, 238, 240
Committee on Eskimo Affairs, 5, 29
Communication failure, 96, 103, 115
Community associations, 12, 21
Community awareness, 18, 30
Community councils, 29
Community development, 21, 28, 59
Community land identification and negotiating teams (CLINTs), 151, 156, 209, 213, 216, 218, 221, 246, 252
Compensation, 79
Comprehensive settlement, 133
Compromise, 254
Concerned Citizens Committee (CCC), 178
Confidentiality, 207
Conflicts, 215, 217
Consolidated Magnorth Oakwood, 142
Consultants, 76, 178, 180, 188, 257
Consultation, 31, 99, 104, 124, 132, 141, 145, 195, 205
Contamination, 175, 189, 198
Control of development, 76
Control of land, 30, 64, 68, 86, 102, 124, 132, 140, 146, 157, 195, 198, 200, 225, 271
Control of mineral lands, 260
Controlled development, 73
Contwoyto Lake, 39, 42, 45, 162, 258, 266
Coolican, Murray, 144, 148, 150
Co-operatives, 21, 28
Committee for Original Peoples' Entitlement (COPE), 52, 64, 69, 77, 79, 129, 138
  Agreement in Principle, 79
  Final Agreement, 138
Copper, 46, 166, 168
Coppermine conference, 51, 61, 63, 81
Coppermine River Ltd., 46
Coppermine, 40, 60, 257
Coral Harbour, 10, 243
Cornwallis Island, 22
Corridors, 215, 220, 241, 248–50, 252, 254–56, 259
Cournoyea, Nellie, 128, 184, 193
Cousins, Mary, 62

Crest, 41
Crnkovich, Mary, 227, 237, 240, 243, 250, 257, 260, 263, 267
Crombie, David, 144
Cross Bay, 247, 253
Crowe, Keith, 80, 132, 139
Crown land, 38
Cullaton Gold Mines, 129, 163
Cullaton Lake, 44, 47, 162
Culture, 31
Cumberland Resources, 188
Cumming, Peter, 52, 60, 63, 69, 81
Cunningham, Michael, 243, 250, 256, 261
Curley, Tagak, 53, 60, 63, 69, 74, 76, 81, 104, 118, 122, 128, 132, 134, 163
  Kiggavik environmental review, 172–75, 183, 188, 200
  land selection negotiations, 243, 249, 250–57
Currie, Walter, 60

D

Dacks, Gurston, 145
Dailey, R. C. and L. A., 9, 17, 20
Davidson, Stanley, 45
Del Tredici, Robert, 175
Demers, Clovis, 133
Dene, 68, 136, 141, 152, 170, 258, 266
Denhez, Marc, 159
Department of Indian and Northern Development (DIAND), 52, 97, 149, 161, 170, 174, 187, 206, 227
Department of Mines and Resources (DMR), 2
Department of National Defence (DND), 216, 258
Department of Northern Affairs (DNA), 16, 58
Depleting, 118
Destitution, 4
Developers, 186
Development, 38, 52, 59, 62, 67, 70, 77, 82, 84, 87, 93, 101, 105, 118, 124, 132–36, 146–48, 159, 163, 173, 189, 198
Devolution, 143, 154
Dew Line, 10, 14, 22, 260

Index · 295

Dewar, Barry, 212, 214, 247, 257
Dia Met, 168
Diamond, 168
Diana River, 253, 256
Dickens, Punch, 35
Disruption of lifestyle, 27
Dogrib Tribal Council, 170
Dominion Explorers, 34
Douglas, Rod, 176
DPA Consulting, 128
Drummie, F. R., 155
Drums, 74
Drury report, 122, 125
Dryborough, John, 5, 35
Dubawnt, 33, 36, 54, 161
Dubawnt sandstone, 49
Dunne, Norman, 62
Dynamic Mining, 49

E

Early-stage liaison, 137
Easton, Andy, 7, 14, 29
Echo Bay Mines, 162
Ecology North, 178
Economic benefits, 143, 259
Economic self-sufficiency, 141, 255
Education, 57
Edwards, Gordon, 175, 199
Eetoolook, James, 131, 155, 247
Eldorado, 47, 166
Elias, Edna, 257, 265
Ellesmere, 40
Elliot Lake, 48
EM-16, 44
Employment, 7, 9, 17, 21, 50, 70, 92, 97, 100, 107, 109, 112, 114, 125–27, 136, 138, 179, 185
Employment contracts, 135
Encroachment, 67
Energy crisis, 140
English, Albert, 89
Ennadai, 23, 26, 34
Ennadai Lake, 252
Ennadai-Rankin, 44
Environment Canada, 200

Environment, 46, 56, 81, 94, 118, 126, 148, 161, 164, 170, 179, 188, 248, 260, 264
Environmental Assessment Review Panel (EARP), 134, 154
Environmental consequences, 125
  perceived impact, 197
Environmental impact compendium (EIC) guidelines, 196
Environmental impact statement (EIS), 176, 182, 186, 193
Environmental monitoring, 183
Environmental review, 132
Environmental Scrutiny, 169
Epp, Hon. Jake, 123, 159
Eqe Bay, 216, 221–24
Equity participation, 151
Erasmus, Bill, 152
Ernerk, Peter, 81, 147, 172, 177, 181, 185, 187, 190–92
Esau, Peter, 51
Eskimo Conference, 60
Eskimo Point. *See also,* 8, 13, 24, 27, Arviat.
Essex Minerals, 83
Etegoyok, Harold, 72
Evaigotailak, Joe Allen, 211, 257, 265
Evaluardjuk, Mark, 155
Evoy, Jim, 195
Exploration garbage, 75
Exploration moratorium, 80, 138, 165
Exploration philosophy, 47
Exposure of workers, 164
Extinguishment, 68, 78, 84, 139, 144, 153, 158

F

Falconbridge Nickel, 40
Family allowances, 1
Farquharson, Graham, 96, 102, 110, 116
Faulkner, Hugh, 79, 82, 139
Faulkner, Neil, 93
Federal concerns, needs, demands, 217, 253, 258
Federal criteria for negotiating, 243
  Exceptional circumstances, 221, 224
  Representative selections of third party lands, 237

Representative share of third party lands, 255
Federal Environmental Assessment Review Office (FEARO), 154, 169, 171, 176, 178, 182, 185, 189, 200
Federal obstruction, 133
Federal Resources Ltd., 47
Female employment, 116
Fenge, Terry, 130, 149, 227, 237, 250, 257, 259, 263, 269
Ferguson Lake, 40
Ferguson River, 33
Ferris, Ed, 188
Fisher, Moses, 43
Flaherty, Martha, 250, 257
Flow through shares, 168
Ford, Fred, 190, 192
Ford, Norman, 75
Fording Coal, xxi
Fort Chimo, 1
Fort Norman, 67
Fort Smith, 36
Foster, Terry, 15
Fragmentation, 220, 222
Fraser, J. A., 40
Free entry system, xix, 38, 150, 261
Free trade, 167
Freehold owners, 209, 261, 268, 271
Freeman, Milton, 65
Freeze, 54, 63, 81, 157, 160
Fuller, W. A., 52
Funding, 61, 133, 162, 178, 184, 195

G

Gamble, Bob, 216, 222, 239
Garry Lake, 16, 23, 27
Gasson, J. E., 61
Geco base metal mine, 43
Genocide, 121, 123
Geological Survey of Canada (GSC), 25, 33, 40, 44, 47, 49, 54, 77, 91, 239, 246
Geonics, 44
Geophysical, 39, 43
Geoscience Forum, 211
Giant Yellowknife, 39
Gibbons, Walter, 208
Gibson, Gordon, 48

Gibson, Robert, 99, 103, 105
Gillies, Bruce, 156, 206, 208, 211, 213, 219, 224, 240, 243, 250, 257, 267, 269
Gingerich, H. M., 163
Ginn, Bob, 226
Goff, Steve, 243
Gold, 39, 165, 188
Golden, Aubrey, 84
Goodarsi, Fari, 246
Government, 107, 137, 140
Graham, William, 84
Great Whale River, 22
Green Plan, 270
Greenland, 187
Griep, J. L., 168
Grise Fiord, 22
Gross, G. A., 43
Guidelines, 178
Gulf Minerals, 49, 160
Gulf Oil, 47
Gunnar Mining, 41, 43

H

Halli, Paul, 224
Hallow, 26
Hammell, Jack, 34
Haqpi, Michael, 149
Harding, Jim, 174
Hay, Keith, 149, 213, 236
Health study, 177
Helicopters, 25, 40, 44, 51, 54
Henik Lake, 23, 25
Heninga Lake, 252
Herchmer, H., 81
Heritage, safeguarding, 136
Hicks, Jack, 172, 174, 177, 180, 186, 189, 198
Hinds, Margery, 3, 90
Hobart, Charles, 112
Hodgson, Stuart, 104
Hoefer, Tom, 226
Hogarth, Don, 227
Holyk, Walter, 91
Horning, Donald, 46
Housing, 116
Houston, James, 23

Index · 297

Hubachek, Peter, 188
Hudson Bay, 34, 142, 148, 184
Hudson Bay Railway, 33, 80
Hudson's Bay Company (HBC), 3, 10
Hunt, Digby, 66, 105
Hurwitz formation, 44

I

Identification, 153, 156
Igloolik, 157, 209
Igyaka, 26
Ihalmiut, 27
Ikuutaq, David, 72
Impact review, 132
Impacts, 71, 82, 124, 127, 138, 142, 171, 176, 183, 186, 188, 190
Imperial Oil, 46
In perpetuity, 153
Indian Eskimo Association (IEA), 57, 60
Indians, 252
Infrastructure, 109, 164, 261
Injunction, 83
Innakatsik, Matthew, 72
Integrated management system, 154
Interdisciplinary System Ltd., 82
Interests of all Canadians, 137
Interim agreements, 219
Interim withdrawal, 157, 227. *See also* land freeze.
International Institute of Conern for Public Health (IICPH), 175
International Nickel Company (INCO), 39, 45
Inuit Circumpolar Conference (ICC), 187, 191
Inuit Cultural Institute, 64, 69
Inuit Development Corp., 118, 126, 129, 163
Inuit Impact and Benefits Agreement (IIBA), 146, 194
Inuit lands, 142
Inuit rights, 60, 124, 130
  offshore, 141, 147, 151
Inuit Settlement Lands, 155
Inuit traditional knowledge, 71, 142, 198
  lifestyle, 64, 101, 123, 137, 174, 179

Inuit Tapirisat of Canada (ITC), 54, 64, 78, 85, 101, 116, 122, 129, 162
Inuk, Joe, 225
Inuk prospector, 265
Inukjuak, 3, 22
Inuktitut, 12, 91, 97, 174, 186, 189
Inuvialuit lands, 80
Inuvialuit Nunangat, 77
Inuvialuit, 67, 77, 130, 138
Inventory, 206, 210
Involvement, 132
Iootna, 24
Iqulik, Toona, 72
Iron formation, 39
Iron ore, 38
Iron Ore Company, 77
Itavia, 16
Ittinuar, Lizzie, 177
Ittinuar, Ollie, 104
Ittinuar, Peter, 125
Ivvalu, Lucassi, 206, 209, 213, 217, 223, 236, 243
Izok, 261–64, 267

J

J.C. Sproule & Associates, 43
James Bay Agreement, 68, 79, 130
Jefferson, Charlie, 238
Jelliss, Arvin, 117
Jenness, Diamond, 90, 107
Jobs, 102, 181, 196
Johan Beetz, 43, 48
Johnston, Dave, 238
Johnston, Oscar, 5, 35
Joint ventures, 136
Jorah, Samson, 176
Joubin, Frank, 135
Judge Sissons Lake, 189

K

Kabloonamiut, 18–20
Kablutsiak, Josiah, 62
Kadjuk, Anthyme, 75
Kadlun, Bob, 131, 142, 147–51, 155, 240, 251, 253, 255, 257, 260, 264

Kakaseak, 91
Kalluak, Mark, 170
Kalluk, Levi, 98, 109
Kaludjak, Paul, 250, 253
Kaminak, 34
Kamingoak, Peter, 266
Kaminuriak, 34
Kamuk, Alvin, 186
Karetak, Rhoda, 177
Karlik, Pierre, 75
Karpen, Bernie, 257
Kasba Lake, 34
Kashla, Bill, 72
Kazan River, 33
Kazan River Falls, 49, 54, 160
Kazan River people, 16, 27
Keane, Tony, 242
Keewatin, 71, 131, 172, 181, 196, 206, 250
Keewatin Chamber of Commerce, 196
Keewatin Inuit Association (KIA), 64, 70, 142
Keewatin Regional Council (KRC), 172, 180
Kerr, Bill, 24
Kerr Addison, 41
Key Lake Inquiry Board, 164
Key Lake Mining Corporation (KLMC), 164, 181
Keyootak, Pauloosie, 155
Keytook, Paul, 214
Kidd Creek Mines, 111
Kiggavik, 121, 161, 167, 169, 171, 175, 185, 189, 193, 195, 200, 247
Kikkik, 26
Kilabuk, Ipeele, 62
Kilabuk, Moe, 149
Killulark, John, 131
King William Island, 40
Kingaryalik Lake, 35
Kinship, 18
Kitikmeot, 131, 206, 210, 257
Kitts-Michelin, 165
Knight, Cyril, 35
"Known and potential mineral deposits", 155, 218
Koe, James, 52
Koneak, George, 29, 59
Koodlooalook, Jimmy, 3
Koonoo, Moses, 223, 225

Kowtiuk, Andy, 254, 257
Kreelak, Martin, 190
Kritterdlik, David, 200
Kudloo, Thomas, 176
Kujarik Park, 256
Kulchyski, Peter, 3, 25, 32
Kupeuna, Jack, 148, 257
Kupsch, Walter, 137
Kusugak, Jose, 252
Kusugak, Mike, 188
Kusugak, Peter, 171

L

Labour, 106, 115
Labow, Larry, 45
Lake Garrow, 169
Lancaster Sound, 142, 167
Land, 67, 124, 147, 166
Land administration, 271
Land claims, 53, 63, 67, 74, 123, 131, 138, 159, 164, 167, 176, 193, 200, 223, 266
Land claims effect, 167
Land freeze, xxiii, 51, 54, 56, 63, 69, 74, 82, 132, 145, 160, 200, 206, 210–12
Land identification, selection, 80, 142, 204, 255
Land ownership, 141, 147, 149
Land ownership negotiations, 152, 157, 269
Land per capita, 139
Land rights, 79
Land use, 65, 170, 200, 247, 249
Land use and occupation, 142
Land use plans, 154
Land use regulations, 54, 56
Language, 31, 59
Lapis lazuli, 227, 237
Laporte, Pierre, 174
Larsen, Henry, 4
Le Moyne, 38
Leaders, 13, 15, 18, 20, 25, 28, 31, 51, 61, 124, 135, 142, 163, 176, 271
Legal surveys, 215, 223, 238
Legislative Assembly, 173, 175, 190
LeMeur, Rev. Robert, 129
Leonard, Bob, 196
Levi, 91

Index · 299

Lidbetter, Anne Padlo, 59
Limited entry regime, 272
Lindsey, Thayer, 48
Local initiative, 29
Lone Gull, 161, 164
Lord, Cliff, 25
Lotz, Jim, 52
Lucky Eagle Mines, 188
Lupin mine, 40, 259, 261, 263
Lush, Don, 188
Lyall, Charlie, 155
Lynn, P., 24
Lynn Lake, 14

M

MacAlpine, C.D.H., 36
Mackenzie Delta, 3
Mackenzie gas pipeline, 53
Mackenzie Valley pipeline, 70, 77
Mackenzie Valley Pipeline Inquiry, 66
MacMillan, Viola, 42
MacPhar geophysics, 39
MacQuarrie, Bob, 176
Maghagak, Allen, 132, 141
Magsagak, John, 153, 266
Maguse, 34
Mahoney, Justice Patrick, 83–86
Mahoney decision, 159
Makkovik, 48
Makpah, Bernadett, 250
Makpah, C., 62
Malani, David, 176
Management, 126, 132
Marble Island, 7, 252
Marshall, David, 176
Marshall, Jim, 24, 110, 116
Mary River, 40, 50, 107
Mary-Rousselier, Father Guy, 100
Masse, Marcel, 145
Matromatteo, Earnest, 176
Mautaritaaq, Michael, 82
McCandlish, Keith, 207
McEwen, Ernest, 60
McInnes, J. W., 91, 138
McInnes, Simon, 138
McKinnon, Ross, 151, 243, 250, 257
McKnight, Bill, 142, 148, 150

McNeith, Lorne, 215
McPhar Geophysics, 45
McPherson, Bob, 204, 214
Meadowbank Lake, 188
Megill, D. K., 50
Meliadine gold trend, 254
Melling, John, 57
Menarik, Elijah, 29, 59
Merritt, John, 212
Metallgesellschaft, 47, 99
Mike, Johnnie, 236
Milne Inlet, 50
Milortuk, Donat, 75, 131, 141, 145, 150, 172, 189, 243
Mineral advisors, 211
Mineral advisory committee, 123, 159
Mineral and Energy Resource Assessment Committee (MERA), 239
Mineral claims, 166, 168, 192, 208, 216, 218, 221, 237
Mineral data compilation, 207
Mineral exploration, 38
Mineral resources, 123, 206
Mineral Resources International (MRI), 92, 94, 99, 109
Mineral rights acquisition, 240
Mineral rights ownership, 69, 125, 130, 269
Mineral rights selection, 242
Mineral value, 141
Mingeriak, Pudloo, 155
Mining Association of Canada (MAC), 161
Misheralak, Tommy, 75
Mistake Bay, 34
Mitchell, Robert, 132, 164
Mitchell, Scott, 249
Mitigation, 183
ML31, 247
Moelaert, John, 172
Mogul Mining Corporation, 7
Mollison, Dick, 91
Molloy, Tom, 132, 151, 155, 157, 214–67, 270
Monture, Gilbert, 58
Moore, Mike, 117
Moral issues, 172, 175, 200
More, Minnie, 59
Moses, Elliot, 59
Mosesee, 50
Motka, 47

Mounik, 24
Mowat, Farley, 2, 5, 11, 27, 29, 33
Mudry, Phil, 205, 211, 249, 251, 254–56
Munro, Hon. John, 58, 122, 125, 130
Murdoch, Peter, 28
Murray Mining, 38
Murray, John, 184, 199
Muskox Intrusion, 40

N

Naden, Bill, 36
Nanisivik mine, xxii, 41, 64, 69, 75, 89, 92, 99, 104, 108, 115, 166, 216, 225, 239
Nashook, Mary, 250
National parks, 143, 158, 216, 220, 222, 236, 238, 247, 260, 269
Nationalistic oil policy, 140
Native claims, 66
Native councils, 12
Native opposition, 165
Neary, Bernard, 58
Negotiations, 86, 123, 132, 147, 150, 191, 226, 250, 255
Neptune Resources, 170
Neuterajuk, 91
New records, 161
Nickel mine, 186
Niego, Joe, 170
Ningeocheak, Raymond, 243
Ningiuk, Laisa, 222
Nininger, Robert, 45
Nipissing Mining, 35
Niptaniatiak, Joe, 266
Nishga'a, 52, 59
Noranda, 54, 83
Norlands Petroleum, 118, 128
Norris, Malcolm, xxi, 59, 136
North Rankin Nickel Mines, 6, 30, 76, 177
Northern Advisory Committee, 125
Northern Anti-Uranium Coalition (NAUC), 174, 177, 180, 189
Northern Development Conference, 123, 138
Northern mineral policy of government, 148, 167
Northern service officers (NSOs), 2, 23, 27

Northwest Territories, Government of the (GNWT), 100, 102, 173, 185, 187, 243
Northwest Territories Chamber of Mines, 171, 188, 191, 200, 209, 226
Northwest Territories Federation of Labour, 195
Nuclear Free North, 169, 184
Nuclear power, 171
Nuclear weapons, 171, 175, 182
Nueltin Lake, 23
Nunamiut, 18, 31
Nunasi Corp., 134
Nunavut, 73, 143, 145, 147, 163, 193, 266, 270
Nunavut Final Agreement, 269
Nunavut impact review board (NIRB), 134, 142, 146, 148, 151, 153, 194
Nunavut planning commission (NPC), 154
Nunavut proposal, 70, 77
Nunavut Settlement Area, 156
Nunavut Trust, 157
Nunavut Tunngavik Inc., xx
Nunavut Water Board (NWB), 154
Nunavut Wildlife Management Board (NWMB), 154
Nungak, Zebedee, 128
Nutter, Dave, 177, 188, 200, 206, 209, 240, 242

O

O'Brien Energy, 162
O'Brien, Chris, 174, 177
Office of Native Claims, 80, 132
Oil and gas development, 64, 67, 70, 135
Oil and gas rights, 77
Oil and gas spill, 123
Okalik, Looee, 250
Okpik, Abraham, 29, 31, 58
Omigluk, Angela, 42
Onalik, Janet, 180
Ontario Securities Commission (OSC), 48
Oohootuk, 24
Oolooyuk, Rosie, 152
Ootuk, 26
Operation Keewatin, 25

Organization of the Petroleum Exporting Countries (OPEC) energy crisis, 52, 139
Orsokitik, Louis, 75
Otokiak, Joe, 211, 213, 257, 262, 267
Outcrop Ltd., 128
Owetaltuk, Jacob, 62
Ownership by Inuit of land, 64, 69, 130, 133, 141, 211
Ownership by Inuit of land in every area, 219
Ownership by Inuit of land in finite areas, 140

P

Padgham, Bill, 161, 165, 168, 208
Padlei, 24, 26, 35
Palitug, Sam, 215
PanArctic Oils, 42, 47, 93, 100, 160, 166
Pan Canadian Petroleum, xxi
Pan Ocean, 54, 83, 160, 164
Panegoosho, Mary, 29
Pangnirtung, 2, 34, 227
Parker Line, 261, 263, 266
Parker, John, 102, 122, 136, 155, 163, 257
Parks Canada, 216, 236, 239
Parre, Mike, 216
Participation of Inuit in decisions, 144, 147
Participation of Inuit in mineral developments, 118, 124, 129, 138, 163, 262
Partnership ventures, 134
Paterk, Joe, 75
Patterson, Dennis, 122, 155, 173, 193
Paulette, François, 53
Pearce, Richard, 36
Pearson, Brian, 178
Pearson, Chris, 130
Pemberton, Roger, 46
Penny, Brian, 117
Perrin, Dave, 236, 250, 257
Peter, Saali, 157
Petition, 176, 180, 190
Petro Canada, 162
Phillips, R.A.J., 58
Pierce, Jon, 206, 208, 212
Pikuyak, Percy, 110, 112

Pilakapsi, Louis, 131, 155, 171, 183, 187, 189, 255, 257
Pine Point, 39–43, 160, 165
Piryuaq, Barnabas, 181
Pissuk, 75
"Place where people find things", 90, 109
Planning development or land use, 132, 180, 200, 212, 272
Plebiscite, 133, 190, 192
PNC Exploration (Canada) Co. Ltd., 166, 192, 248
Pokiak, Bertram, 65
Polar Gas pipeline, 69, 71, 124, 169
Polaris mine, 107, 116, 118, 126, 165, 169, 226, 238, 242
Policy, 145, 147, 149, 162, 207, 218, 258, 260
Polluters, 76
Pond Inlet, 22, 213
Price, Ray, 48
Prima, Dan, 169
Prince of Wales, 40
Private agreement with developer, 170, 194
Production of mineral concentrates, 110
Proportionate share, 263
Prospecting, 25, 38, 41, 55, 238
Prospecting permits, 39, 127, 160, 166, 217, 221, 262
Prospective areas, 214
Prospectors, 22, 24, 37, 40, 60, 66, 72, 147, 162, 206, 226, 265
Prospectors and Developers Association of Canada (PDAC), xx, 41, 161, 226
Protection zone, 182
Protest movements, 21
Prudhoe Bay, 64
Pyke, Murray, 205, 207, 211, 227, 243, 247, 249, 257, 260, 262, 271
Pyramid Mining Company, 41

Q

Qanatsiag, Solomon, 216, 223
Qaunaq, Sakiassie, 112
Quantum, 150, 152, 156, 207, 210, 213, 217, 220–25, 236, 243, 252, 256, 264, 266

Quassa, Paul, 143, 146, 151, 157, 209, 214, 218, 220–69
Qumak, Noah, 62, 81

R

Rabbit Lake, 49, 166, 181
Raddi, Sam, 66, 69, 79, 130
Radioactivity, 171, 175, 177, 181, 183, 200
Raglan Nickel Mines, 38
Rankin Inlet, 9, 15, 20, 34, 48, 71, 75, 175, 184, 249, 255
Rankin Inlet Nickel Mines, xxii, 6
Rankin-Ennadai, 167
Rasky, Frank, 50
Ratification, 157, 203, 236
Rayrock, 40
Recession, 165
Referendum, 178
Regional associations, 131
Relocation, 15, 22, 27
Renaud, Father André, 57
Repulse Bay, 8, 13
Resolute Bay, 22, 126, 128, 214, 237
Resource management, 67, 195
Revenue sharing, 219
Rexspar, 165
Riewe, Rick, 149
Rio Alto, 54
Risk capital, 48
Risks, 172
Roberts Mining Company, 42
Robertson, Gordon, 30
Ronka, 44
Rotation of work, 115, 117, 126, 128
Rowley, Graham, 50, 163
Royal Canadian Mounted Police (RCMP), 2–4, 22, 24, 26
Royalty, 39, 68, 72, 78, 80, 94, 123, 130, 132, 146, 157, 164, 169, 241, 251, 263
Rudnicki, Walter, 22, 27

S

Safety concerns, 193, 200
St. Simeon, 43

Salluviniq, Allie, 214, 217, 224, 226, 241
Sandy, Tongola, 214, 223, 243, 256
Sanikiluaq, 156
Saskatchewan, 117, 164, 181
Saunders, Douglas, 60
Schiller, Ed, 179, 182
Science Council of Canada, 103
Scintrex Ltd., 49
Scoping workshops, 176, 180, 182, 184
Scott, Frank, 58
Scott, Paul, 169, 176, 188
Scottie, Joan, 176, 178, 180, 186
Searle, David, 137, 170
Seaton, John, 209
Seismic, 51, 60
Selco Exploration, 40, 44, 47
Selection of land, 78, 141, 149, 215
Self-determination, 1
Self-expression, 22
Self-government, 32
Self-sufficiency, 252
Semmler, Agnes, 52, 81
Settlement council, 73, 95, 98, 109
Settlement lands, 155, 203
Shappa, Kik, 131
Sharp, Bob, 242
Sherman, Patricia, 199
Sherritt, Gordon, 7, 14, 24, 36
Shilts, W., 54
Shingituk, 8, 29
Shirley, Jim, 173
Sibbeston, Nick, 143
Sidden, Thomas, 155
Sidney, Peter, 51
Silver, Ray, 178
Simailak, David, 177, 179, 185
Simmons, Mike, 175
Simonee, 50
Siniktarvik Ltd., 123
Skilled jobs, 16
Skills, 114, 117, 179
Smith, Garry, 190, 192
Smith, Robert, 272
Snake River, 41
Social assistance, 15
Social consequences, cost, 113, 124
Social disruption, disintegration, 21, 73
Social equality, 12
Social impact, 94, 105, 127, 173, 193, 197

Index · 303

Social problems, 73
Society Cliffs formation, 91, 208, 239
Socio-economic impact study, 128
Sogepet, 135
Somerset Island, 40
Soper River Territorial Park, 237
Sovereignty, 218
Sposs, Werner, 182, 184, 193
Squatters, 93
Staking, 38, 168, 201, 221, 249
Starvation, 23, 27
Steen, Vince, 67
Stephen, Bill, 174, 187, 189
Steuart, Senator Davie, 130
Stoney Rapids, 36
Strathcona Sound, 89
Stuart, Mick, 169, 171, 183, 186, 194
Subsurface lands, 141, 217, 221, 223, 236, 247, 251, 254, 258, 261
Subsurface quantum, 130, 208, 250
Subsurface rights, 69, 80, 139
Subsurface selection, 254, 256, 263
Suluk, Luke, 149, 252
Suluk, Thomas, 86, 121, 129, 132, 134, 250
Superior Oil, 46
Surface Rights Tribunal (SRT), 154
Survival and prosperity of Nunavut, 272
Suspend proposal for mining, 193
Sverdrup Basin, 156, 209, 214, 224
Swanson, Al, 49
Symons, T.H.B., 61

T

Tagoona, Armand, 73
Tagoona, David, 73
Tagoona, Erik, 82, 124
Tagoona, William, 73
Tailings, 100, 105, 118, 169, 172, 174, 177, 179, 181, 183, 188, 191, 199, 249
Taipana, Simon, 133
Taparjuk, Louis, 131
Taparti, Lucien, 81
Tapatai, Peter, 82
Tatty, Annie, 250
Tatty, Robert, 252
Tautu, Andre, 70, 247
Tavani, 34

Technology of exploration, 38
Tellier, Paul, 126
Tenure system, 271
Term Island, Point, 6, 35
Territorial division, 137
"Testament to good faith", 270
T'Seleie, Frank, 121, 123
Tester, Frank, 3, 24, 32, 70, 76, 86
Texasgulf, 41, 91, 93
Theft, 24
Thelon, 26, 33, 167
Third parties, 85, 87, 143–53, 169, 212–21, 226, 237, 240, 243, 256, 270
Thompson, Andrew, 69
Thompson, J. E., 43
Thrasher, Peter, 65
Thurber Consultants, 184
Tibbett, J. F., 91
Tiktak, Thomas, 82, 181
Tiktalik, Mona, 257
Todd, John, 123
Tookoome, Simon, xiii, 55
Topilak, Walter, 256
Townsite, 93, 97, 116
Towtongie, Cathy, 177
Trades training, 128
Traditional land, 84
Treminco Resources, 170
Trinka, Joe, 258
Trudeau, Dennis, 170
Tuberculosis, 31
Tukturdjuk, David, 155
Tundra Conference, 61
Tungavik Federation of Nunavut (TFN), 131, 141, 143, 146, 149, 152, 168, 173, 186, 189, 201, 204
Tunguaq, Luke, 74
Turner, George, 42
Turnover, 113, 117
Tyrell, J. B., 5, 25, 33, 37

U

United effort sought by Inuit, 129
Uranerz, 54, 160
Urangesellschaft (UG), xxii, 160, 164, 166
Uranium, 41, 47, 50, 54, 81, 124, 136, 165, 176, 178, 181, 189, 210, 248

Uranium inquiry, 171
Use and occupancy recognized by government, 140
Usher, Peter, 65, 107
Uttak, Andre, 149

V

Valentine, Vic, 58
Vallee, Frank, 16, 18, 20, 28, 58
Van Loon, Rick, 150
Vangorda Creek, 41
Victoria Island, 46, 166
Vocational training, 128
Voisey, Louis, 24, 28

W

Wabush mine, 41
Waddell, Ian, 145
Wage economy, 30, 114
Wage employment, 252
Wage labour economy, 174, 177
Wages, 9, 11, 13, 76, 113
Wagner, Wayne, 239
Wah-Shee, James, 67, 84
Wakeham Bay, 37
Water bodies requiring land corridors, 141
Water claims, 252
Watts, Griffis and McQuat, 94
Watts, Murray, 5, 38, 40, 46–50, 216
Webber, W. W., 7
Weeks, Lud, 25, 34
Weetaltuk, John, 112

Western Arctic Region, 77
Western Canadian Airways, 35
Western Mines, 83
Whale Cove, 16, 27, 35
White bosses, 17, 20
White foxes, 4, 76, 247
White Paper on Native Policy, 60
Whitehorse Mining Initiative, 219, 270
Wiebe, Nette, 175
Wilkinson, Doug, 30
Williams, Norman, 214, 221, 240, 247, 250, 257, 259
Williamson, Peter, 178
Williamson, Robert, 10, 14, 21, 25, 178
Wilson, Kenneth, 6
Wilson, Meeka, 53
Windfall affair, 42
Wives, 113
Wolf, Paul, 192
Wollex, 205, 212
Woodford, James, 51
Wray, Gordon, 83, 178, 181, 185, 192

Y

Yahah, 26
Yathkyed Lake, 55
Yellowknife, 15, 48
Yukon, 53

Z

Zeemel, Albert, 41
Zinc, 41

OCT - - 2019

OCT - - 2019